HUMAN FACTORS OF VISUAL AND COGNITIVE PERFORMANCE IN DRIVING

HUMAN FACTORS OF VISUAL AND COGNITIVE PERFORMANCE IN DRIVING

EDITED BY
CÁNDIDA CASTRO

CRC Press is an imprint of the
Taylor & Francis Group, an **informa** business

CRC Press
Taylor & Francis Group
6000 Broken Sound Parkway NW, Suite 300
Boca Raton, FL 33487-2742

First issued in paperback 2019

ISBN-13: 978-1-4200-5530-6 (hbk)
ISBN-13: 978-0-367-38635-1 (pbk)

Library of Congress Cataloging-in-Publication Data

Human factors of visual and cognitive performance in driving / editor, Candida Castro.
p. cm.
"A CRC title."
Includes bibliographical references and index.
ISBN 978-1-4200-5530-6
1. Automobile drivers--Mental health. 2. Automobile driving--Physiological aspects. 3. Visual perception. 4. Cognitive psychology. I. Castro, Cándida. II. Title.

TL152.35.H86 2009
629.28'3--dc22 2008020511

Visit the Taylor & Francis Web site at
http://www.taylorandfrancis.com

and the CRC Press Web site at
http://www.crcpress.com

Dedication

To Irene, my youngest daughter, who was born at the same time as this book.

To Pedro, David, and Laura, my sons and daughter too.

Contents

Preface ix
Acknowledgments xiii
The Editor xv
List of Contributors xvii
Contributors xix

Chapter 1 Visual Demands and Driving 1

Cándida Castro

Chapter 2 Visual Requirements of Vehicular Guidance 31

Frank Schieber, Ben Schlorholtz, and Robert McCall

Chapter 3 Estimations in Driving 51

Justin F. Morgan and Peter A. Hancock

Chapter 4 A Two-Dimensional Framework for Understanding the Role of Attentional Selection in Driving 63

Lana M. Trick and James T. Enns

Chapter 5 Driver Distractions 75

Miguel A. Recarte and Luis M. Nunes

Chapter 6 Experience and Visual Attention in Driving 89

Geoffrey Underwood, Peter Chapman, and David Crundall

Chapter 7 The Environment: Roadway Design, Environmental Factors, and Conflicts 117

Marieke H. Martens, Rino F.T. Brouwer, and A. Richard A. van der Horst

Chapter 8 On Allocating the Eyes: Visual Attention and In-Vehicle Technologies 151

William J. Horrey

Chapter 9 **Enhancing Safety by Augmenting Information Acquisition in the Driving Environment** 167

Monica N. Lees and John D. Lee

Chapter 10 **Crossmodal Information Processing in Driving** 187

Charles Spence and Cristy Ho

Chapter 11 **Interventions to Reduce Road Trauma** 201

Narelle Haworth

Chapter 12 **On Not Getting Hit: The Science of Avoiding Collisions and the Failures Involved in That Endeavor** 223

Peter A. Hancock

Index 253

Preface

I find it difficult to separate my reflections on my chapter from my general reflections on the book as a whole. The contents of this book are more than just academic information about human factors of visual and cognitive performance in driving. Nowadays, we are still asking ourselves the same crucial question concerning road accidents proposed by Gibson nearly a century ago: *"What goes on when a man drives an automobile?"* In fact, no other machine by itself leaves so many people have access to such a high-risk system in which their own lives are critically at risk. "Of all the skills demanded by contemporary civilization, the one of driving an automobile is certainly, the most important to the individual, in the sense, at least, that a defect in it is the greatest threat to his life" (Gibson and Crooks, 1938).

Driving is an active search process through which information is selected and transformed. However, this process is complex. Road users are exposed on their journeys to a multitude of stimuli that are mainly visual. They must make a choice based on these stimuli that will in turn determine their behavior. Within this concept, the driver is an information processor. However, humans are inherently creatures of restricted capacity; that is, we are able to process only a limited amount of sensory information at any given time. *Driver inattention* represents one of the leading causes of car accidents. The advent of new, complex in-vehicle technologies, such as satellite-based global positioning systems (GPS), cellular phones, and so forth, and the increased power of computers are in the process of revolutionizing many aspects of transportation, but, at the same time, they are exacerbating the driver's attentional limitations.

The driver should interpret the available information continuously, so as to predict how a situation might evolve without him taking any action and how it might change depending on his decisions. The possible consequences of all the options open to him have to be estimated and considered all the time. The driver is constantly making decisions related to his journey, based on his interpretation of the situation and his forecast of the future state of the increasingly complex system he is controlling. These decisions lead to actions.

This definition of the complex behavior of driving can be enriched by an analysis from different perspectives, such as that of cognitive and ergonomic psychology. This work is undoubtedly a step in that direction. A good understanding of it will depend on:

- Allowing for the driver's information processing limitations.
- Being able to explain how the psychological processes involved in driving work: perception, memory, attention, decision making, learning, motivation, and motor responses.
- Analyzing person–machine or person–vehicle interaction, assessing the degree to which driving responds to well-known ergonomic criteria, comfort, safety and productivity, staff selection and training, and so forth.

- Understanding in greater depth the subtasks involved in driving: those carried out continuously such as tracking, by which the vehicle adapts its displacement to the bends in the road; and others that are carried out simultaneously and intermittently, such as changing gears, and which in part become automatic reactions. No less important are tasks such as information acquisition from the road environment, traffic devices, or from inside the vehicle.
- Considering the changes that take place regarding navigation tasks. Those tasks that include geographic and visual–spatial elements also become important, especially when driving in unfamiliar areas. The onset of new technologies, such as Global Positioning Systems (GPS) by satellite and the growing power of computers, is revolutionizing many aspects of both land and air transport. However, it is not all positive, and of all the machines that man handles and interacts with, the vehicle brings with it the assumption of great risk to the driver's and others' lives. This trade-off is just one among the many found in human performance, such as the well-known one between reaction time and accuracy whereby we can make a rapid response but at the cost of diminished accuracy. There is no easy solution to the question of which devices are helpful and which are detrimental. Yet, it is possible to produce intelligent and creative designs based on the knowledge we have about human processing limitations.
- Taking the speed factor into account, as it multiplies the number of accidents. Given the fact that traveling happens at great speed, intelligent information systems affect transport safety and are without doubt determining factors in the greater or lower probability of accidents happening.
- Understanding changes in the driving environment, which happen dramatically even on the same day, whether it be daylight or nighttime, a wet or a sunny day, or whether we are driving in dense or light traffic. Without a doubt, all these aspects produce significant changes in human interaction with transport systems.
- Analyzing different driving situations. In order to formulate new ideas and obtain guidelines from cognitive psychology and human factors or ergonomics, it is also vital to analyze the specific driving skills required in each driving situation, such as overtaking, parking, and driving on motorways or on minor roads. This analysis will determine which psychological processes are necessary for input, information handling, output, or metacognition for every driving situation.

It is hoped that, in the meantime, this work can make recommendations and suggest improvements in the design and rethinking of how we define driving and of vehicles and traffic environments, taking into account information from cognitive psychology and ergonomics, which have sometimes been considered very separate from each other.

Cándida Castro

REFERENCE

Gibson, J.J., and Crooks, L.E. (1938). A theoretical field-analysis of automobile driving. *The American Journal of Psychology*, 51(3), 453–471.

Acknowledgments

Preparation of this book was supported by grants: SEJ-2007-61843 from the Spanish Ministry for Education and Sciences, and the Excellence Research Grant P07-SEJ-02613 from the Junta de Andalucia, Spain.

The Editor

Cándida Castro holds a PhD in psychology. She is a senior lecturer at the Faculty of Psychology, University of Granada, Spain. Her areas of specialization are cognitive psychology and traffic, especially attention and perception while driving, and she has been an author of papers in these areas. Her PhD thesis was concerned with the arrangement of traffic signs in road environments. She has carried out several overseas academic working visits, including the psychology department, Royal Holloway & Bedford New College, University of London, in 1994; the Institute of Behavioral Sciences, University of Derby, in 1998; and the Accident Research Unit, University of Nottingham, in 2007. Dr. Castro previously edited the book entitled: Human Factors of Transport Signs, www.vgr.es//local/segvia. She can be reached at the Departamento de Psicología Experimental y Fisiología del Comportamiento, Facultad de Psicología, Campus Cartuja, s/n, 18071, Granada; e-mail: candida@ugr.es.

List of Contributors

Rino F.T. Brouwer
TNO Human Factors
Soesterberg, Netherlands
Rino.Brouwer@tno.nl

Cándida Castro
Department of Experimental Psychology and Behavioural Physiology
Faculty of Psychology
Campus Cartuja
University of Granada
Granada, Spain
candida@ugr.es

Peter Chapman
Accident Research Unit, School of Psychology
University of Nottingham
Nottingham, United Kingdom
peter.chapman@nottingham.ac.uk

David Crundall
Accident Research Unit, School of Psychology
University of Nottingham
Nottingham, United Kingdom
david.crundall@nottingham.ac.uk

James T. Enns
Department of Psychology
University of British Columbia
Vancouver, British Columbia, Canada
jenns@psych.ubc.ca

Peter A. Hancock
Department of Psychology
University of Central Florida
Orlando, Florida, United States
phancock@pegasus.cc.ucf.edu

Narelle Haworth
Faculty of Health, CARRS-Q School of Psychology and Counselling
Queensland University of Technology
Queensland, Australia
n.haworth@qut.edu.au

Cristy Ho
Department of Experimental Psychology
University of Oxford
South Parks Road, Oxford, United Kingdom
cristy.ho@psy.ox.ac.uk

William J. Horrey
Liberty Mutual Research Institute for Safety
Hopkinton, Massachusetts, United States
william.horrey@libertymutual.com

John D. Lee
Cognitive Systems Laboratory, Department of Mechanical and Industrial Engineering
University of Iowa
Iowa City, Iowa, United States
jdlee@engineering.uiowa.edu

Monica N. Lees
Cognitive Systems Laboratory, Department of Mechanical and Industrial Engineering
University of Iowa
Iowa City, Iowa, United States
mnlees@engineering.uiowa.edu

Marieke H. Martens
TNO Human Factors
Soesterberg, Netherlands
marieke.martens@tno.nl

Robert McCall
Department of Psychology
University of South Dakota
Vermillion, South Dakota, United States
Robert.Mccall@usd.edu

Justin F. Morgan
Virginia Tech Transportation Institute
Blacksburg, Virginia, United States
jfmorgan@gmail.com

Luis M. Nunes
Faculty of Psychology
Universidad Complutense de Madrid
Campus de Somosaguas
Madrid, Spain
luisnunesg@gmail.com

Miguel A. Recarte
Faculty of Psychology
Universidad Complutense de Madrid
Campus de Somosaguas
Madrid, Spain
marecarte@psi.ucm.es

Frank Schieber
Department of Psychology
University of South Dakota
Vermillion, South Dakota, United States
Frank.Schieber@usd.edu

Ben Schlorholtz
Department of Psychology
University of South Dakota
Vermillion, South Dakota, United States
Ben.Schlorholtz@usd.edu

Charles Spence
Department of Experimental Psychology
University of Oxford
Oxford, United Kingdom
charles.spence@psy.ox.ac.uk

Lana M. Trick
Department of Psychology
University of Guelph
Guelph, Ontario, Canada
trick@psy.uoguelph.ca

Geoffrey Underwood
Accident Research Unit, School of Psychology
University of Nottingham School
Nottingham, United Kingdom
geoff.underwood@nottingham.ac.uk

A. Richard A. van der Horst
TNO Human Factors
Soesterberg, Netherlands
Richard.vanderhorst@tno.nl

Contributors

The Psychology Department of the University of Leiden provided an inspiring setting for **Rino F.T. Brouwer** to study psychonomy and to receive his PhD with the thesis "One Feature to Rule Them All: The Dependencies between Position, Colour, and Form." In this highly fundamental piece of work he investigated the relation between different object features in visual information processing. This work strongly related to selective attention in vision. In 1998 he started working with the Traffic Behaviour Group of TNO Human Factors in the Netherlands. He has been involved in different national and international projects on a wide range of topics. His main interest lies at detection of impaired driving, the role of visual attention in driving, relating driving behavior and traffic safety, and the relation between driver support systems and driving behavior.

Peter Chapman is a lecturer in the School of Psychology at the University of Nottingham and a member of the Accident Research Unit and the Cognition and Language Group. He does research in applied cognitive psychology. His main area of application is the psychology of driving, while his more theoretical interests are in vision and memory. Some of his research actually fuses all three of these areas, that is, where do drivers look and what do they remember after they have looked there? Some examples of the research Chapman is involved with are visual search in novice and experienced drivers, eye movements in dangerous driving situations, memory for accidents and near accidents, attention and memory failures in routine tasks, and traffic accident liability.

David Crundall is a member of the Accident Research Unit (ARU); the Cognition and Language Research Group; and the Risk, Social Processes and Health Group (RaSPH) at the University of Nottingham. His current research is concerned with the allocation of visual attention in both theoretical and applied domains. The applied nature of his work tends to focus upon driving and driving-related behavior within the Accident Research Unit, though not exclusively. His theoretical research is currently focused on a number of key concepts in the attention literature, though often the theoretical and applied strands of research blend in a reciprocal manner. His research and current research projects are concerned with hazard perception in learner drivers, profiling risky road users, visual attention and line tracing, eye movements during map reading, and facial emotion and the attentional blink.

James T. Enns is a distinguished university professor in the Department of Psychology and the Graduate Program in Neuroscience at the University of British Columbia, and is a fellow of the Royal Society of Canada. He received his bachelor of arts (with honors) at the University of Winnipeg and his master's and PhD at Princeton University. His research is on how the visual world is represented inside and outside the focus of attention and how attention changes the contents of consciousness. He has investigated preattentive vision and attentional capture, perceptual organization, reentrant cortical processing, and transfer and competition between the cortical hemispheres. His longstanding interest in the lifespan development of attention has also led him to the study of dyslexia and autism. His research on vision and attention lends itself quite naturally to applications in related fields, and has resulted in collaborative projects on multidimensional data visualization and more recently, driving. He served as associated editor for *Psychological Science* and *Visual Cognition* and has authored or coauthored several books (see http://www.psych.ubc.ca/~ennslab/people/jim_index.html).

Peter A. Hancock is a Provost's Distinguished Research Professor in the Department of Psychology and the Institute for Simulation & Training and Department of Civil and Environmental Engineering, University of Central Florida. He holds a DSc from Loughborough University, Loughborough, England, in human–machine systems, 2001; a PhD from University of Illinois, Champaign, Illinois, in motor performance, 1983; an MSc from Loughborough University, Loughborough, England, in human biology, 1978; a BEd from Loughborough University, Loughborough, England, in anatomy and physiology, 1976; and a certificate of education from Loughborough University, Loughborough College of Education, England, 1975. Hancock was the principal investigator on the Multi-Disciplinary University Research Initiative, in which he oversaw $5 million of funded research on stress, workload, and performance. His current experimental work concerns the evaluation of behavioral response to high-stress conditions. His theoretical works concern human relations with technology and the possible futures of this symbiosis. Additional information concerning Dr. Hancock and his research program can be found at www.peterhancock.ucf.edu.

Narelle Haworth was appointed as professor in injury prevention and rehabilitation at the Centre for Accident Research and Road Safety, Queensland, Australia, in January 2006. In her earlier employment at the Monash University Accident Research Centre, she conducted research that spanned the breadth of human factors issues across all modes. These projects have provided her with knowledge of the interplay of technical, human, and organizational factors that influence transport safety. Haworth has conducted extensive research in road safety, including studies of fatigue in driving, seat-belt wearing by truck drivers, road-user behavior in developing countries, development of data collection methodologies, driver training and licensing, coin-operated breath testing, motorcycle safety, and single-vehicle crashes. Recently, Haworth's work in road safety took on a much more systemic approach, assisting states and large organizations in the development of road safety strategies and providing advice and monitoring of their implementation. In rail research,

Haworth has examined the effectiveness of deadman and vigilance control systems, managed a simulator study of potential distraction of train drivers, identified factors contributing to the occurrence of collisions, and studied the characteristics of suicides in the rail network. Other past research has examined the human factors aspects of ship evacuation and development of improved data systems for aviation incidents.

Cristy Ho is a postdoctoral research scientist at the Crossmodal Research Laboratory, University of Oxford, United Kingdom. She received her DPhil in experimental psychology from the University of Oxford in June 2006. Her research has focused on investigating the effectiveness of multisensory warning signals in driving. In 2006 Ho was awarded the American Psychological Association's New Investigator Award in Experimental Psychology: Applied. This award is given for the most outstanding empirical paper authored by a young scholar published in the *Journal of Experimental Psychology: Applied.*

William J. Horrey is currently a research scientist at the Liberty Mutual Research Institute for Safety in Hopkinton, Massachusetts, United States. He joined the institute in 2005. He earned his PhD and MA in psychology from the University of Illinois and his BSc (with honors) in psychology from the University of Calgary (Alberta, Canada). His fields of professional interest are applied experimental psychology, transportation human factors, driver distraction, visual attention, perception, and system automation. Currently, his research focuses on drivers' awareness of performance decrements due to distraction as well as driver decision making with respect to self-initiated distractions. He is also interested in modeling drivers' visual attention while interacting with in-vehicle devices, and determining the associated performance decrements. A member of the Human Factors and Ergonomics Society (HFES), the American Psychological Association, the Association for Psychological Science, and the Transportation Research Board (TRB), he currently serves on the HFES Publications Committee and the TRB Vehicle User Characteristics committee.

John D. Lee is a professor in the department of mechanical and industrial engineering at the University of Iowa, and is the director of human factors research at the National Advanced Driving Simulator. He is also affiliated with the Department of Neurology, the Public Policy Center, the Injury Prevention Research Center, and the Center for Computer-Aided Design. His research focuses on the safety and acceptance of complex human–machine systems by considering how technology mediates attention. Specific research interests include trust in technology, advanced driver assistance systems, and driver distraction. He is a coauthor of *An Introduction to Human Factors Engineering*, a textbook, and is the author or coauthor of over 220 articles. He received the Ely Award for best paper in the journal *Human Factors* (2002), the best paper award for the journal *Ergonomics* (2005), and a Donald E. Bently Faculty Fellow. He is a member of the National Academy of Sciences committee on human factors and has served on several other committees for the National Academy of Sciences. Dr. Lee serves on the editorial board of *Cognitive Engineering and Decision Making*; *Cognition, Technology and Work*; *International Journal of Human Factors Modeling and Simulation*; and is the associate editor for the journal *Human Factors and IEEE-Systems, Man, and Cybernetics.*

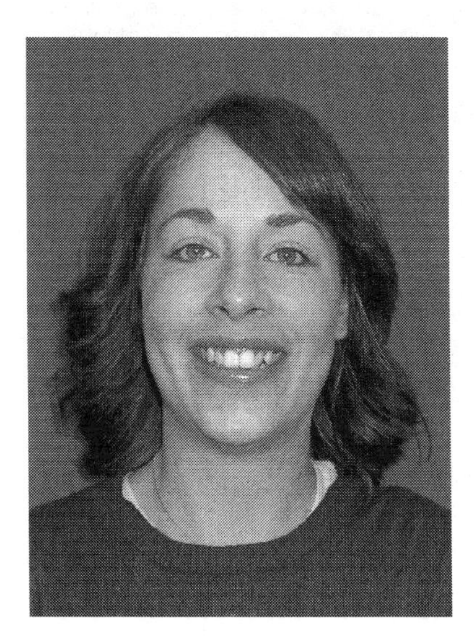

Monica N. Lees is a PhD candidate in industrial engineering at the University of Iowa. She earned an MS in industrial engineering at the University of Iowa in 2005 and a BA in psychology, University of Calgary, 2003. Her areas of interest are interface design and driver support, automation failures, attitudes toward automation, and driver distraction. Currently, her research focuses on developing and evaluating in-vehicle technology to aid attention impaired drivers (NIH). Her dissertation-related work focuses on how different alarm failures influence driver behavior and acceptance in collision warning systems.

Marieke H. Martens studied experimental psychology and ergonomics at the Free University of Amsterdam. Since 1996 she has been working at TNO Human Factors in the Netherlands, with the Traffic Behaviour Group. She is a researcher in the area of road design, driver support systems, tunnels, dynamic information, eye movements, visual attention, workload, and driver state. Marieke finished her doctoral dissertation in 2007. Her doctoral thesis, titled "The Failure to Act Upon Important Information: Where Do Things Go Wrong?," deals with the phenomenon that sometimes drivers fail to respond to information that is clearly visible and very relevant for the driving task. Martens has worked on many national and international projects, and has been the coordinator of the Traffic Behaviour Group since 2002.

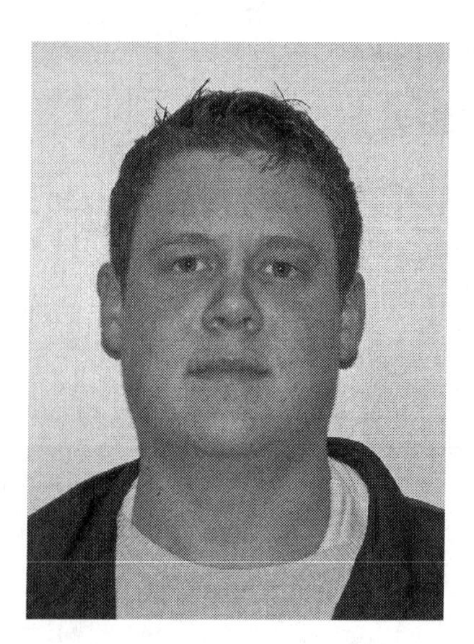

Robert McCall was born in Fort Worth, Texas, on December 21, 1982. He received his BA degree in psychology from Texas Tech University in 2006 and is currently working toward his doctoral degree in human factors psychology at the University of South Dakota. He has a broad interest in transportation psychology but has become especially interested in the role of attention in peripheral vision as a mediator of age-related driving problems.

Justin F. Morgan is a research associate at the Virginia Tech Transportation Institute. As a researcher, he has worked on a variety of projects examining human information processing and decision making in the driving task, including issues of highway design standards, in-vehicle displays, and multimodal communication. He holds a BA in psychology from the University of North Carolina at Asheville, and an MS and PhD from the University of Central Florida.

Luis M. Nunes, psychologist, graduated from the Universidad Autónoma de Madrid, and during the last fifteen years has developed his activity as researcher in the Direction General de Tráfico–Madrid, responsible for the Argos Research Program. The program includes the development of the instrumented vehicle, with the collaboration of the Computer Architecture Department of the Polytechnic University of Madrid. He collaborated with Miguel Angel Recarte Goldaracena in most of his research activity and also contributed as an associate professor at the Faculty of Psychology of the Universidad Complutense de Madrid.

Miguel A. Recarte, PhD, Universidad Complutense de Madrid, is presently a professor of psychology of attention and driving psychology in the Faculty of Psychology of the same university. During the last fifteen years his research activity has focused on perceptual and attentional processes in driving, in collaboration with Luis M. Nunes and with the funding of the Direction General de Tráfico–Madrid.

Frank Schieber was born in Philadelphia, Pennsylvania, on July 24, 1954. He received his PhD in experimental psychology from the University of Notre Dame in 1985. He is a professor in the Heimstra Human Factors Laboratories of the University of South Dakota. His research interests center on the converging areas of vision, aging, and driving behavior. He recently completed three terms as chair of the Transportation Research Board's Visibility Committee and serves on the editorial board of *Human Factors*. Currently, he is exploring how eye movement data and models can be employed to better understand the visual requirements of automobile driving.

Ben Schlorholtz was born in Mason City, Iowa, on November 16, 1979. He received his BA degree in psychology from Kansas State University in 2002 and is currently a doctoral candidate in human factors psychology at the University of South Dakota. His research interests include multiple-resource theory as well as the information processing requirements of driving an automobile. He recently joined the staff of RED-Inc. in Lexington Park, Maryland.

Charles Spence is the head of the Crossmodal Research Laboratory based in the Department of Experimental Psychology, Oxford University (http://www.psy.ox.ac.uk/xmodal/default.htm). He is interested in how people perceive the world around them. In particular, how our brains manage to process the information from each of our different senses (i.e., smell, taste, sight, hearing, and touch) to form the extraordinarily rich multisensory experiences that fill our daily lives. His research focuses on how a better understanding of the human mind will lead to the better design of multisensory foods, products, interfaces, and environments in the future. His research calls for a radical new way of examining and understanding the senses that has major implications for the way in which we design everything from household products to mobile phones, and from the food we eat to the places in which we work and live. Spence is currently a consultant for a number of multinational companies advising them on various aspects of multisensory design. He has also conducted research on human–computer interaction issues on the crew workstation of the European space shuttle, and currently works on problems associated with the design of foods that maximally stimulate the senses, and with the effect of the indoor environment on mood, well-being, and performance. Spence has published more than 250 articles in topflight scientific journals over the last decade. He has been awarded the 10th Experimental Psychology Society Prize; the British Psychology Society, Cognitive Section Award; the Paul Bertelson Award, recognizing him as the young European Cognitive Psychologist of the Year; and, most recently, the prestigious Friedrich Wilhelm Bessel Research Award from the Alexander von Humboldt Foundation in Germany.

Lana M. Trick is an associate professor in the Department of Psychology in the area of neuroscience and applied cognitive science and is codirector of the University of Guelph DRIVE lab (http://www.uoguelph.ca/drive). She received her bachelor's of science (with honors) at the University of Calgary, and then a master's and PhD at the University of Western Ontario in London. Her doctoral dissertation was on the relationship between multiple-object tracking and subitizing (the ability to know at a glance the number of items when there are fewer than four or five items), and it was done under the supervision of Zenon Pylyshyn, best known for his work on computational theory of mind. Upon graduation, she was awarded a Killam Post-Doctoral Fellowship to work with James Enns at the University of British Columbia. It was there that she began research on how attentional abilities change across the lifespan, from childhood to old age. She became interested in researching driving as a result of a project done in collaboration with the Insurance Corporation of British Columbia. She has done a variety of studies on driving, some looking at the effects of age and experience, and others looking at the impact of in-vehicle technologies such as navigation systems, cellular phones, and a system designed to warn of the presence of large animals on the road (the University of Guelph "moose detector"). Recently she has become involved in a study of the effects of medication on driving performance in adolescents diagnosed with attention deficit hyperactivity disorder, as well as a project investigating the use of vestibular stimulation as a means of overcoming simulator adaptation syndrome (also known as simulator sickness), a common methodological problem in research involving driving simulators.

Geoffrey Underwood holds a BSc, psychology, Bedford College, University of London; and a PhD, University of Sheffield (Department of Psychology). His thesis presented experiments on selective attention to spoken messages (supervisors were Neville Moray and Harry Kay). He also holds a DSc, University of London. Underwood is currently a professor of cognitive psychology at the University of Nottingham. Previously he was an associate professor at the University of Guelph, Ontario, Canada; and an assistant professor at the University of Waterloo, Ontario, Canada. His ongoing research is carried out at the Eyetracker labs at Nottingham University. Currently in use on a number of related projects that investigate the relationship between visual attention and skill, his NAC EyeMark VII eyetracker is being used to investigate visual attention in drivers. This project compares novice drivers who are at high or at low accident risk, using a range of measures that include eye fixations on critical objects while driving an instrumented vehicle. His SR Research EyeLink eyetracker is being used for investigations of the cognitive processes used in reading, with particular interest in the distribution of attention between text that is fixated and text that is available in parafoveal vision.

A. Richard A. van der Horst holds an MSc degree in electrical engineering (1973) and a PhD (1990) from the Delft University of Technology (PhD thesis: "A Time-Based Analysis of Road User Behaviour in Normal and Critical Encounters"). Van der Horst is both business developer and senior research scientist in the Traffic and Transport area at TNO Human Factors in the Netherlands. He conducted basic research on the direct use of time-related measures such as time-to-collision and time-to-intersection in road users' decision making in traffic. He has been working in the field of human factors in road traffic for more than thirty-three years with a main focus on evaluation and assessment of road design, traffic management, and road user support systems in terms of road safety and road user behavior. Driver modeling and development of tools and techniques for evaluating ITS applications have his special interest. He is a member of several national and international committees on road design and human factors-related topics.

1 Visual Demands and Driving

Cándida Castro

CONTENTS

Reflection 2
1.1 Introduction 2
1.2 Models of Driving Information Acquisition 3
 1.2.1 Information Theory 3
 1.2.2 Gibson's Driving Information Acquisition Model 7
 1.2.3 Sign Detection Theory 8
 1.2.4 Other Models Explaining Perception in Driving 9
1.3 Methodology Used to Study Driving Information Acquisition 11
1.4 Driving Simulators 13
1.5 Research Methods 15
 1.5.1 Observation 15
 1.5.2 Testing 16
1.6 Recording Techniques: How to Measure the Driving Dependent Variable 17
 1.6.1 Eye Movement Recording 17
 1.6.2 Drivers' Self-Assessments 20
 1.6.3 Memory 21
 1.6.4 Sign Recognition 22
 1.6.5 Traffic Accidents 22
 1.6.6 Reaction Time and Other Measurements of Driving Performance 22
 1.6.7 Psychophysiological Techniques 24
1.7 Adaptation to Driving: A Challenge Overcome or Still a Challenge? 24
References 26

REFLECTION

The main purpose of this book is to draw together knowledge about how drivers acquire information, from a human factors psychology standpoint, in order to help provide deeper insights into the way road users behave.

Driving is inherently more hazardous than other everyday activities. Therefore, the application of human factors knowledge to the study of transport information acquisition could not only help improve drivers' behavior, but also help prevent accidents.

The driver's statement "looked but failed to see" after any car accident is the best theoretical justification for the in-depth study of visual perception and attention while driving. The driver is not only a passive receptor of the traffic scene (bottom-up processing) but also, and mainly, an active processor who continuously selects and transforms information from the driving environment. This search is biased by the motivation, interests, and expectations the driver imposes on the environment and is driven by his/her attention (top-down processing).

Overall findings show that many well-established principles of engineering psychology, cognitive science, social psychology, visual design, and road-user behavior are applicable and should be taken into account by engineers and other transport specialists or professionals in order to be realistic and usable. This is a very fertile research area for human factors professionals. To reduce traffic accidents, it is essential to consider human strengths and weaknesses when processing information as well as the visual demands of the driving task.

1.1 INTRODUCTION

Three figures are all that is necessary to understand the importance of studying perception and attention processes in the road safety context: (1) More than 90% of traffic accidents are due to human error (Fell, 1976); (2) more than 90% of these are due to visual information acquisition problems (Hills, 1980; Olson, 1993; see also Sivak, 1996, for more details); and (3) the majority of explanations given by the drivers are of the "I looked, but I didn't see it" type.

An example is provided by Langham, Hole, Edwards, and O'Neil (2002) when they reviewed accident reports that involved drivers crashing into a very noticeable vehicle (a police car) parked on the hard shoulder. The drivers stated that they had not seen the police car prior to colliding with it. Previous research has tried to prevent the so-called "I looked but I didn't see it" accidents by making vehicles, traffic signs, and elements in the roadway environment more noticeable and conspicuous. However, these authors consider that other reasons, such as an error in vigilance, could be the most direct cause of this type of accident and not that these types of vehicles are actually difficult to see.

This statement of "looking but not seeing" in spite of having the stimulus before our very eyes or even behind them, reflected off our retina, provides the best theoretical justification for the need to study visual perception and attention in the road safety context.

Perception was traditionally seen as the structuralization of what the driver sees around him or her in an exclusively passive way, by receiving luminance parameters that are reflected on the retina. Yet it must also, even rather, be understood as an active phenomenon based on continuous information selection and processing. This active process is steered, biased, and distorted by the functioning of other cognitive processes (memory, motivation, interests and expectations, etc.) that the driver imposes on the context and is directed by our attention through what has become known as top-down processing. This way of understanding perception as a complex process that involves processing guided by data as well as conceptual issues is upheld by many classic statements in the field of road safety. Such statements underline the importance of perceiving the driver as an active information processor. It also seems essential to highlight the adaptive and functional purpose of using perception and attention systems in an everyday situation, such as driving.

Thanks to the functioning of the perception system, we can survive by being able to detect stimuli quickly and efficiently. The operation of this system enables us to detect errors between our predictions or expectations and reality. However, over the last few centuries we humans have abandoned our natural environments and started to take on new activities, such as driving cars in artificial environments like modern roads. In these situations, our perceptual systems have become partially inefficient. Typical examples that illustrate the shortcomings in human adaptation to these environments are the delay or even failure in detection when driving at night or in seeing vehicles approaching the limits of our field of vision.

Nevertheless, it would be difficult to imagine life without vehicles or the services they provide to us. We use them, among other things, to go to work, go food shopping, take our children to school, go to the movies, or go on holiday. We use them to move around and avoid being tied to one place. They are without a doubt one of the elements that drive our society, economy, and lives.

However, the use of cars not only has advantages but also serious drawbacks. Gibson stated as long ago as 1938 that one in every twenty drivers was involved in an accident resulting in death or injury. This problem continues to be equally or even more serious today. It is extremely important to know the nature of driving acquisition and performance, and to optimize the devices used. For this reason, Gibson and Crooks (1938) also argued that "of all the abilities that contemporary civilization requires of us, driving is the most important for individuals in the sense that errors in this ability translate into the greatest threat to human life."

Several models of driving information acquisition have been specified to better understand driving behavior.

1.2 MODELS OF DRIVING INFORMATION ACQUISITION

What is information? Sivak (1996) states that it is difficult to find consensus on the concept of information and mentions some of the attempts made to describe it.

1.2.1 Information Theory

According to Shanon and Weaver (1949), the creators of information theory, information can be defined as a reduction in uncertainty. Unlikely events are more informative than those more likely to occur. Furthermore, because there are few chances

TABLE 1.1
Events with Different Occurrence Probability

	A	B	C	D
Pi	0.5	0.25	0.125	0.125
1/Pi	2	4	8	8
$[\log_2(\frac{1}{Pi})]$	1	2	3	3

$$H_{average}' = \sum_{i=1}^{n} Pi[\log_2(\frac{1}{Pi})] = 0.5 + 0.5 + .0.375 + 0.375 = 1.75 Bits$$

of them happening, they contribute little to the final calculation of the average information transmitted by a set of events. According to Miller (1956), a bit of information is the amount of information needed to decide between two equally probable alternatives. However, some authors consider that this definition is difficult to apply to the driving context where the degree of uncertainty varies from one driver to another and from one moment to the next.

As mentioned, the first factor to have a bearing on the amount of information transmitted is occurrence probability (Fitts and Posner, 1968; see Tables 1.1 and 1.2):

$$H_s = \log_2(\frac{1}{Pi})$$

where Pi is the occurrence probability of an event and Hs is the information from the stimulus.

The short-term context in which information is introduced also affects information transmission. Even if an event has a generally low occurrence probability, it may happen that in a particular context the probability increases, which makes it less informative in that case. For example, a sign warning us of a holdup (delay) in 3

TABLE 1.2
Events with Equal Occurrence Probability

	A	B	C	D
Pi	0.25	0.25	0.25	0.25
1/Pi	4	4	4	4
$[\log_2(\frac{1}{Pi})]$	2	2	2	2

$$H_{average} = \sum_{i=1}^{n} Pi[\log_2(\frac{1}{Pi})] = 0.5 + 0.5 + .0.5 + 0.5 = 2 Bits$$

kilometers is not as informative as a signal to stop given by a policeman upon arrival at the actual holdup.

The sequence of events can also have that effect. For example, if we were presented for the first time with an ABABABABAB sequence of signs in which the probability of both A and B was initially 50% but where, if we were asked which would be the next stimulus after A, we answered B, then the information transmitted by the sign would be zero bits.

In both cases, the contingent probability in a specific context can be calculated (Pi/X) and will be the same as the probability of an informative event happening since X has occurred, where X is the context.

Another important concept in information theory is redundancy or loss of potential information. It depends on the three variables mentioned earlier: the probability of an event happening (occurrence probability), the effects of context, and the sequence in which information appears.

$$\% \text{ Redundancy} = (1 - H_{real}/H_{max}) \times 100$$

where H_{real} or average is equal to the present average for information transmitted and H_{max} is equal to the maximum amount of possible information that can be provided by the alternatives if they all have the same probability, such as the letters of the alphabet. For example,

$$H_{\text{Real or average}} = 1.5 \text{ bits per letter of the alphabet}$$

$$H_{Max} = \log_2 26 = 4.7 \text{ bits}$$

$$\% \text{ Redundancy} = (1 - 1.5/4.7) \times 100 = 6.8\%$$

It is sometimes important to maximize the efficiency of the information transmission channel by making the redundancy value low. At other times, however, an attempt is made to guarantee the safety of the transmission and the channel even though the redundancy values may be higher.

In discrete signal transmission, it is not only the information presented that is important but also the information the operator succeeds in transmitting and what arrives at the response level.

Some important concepts related to signal transmission are system capacity, transmission speed, and bandwidth, which are all useful for quantifying information transmission (see Figure 1.1). Channel bandwidth is the amount of information transmitted (H_T) per unit of time:

$$\text{Channel bandwidth} = H_T/\text{reaction time} = (\text{Bits/sec})$$

Small events need to be defined in order to quantify information transmitted by continuous signals, such as bends in the road as we drive, the speedometer needle

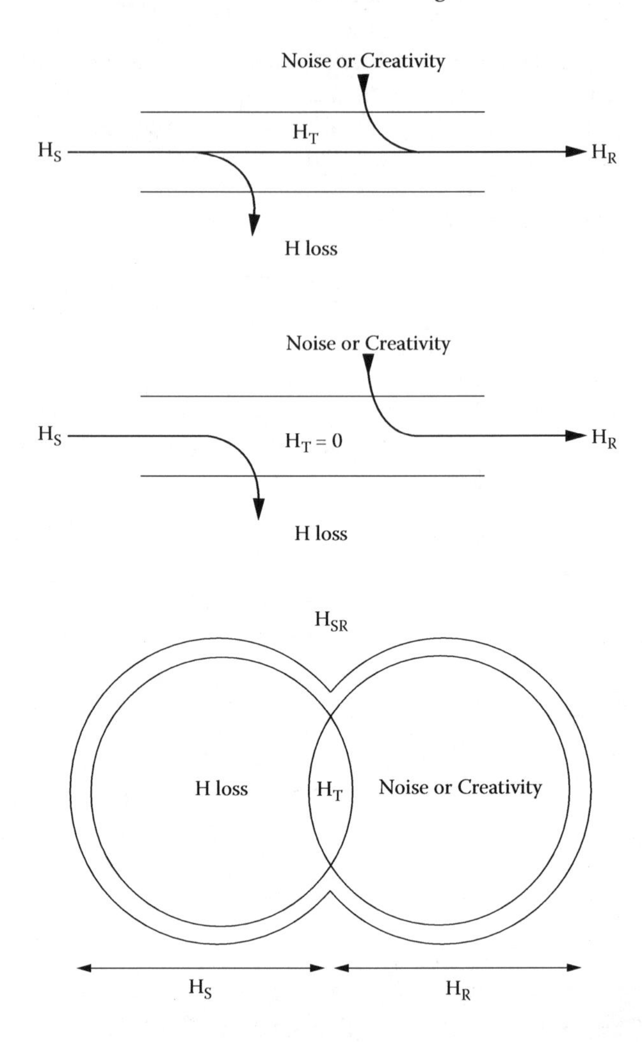

FIGURE 1.1 Different diagrams showing the information transmission channel, where H_S is information from the stimulus, H_T is the transmitted information, and H_R is the information that reaches the response. The top diagram represents information transmitted through the channel; in the middle one, no information is transmitted; in the bottom one, transmitted information is shown by Euler/Venn diagrams.

moving up and down, or general tracking tasks which do not provide discrete information although they do provide the person with information about the surrounding world.

A continuous signal can be divided into discrete levels. This happens when possible changes are known in advance, for example, a change in the direction of the needle or a bend. Information provided by each movement provides 1 bit because there are two positions or locations toward which a change can go, for example, up/down or right/left (bit/changes = 0.5). However, sometimes the information can adopt an infinite number of possible states, making it more difficult to define the accuracy with which the operator can find the signal position.

Information theory also has its limitations. One of them is that HT only measures responses consistently associated with the matrix stimuli and does not measure whether they are correctly associated. Moreover, this measurement does not take into account the size of the error, which can sometimes be substantial. Bigger mistakes should be penalized more than smaller ones. That is why a correlation coefficient or some measurement that includes the mistake over time should be used.

Despite the fact that information theory can be difficult to apply to the driving context due to varying degrees of uncertainty from one driver to another and from one moment to another, it can provide us with useful references to describe information provided by traffic signs and other traffic devices. The message redundancy of a sign, for example, should be estimated by considering whether we want to convey a lot of information or less information but in a safer way. The number of times a sign is seen may influence the amount of information conveyed. It has been demonstrated, for example, that the effect of priming repetition produces faster reactions to the repeated sign (Crundall and Underwood, 2001; Castro, Horberry, Tornay, Martinez, and Martos, 2003). However, the amount of information transmitted by the second sign is much less than that transmitted by the first. Undoubtedly, this analysis both enriches and limits the conclusions reached using repetition priming as the paradigm for the study of signs.

It is for all these reasons that we believe a description of the road environment in information theory terms—a theory in decline over the past decades—is enriching and useful in order to understand more about driver limitations. It enables us to quantify and assess the usefulness of traffic elements with regard to the amount of information they can transmit, so that we can estimate the guarantees of such information being received.

1.2.2 Gibson's Driving Information Acquisition Model

Gibson (1979/1986) disagreed with the idea that information can be defined in terms of bits when talking about perception. In his opinion, this may be valid for defining communication but not perception. Gibson defines the term *information* as patterns of energies such as degrees of texture, direction of contours, or Fourier spatial frequency analysis.

According to Gibson and Crooks (1938), changes and constancies of such patterns over time and space will give rise to the *affordances* (action bids) available to a driver. These affordances guarantee what Gibson calls the *field of safe travel* to the driver at a given time (Figure 1.2). This consists of the field of possible courses a car can take without having its way impeded by objects found in the area. The field of safe travel can be physically described as a tongue extending along the road, whose limits are determined by objects or characteristic features of the road. Valency is the feature by virtue of which we move toward or away from certain objects. Such obstacles will have a negative valency, while the average profile of the field of safe travel has a positive valency.

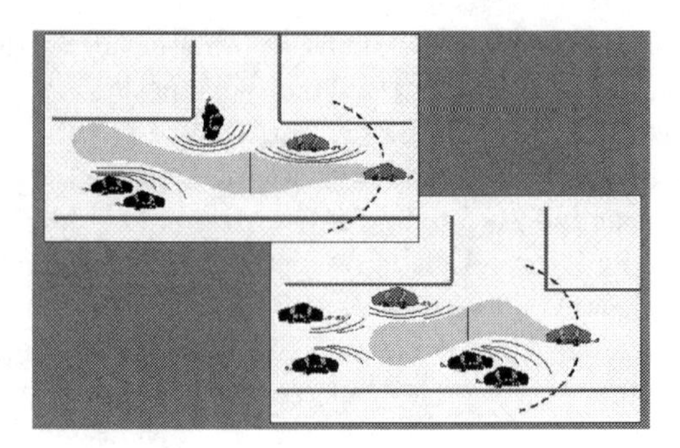

FIGURE 1.2 The driver's field of safe travel and the minimum stopping distance in the traffic environment, at two consecutive moments in time.

The field of safe travel has to be defined at each specific moment. It is a spatial but not a stationary field. The car moves through the field when it crosses the space, and the driver is the point of reference. The field of safe travel changes continuously, turning and becoming longer, smaller, wider or narrower depending on the presence of obstacles on the road limiting its borders. However, it does not involve the driver's subjective experience since it is possible to make an objective assessment of which field a car could safely maneuver (Figure 1.3).

Driving a car can therefore be defined as an activity that is governed by a series of driver reactions in order to keep the vehicle within the average lines of the field of safe travel (Gibson and Crooks, 1938).

1.2.3 Sign Detection Theory

It is also possible to define information by referring to its expression depending on the existing proportion of signs to noise (Tanner and Swets, 1954). Signs are understood

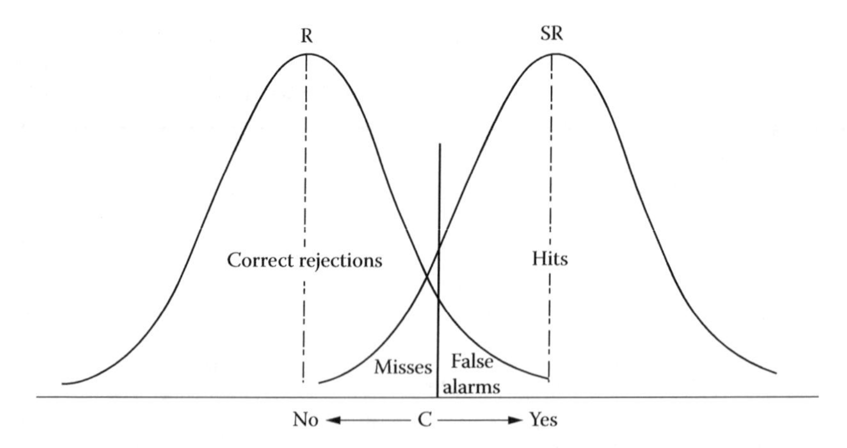

FIGURE 1.3 Graph showing success and false alarm (F.A.) rates generated depending on criteria (C) followed by the observer, given a certain amount of discrimination or distance between noise distribution (R) and sign distribution (SR).

here as the stimulus or stimuli we have to detect when carrying out a particular task, and noise as those stimuli that are not signs but may be confused with them because they can produce certain sensations characteristically produced by signs.

Sign detection theory assumes that two processes are involved, one of which is related to the senses, while the other is related to decision. The sensory process is always activated to a greater or lesser extent. The decision as to whether the sign is detected or not is made by the observer, and is partly dependent on the amount of sensation being felt and on the decision criteria followed. This may be laxer or stricter depending on the motivation, interests, instructions, and so forth affecting the observer (Figure 1.3).

This definition is useful for describing situations where objects around us are detected discretely, taking into account the bottom-up process (sensory entry, a data-guided process) and the top-to-bottom process (decision criteria, a concept-guided process that includes important motivation factors in the detection process as traffic signs, for example). However, driving also clearly implies ongoing information acquisition to keep the vehicle on course, and this is difficult to describe in terms of this theory.

1.2.4 Other Models Explaining Perception in Driving

Other models agree in highlighting the relevance and complexity of perception and attention processes as determining factors in driving (see Moore, 1969; Rumar, 1982). Both Moore (Figure 1.4) and Rumar (see Figure 1.5) view the driver as a traffic system information processor. They underline the importance of search and selection processing of information that may be useful to us, taken from around us by the actual vehicle or driver. Correct receiving and processing of information enables us to make decisions and maneuver appropriately according to each traffic situation.

The difficulty lies in selecting the stimuli providing us with the most relevant information from the wide variety of stimuli all around us. The choice is not only governed by the conditions of each situation and by mechanical devices, but also by the driver's mental functions. Emotional and motivational states and experience in the task influence the driver's expectations and attention processes, which in some way control sensory processes and perception structure. The influence that these processes have on information processing was shown by Rumar (1982) in terms of perceptive, sensory, and cognitive filters.

A more recent model that can be applied to driving is Endsley's (1995). This author highlights the need to take into account the great variability of information being processed as a basic feature of driving. Environmental and task conditions change constantly while driving, resulting in drivers having to carry out continuous decision making based on those variable stimuli conditions. Situations have to be continuously assessed and immediate changes must be anticipated in order to make the right decisions. Endsley refers to this as *situation awareness*, defining it as the "perception of environmental elements in terms of time and spatial measurements, understanding their meaning and foreseeing their state in the immediate future" (1995, p. 36). According to his model, three different levels can be found in situation awareness (Figure 1.6). The first level refers to the perception of elements from the environment. A driver should above all perceive road, traffic, and vehicle conditions,

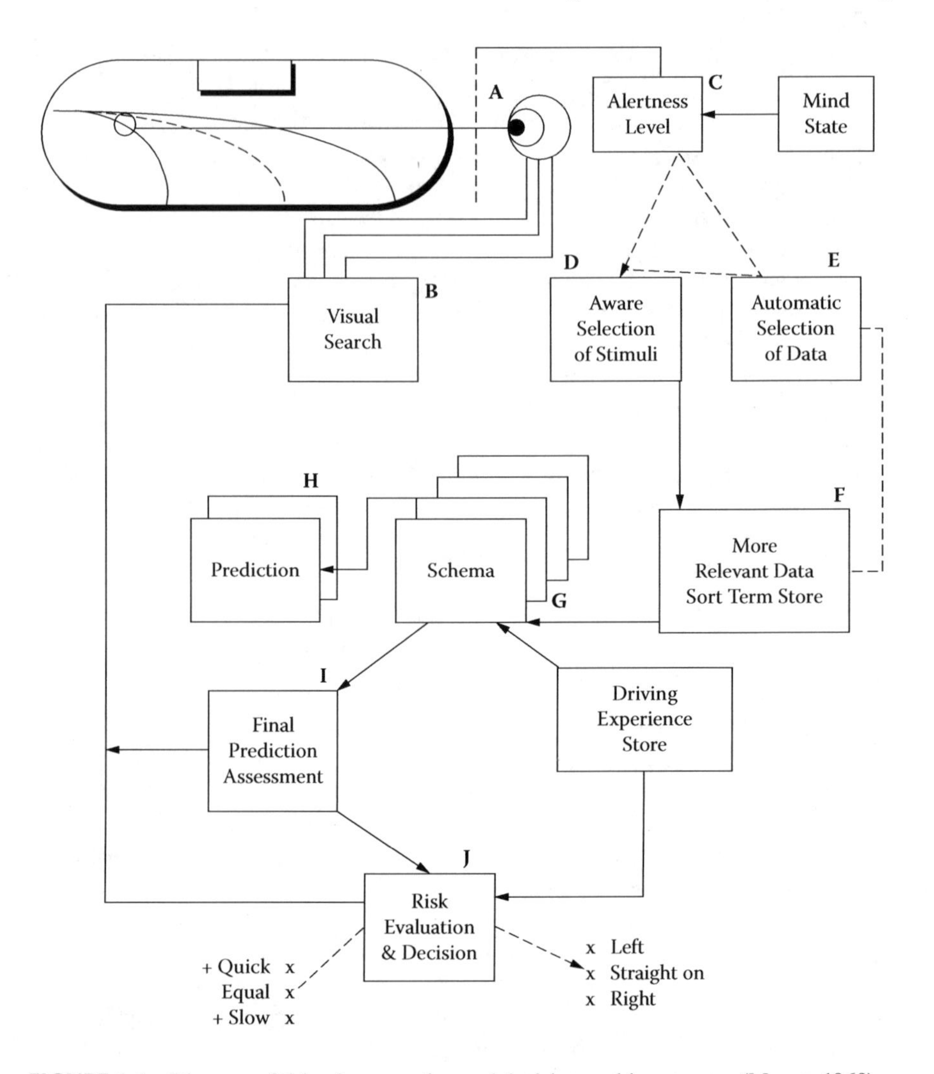

FIGURE 1.4 Diagram of driver's perception and decision-making process (Moore, 1969).

which are continuously changing. The second level is understanding the current situation. As is true of other facets of processing, it is not enough to just perceive these situations; they have to be given meaning. In the case of driving, it is not enough to simply perceive a change in the intensity of the rear red light of the car in front of us; you also have to know that this indicates the car is braking. The third level, foreseeing the near future, is the highest level component in the model. It is reached by knowing the state and dynamics of the elements around us and by integrating the two previous levels. For example, we have to be able to predict a driver's behavior when braking and make decisions about our behavior accordingly to avoid a potentially dangerous situation.

Being aware of a situation depends on much more than the simple perception of information from around us. It includes understanding what the information means,

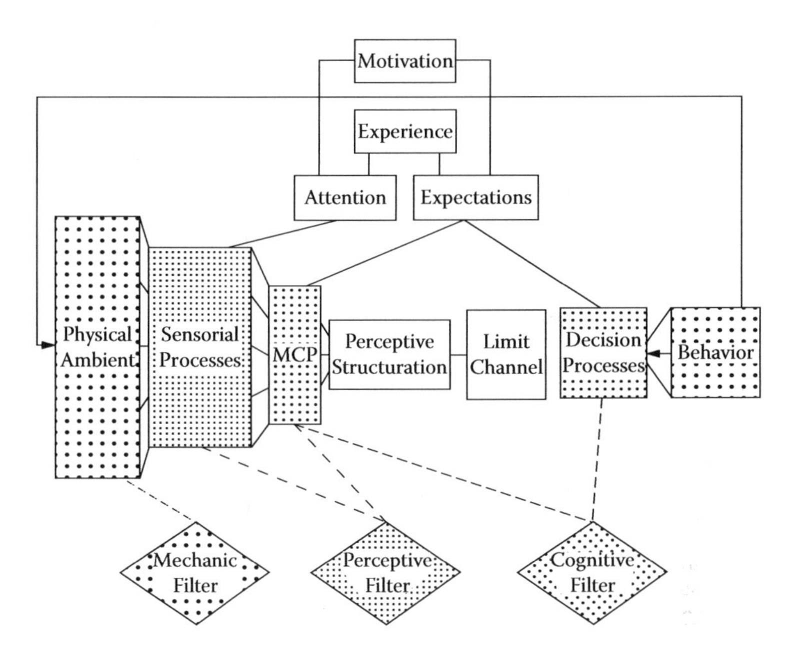

FIGURE 1.5 Diagram showing functions that determine driver information acquisition and processing (Rumar, 1982).

taking everything into account, comparing it with the driver's goals and providing information about possible future states of the surroundings, and considering all this in order to make decisions. At the same time, situation awareness is affected by other factors such as previous objectives, preconceptions and expectations, previous experience, and the driver's abilities.

1.3 METHODOLOGY USED TO STUDY DRIVING INFORMATION ACQUISITION

Seeing drivers as information processors highlights the importance of perceiving the whole traffic situation they find themselves in. Information acquisition while driving has been researched using different types of study: field research, laboratory research, and simulations; different measurements, such as eye movements, reaction time, accuracy, correct memory, and estimation errors; and different tasks, from complex ones such as actual driving to simple ones concerning perception ability or dual tasks. Because of this, results are not always easy to compare. In 1965, Häkkinen proposed a classification, also not exempt from overlapping, in which he included accident statistics, case studies, critical or conflictive situations, perception studies, interviews and questionnaires, experimental studies, theory and model development, and research studies. The group of studies quoted includes those dealing with perception, interviews, and questionnaires, and those experimental methods used most frequently for research into how drivers acquire information.

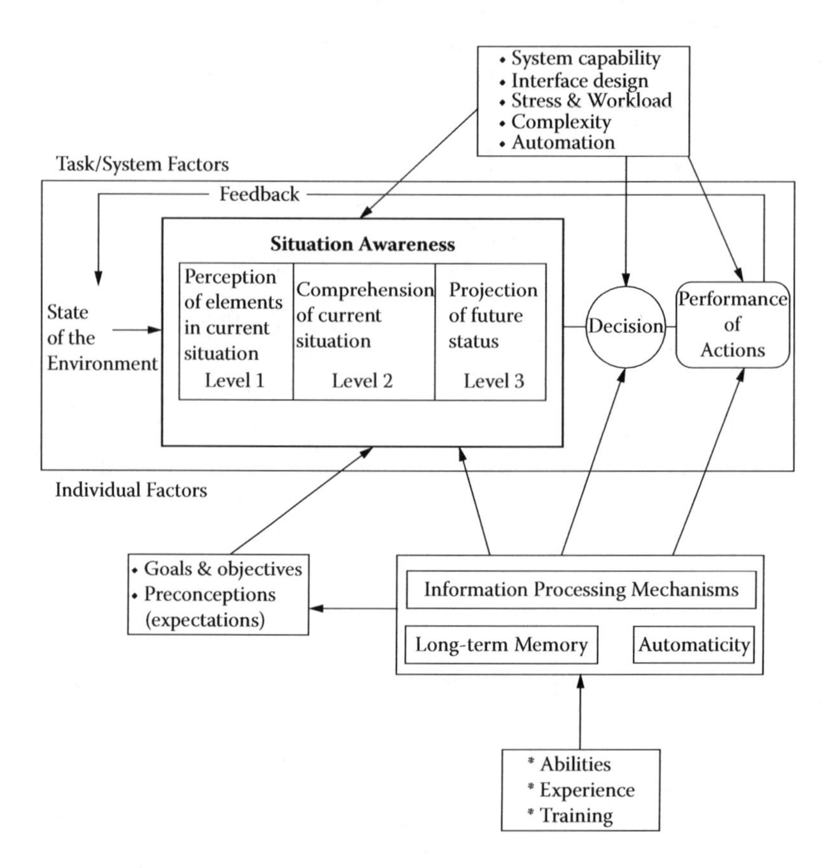

FIGURE 1.6 Model of situation awareness in dynamic decision making proposed by Endsley (1995).

Another classification, drawn up by Luoma (1986), emphasized the difference between two study types: field research and laboratory research. Both of these have advantages and disadvantages. In field research there is good external validity and it is easier to generalize the results, although it involves working in a situation with little control over strange variables. The case of laboratory studies is the opposite. One of the main disadvantages of laboratory studies is the difficulty of simulating situations realistically. De Waard and Brookhuis (1991) and Brookhuis and De Waard (1994) are in favor of using experimental situations in a real environment, for example, to assess changes in the driver's state due to the amount of time he or she has been driving or the effects on driving of external factors such as alcohol or certain drugs. Rothengatter (1997) considers that traffic psychology needs mostly field research and criticizes the use of laboratory experiments to provide information about the relationship between driver behavior and traffic accidents. Rothengatter defends the use of instruments that provide subjective information about the driver's psychological experiences.

Along the same lines, Recarte, Nunes, and Conchillo (1998) emphasized in their study of drivers' attention to traffic signs that laboratory research has focused on the structural and content conditions that signs need to fulfill in order to reach maximum information. However, the fact is that in real traffic contexts, attention span can be focused or divided and is certainly ever-changing making it difficult to generalize laboratory results in traffic systems, which is where they need to prove their effectiveness. In field research, this complexity of context is what makes it difficult to control the multiplicity of factors that can influence some results and consequently lead to an incorrect interpretation of them, especially if explanations about the causes are sought rather than just correlative data. However, some studies have produced similar results when carried out in the laboratory and in a real situation (Lajunen, Hakkarainen, and Summala, 1996). Some of these laboratory studies have aimed at understanding how drivers go about extracting information from traffic signs (Jacobs, Johnston, and Cole, 1975; Ells and Dewar, 1979; Whitaker and Sommer, 1986; Horberry, Halliday, Gale, and Miles, 1998; Castro and Martos, 1995, 1998; Wogalter, Kalsher, Frederick, Magurno, and Brewster, 1998; Luna-Blanco and Ruiz-Soler, 2001; Drakopoulos and Lyles, 2001, to mention just a few). For example, if we only focus on signs conveying danger, Wogalter and Laughery (1996) argue that controlled studies are useful to assess the effectiveness of different designs of danger signs before putting them on roads. The basic problem is certainly the complexity of traffic environments. If one wants to carry out research in real contexts, it is practically impossible to control all the factors affecting the process under scrutiny.

An option somewhere in the middle is to simulate driving, which is economically viable because it saves time and cost. Moreover, it also has some additional advantages. It is possible to predict the effects of a particular measure, or to easily repeat the conditions selected by the researcher and remain faithful to those conditions. Nevertheless, it is not exempt from disadvantages. One of them is the extent to which the simulated situation can be considered like the real one. The validity of a simulation can be assessed by looking at two basic correspondence criteria: similarity to the behavior in question and similarity to the physical situation. Psychologically speaking, it would suffice to obtain the first type of correspondence. However, many attempts have been made to produce excellent and very expensive physical simulations in which the behavioral side has been jeopardized. Another problem emerges when trying to simulate situations involving risk, such as traffic accidents. This, of course, renders simulations unviable if the aim is to achieve a high degree of realism.

1.4 DRIVING SIMULATORS

In the history of traffic psychology, many tasks, tests, and recording instruments have been designed, built, and used to collect objective information about vehicle control and relevant cognitive, motor, and physiological activity produced when driving. One of the oldest tasks designed for these purposes was the one developed at the beginning of the twentieth century by Münsterberg to assess the skills needed to drive a tram safely. According to this researcher, it was necessary to be able to deal continuously with the numerous objects the driver might come across en route and to anticipate both the speed and direction of the movements of these objects. For this purpose he designed some devices, described in detail in Caparrós (1985), by means of which he tried to schematically represent the driving task of a tram along a route where pedestrians and other vehicles could interfere. Serving as performance measurements of these tasks, the total time taken by the driver to perform them was obtained, as well as the number of times he ignored interfering stimuli along the route and the times he incorrectly estimated that a stimulus was interfering.

Experimental vehicles and simulators are tools we can use to obtain objective information in controlled conditions. The vehicles are apparently normal but have been adapted to record data to provide information about different aspects of the driving task. This kind of vehicle has been used in real contexts as well as in circuits closed off to traffic. Government agencies in charge of traffic and road safety usually have circuits available for research. In Spain, for example, the Traffic Department has one at the *Instituto Nacional de Técnica Aeroespacial* (INTA; Spanish National Institute for Aerospace Technology) in a banked ring shape, 455 m in radius and 2859 m long. Another example is the Satolas test circuit in Lyon, France, where research studies are carried out for the *Institut National de Recherche sur les Transports et leur Sécurité* (INRETS; French National Research Institute for Transport and Safety Research). Information about this circuit is available at http://web.inrets.fr/labos/equipement.e.html.

Driving simulators are machines with which individuals attempt to behave as though they were driving in a real context. The use of simulators for the psychological study of driving goes back as far as World War I. Around that time, Moerde and Piorkowski designed a simulator in Berlin to assess the participants' reactions to lit-up auditory stimuli similar to those in a real context (Caparrós, 1985). Consequently, those first simulators were basically a replica of the classic reaction time experiment model (Barjonet and Tortosa, 2000). Nowadays, driving simulators reproduce dynamic fields of vision. In other words, they can make static observers perceive a visual scene in the same way they would perceive it in motion. These simulators try to induce illusory sensations of movement in participants by showing sequences of visual scenes in which objects move following typical patterns of a vehicle in motion. They may have either a static or a mobile base.

Driving simulators have been used to study drivers' actions in different traffic situations, with the aim of clarifying this behavior and inferring the characteristics of the underlying psychological processes. Speed, accuracy, and adequacy in the tasks involved in driving are measured with various indicators. Apart from their use in research, driving simulators are used to teach the abilities and skills needed to drive—although there are less than twenty models on the market for this purpose—and, to a lesser extent, to assess drivers' skills.

In any case, the advantages of using a simulator are the elimination of the risk involved in real controlled driving and the possibility of replicating traffic situations accurately by controlling relevant conditions. These advantages are of particular interest to study the efficiency of strategies for improving road safety and to explore driving impairment as the result of different factors (medication, drugs, lack of sleep, thick fog, and so on).

The variables usually measured by these simulators are vehicle speed, vehicle position in relation to road markings, distance from the vehicle in front, angle of the steering wheel position, and amount of pressure applied to the brake pedal. From these data it is possible to obtain average indicators of vehicle control (operative level) and of the driver's ability to perform basic maneuvers (tactical level).

The use of simulators also has its disadvantages. Both the validity and reliability of these instruments have been questioned (Brown, 1997). The same observations had already been made by Münsterberg when he showed that these instruments were incapable of evoking the state of mind under study in the participants. The ideas, feelings, or choices that arise in real conditions cannot be generated by using mere reproductions of isolated events that happen in real contexts.

Given these limitations, the validity of driving simulators needs to be analyzed. This implies assessing the correspondence between driver behavior in the real context and behavior using a simulator (McLane and Wierwile, 1975; Blaauw, 1982; Reed and Green, 1999) comparing different types of measurement: (1) measurements of the performance of driving-related tasks (e.g., variability in the side position of the vehicle); (2) physiological measurements (e.g., heart rate); and (3) subjective measurements.

Despite the disadvantages of driving simulators, most traffic research centers have them available for use, which are satisfactory to a greater or lesser extent. These simulators are often the result of the joint collaboration of engineers, computer scientists, and electrical engineers. Among others, the following simulators stand out as worth mentioning: the one at the University of Leeds in the United Kingdom (http://www.its.leeds.ac.uk/facilities/uolds/index.php); the one at the Transport Research Laboratory also in the United Kingdom (http://www.trl.co.uk/content/main.asp?pid=70); the one at the University of Groningen in Holland (http://www.stsoftware.nl/simulator.html); and those referenced by INRETS (http://web.inrets.fr/ur/sara/Pg_simus_e.html) in France. There are excellent simulators in North America, for example, at the University of Michigan at the Transportation Research Institute (http://www.umich.edu/local/driving/sim.html). Others can be found in Australia at the University of Monash Accident Research Center (http://www.general.monash.edu.au/muarc), and in private centers, motoring property, and businesses.

1.5 RESEARCH METHODS

1.5.1 Observation

The main advantage of using observation as a method of exploring driving is that behavior can be studied and analyzed in a nonobtrusive way. The occurrence or not of a dependent variable—the individual's behavior—is recorded. This usually leads to results that are valid and can be generalized to the population the participants belong to. It is a relatively low-cost method. Its main disadvantage, however, is that the observers may be biased when running the tests or interpreting what they observe. Therefore, the interobserver reliability index should be calculated. Another limitation of this method is that the associations and conclusions established from the results can only be considered in a comparative way. For instance, it can only be stated that one driver runs red lights more than another driver, but we don't know the reason for this behavior. It is not possible to extract causality relations or correlations from the observation results. An example of the use of this methodology is Aberg's study (1995); the head movements of 1999 drivers were analyzed when they were approaching a level crossing with traffic lights, with the collaboration of specially trained observers for the job. With very different aims, Steyvers and De Waard (2000) video-recorded vehicles driving along country roads with little traffic to assess their lateral positions depending on whether there were road markings on the sides of the road.

1.5.2 Testing

This involves the use of at least two variables, an independent variable (IV) controlled by the researcher and a dependent variable (DV), which is the individual's behavior. The essence of this method consists of two additional characteristics: First, the IV must be controlled explicitly by the researcher, that is to say this variable can only take on values that have been preset by the researcher. Second, any other variables that may influence the person's behavior—the DV—must be controlled thoroughly.

An advantage of the experimental method is that it can establish causal connections between the IV and the DV. The IV is the cause of behavioral changes (DV), although the relation between cause and effect can only ever be probable because it is possible that not all the variables are controlled. An allowance is always made for a margin of error, although it is usually very small. In addition, this experimental method is economical. With observation and correlation methods, the researcher has to wait for the participant's behavior. In the experimental method, however, the participant is forced to carry out the task, that is, to receive different levels of the independent variable. This saves time and energy. To understand the disadvantages it is vital to first define the concepts of internal and external validity.

Internal validity is a function of the level of control over all the variables that are not of interest in a given experiment. It is an index of the degree of certainty with which it is possible to determine the causal relationship between the IV and the DV. External validity refers to the possibility of generalizing a causal relationship between the IV and the DV, that is, the data obtained. Generalizing means applying the causal relationship to participants other than those being studied.

Experimental methods have high internal validity but low external validity because restricted study conditions are a limiting factor when it comes to generalizing the results. The greater the control over the variables, the higher the internal validity but the lower the external validity.

Different attention tasks such as perceptive (mainly visual and auditory), memory, learning, and thinking tasks have been used to explore the understanding of meaning, free recall, cued recall, the relationship between associated pairs, other association tasks, the classification of signs according to whether they are understood to a greater or a lesser extent, the design or rough outline of the type of sign wanted, and an estimation of how familiar it is.

Sometimes the stimuli used are already part of the traffic context or of the vehicle. They are usually visual, such as traffic signs, vehicles, and other elements, especially when studying the attention paid to the traffic context, perception of the aforementioned stimuli, and the immediate recall of traffic signs. On other occasions, tasks are set up with additional stimuli not normally present in the traffic context or in

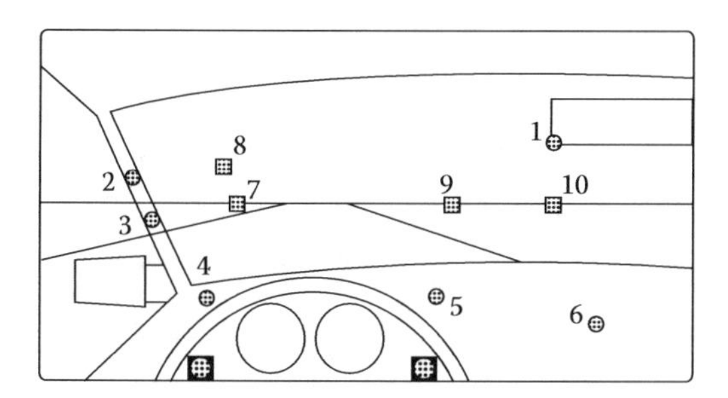

FIGURE 1.7 Descriptive diagram of stimuli introduced for the driver in the traffic environment itself or the vehicle. Adapted from Nunes and Recarte (2002).

the vehicle. These stimuli are as much visual (ranging from simple light flickering to graphics and written messages; see Figure 1.7 as an example) as auditory (from nonverbal sounds to verbal messages). Depending on their nature, added stimuli are introduced via earphones or small speakers installed in the vehicle, on small screens set into the dashboard within the driver's field of vision, or even by projecting the image on the front windscreen of the vehicle. For example, the driver's state of alertness to the traffic environment has been assessed in some studies by examining the way the driver detects sign tasks that are provided aurally.

Additional tasks to those inherently involved in driving are often introduced with the aim of finding out not so much how the driver carries out these extra tasks but rather how these tasks affect information processing and the driving task. These tasks and tests are usually introduced as secondary to the driving task. However, in some cases participants have been asked to pay full attention to them, obviously being careful with the experimental conditions so as to avoid accidents. Examples of this kind of study are those of Summala, Nieminen, and Punto (1996) and of Summala (1998), where participants were required to try to drive along a straight piece of road closed off to traffic without looking at the road, for as long as they could. They were asked to simultaneously carry out certain simple tasks, which required continuous visual attention to information provided for them in a device installed inside the vehicle, in different positions (on the dashboard above the radio, for example). The main purpose of this study was to find out if the drivers' experience showed significant differences when carrying out the task of steering the vehicle using their peripheral vision.

1.6 RECORDING TECHNIQUES: HOW TO MEASURE THE DRIVING DEPENDENT VARIABLE

1.6.1 Eye Movement Recording

One of the most common means to find out how drivers acquire information is eye movement recording, which provides information about visual search patterns. Movements are usually observed using automated recording instruments. Figure 1.8A and Figure 1.8B show examples of a recent recording system and an older one, respectively.

FIGURE 1.8A Adaptation of the current eye movement recording system used by Nunes and Recarte (2002).

FIGURE 1.8B Adaptation of the cumbersome eye movement recording systems used by Mourant and Rockwell (1970).

The study of visual search behavior requires sophisticated systems and powerful analysis instruments capable of processing large quantities of data obtained from each record, because our eyes move in a subtle and rapid way. Current technology has provided a great improvement in eye movement recording. Until not so long ago, this had to be done with cumbersome systems using helmets or glasses, which were uncomfortable for the observer, made movement almost impossible, and led to results that were not very reliable. This made it very difficult to record the driver's eye movements, which is why so few studies have recorded eye movements in real driving situations.

Over recent decades, however, it has become possible to record drivers' eye patterns in real situations without having to use cumbersome pieces of equipment and allowing virtually normal driving. The technology used can vary, but the one that is currently used most also takes advantage of the reflection of light rays off the cornea—a concave mirror in itself—without being intrusive, since the recording device is not placed in a helmet or glasses. Instead, a set of small devices is placed in front of the subject (on the dashboard when driving) and these record the reflection of light given off by the cornea.

This type of recording system has many advantages over the others. Drivers can drive normally toward their destination while the device records their eye movements, without the need to carry out any additional tasks. Attention is normally focused in the same direction as the eyes. The main advantage is that eye movements are relatively free of the bias caused by the instructions because they are involuntary to a large extent.

The disadvantage of these systems is that they provide information about eye fixation but not about subsequent processing of information. Moreover, the role played by peripheral vision in the recording is not clear. Another drawback is that only a small number of participants can be recorded. To avoid some of the criticism, the direction in which the eyes are looking and the field of vision explored by them are normally recorded simultaneously. The devices used include two information channels: one of them records the potential area perceived by the participant, usually by means of a very small camera attached to his or her head, and the other records the participant's eye movements.

Various other factors relating to traffic situations and the driver have also been observed using an eye movement recording system. Some studies have explored the kind of information that attracts the driver's attention, also recording the amount of time the eyes are fixed on the stimulus, how peripheral vision is used, the effects of a vehicle ahead, road and traffic conditions, oncoming bends, eye fixations on signs and other traffic or road devices, the effect of advertising boards on the perception of traffic signs, the effects of different light conditions and types of lighting, and so on. Other studies have compared the varying eye movement patterns of drivers involved in accidents to those of other drivers, looked at information search strategies used by experienced and new drivers, and examined the effects of fatigue, alcohol, and stress, among other influences. Other studies have recorded eye movements and taken other dependent measurements at the same time. For example, they have measured the speed of the vehicles while asking the drivers about the last speed limit sign they had seen and recorded eye fixations on the traffic signs. More ambitious studies have taken four measurements: eye fixations on the sign, head movements, sign recollection, and speed at the time of passing the sign.

The study of eye fixations and movements during driving started in the 1960s. Mourant and Rockwell (1970, 1972) were pioneers in these studies. They paid special attention to the characteristics of new drivers' visual exploration compared to that of experienced drivers. Until recently, eye movement recording involved great difficulties. One of them was that it was not possible to record an observer's eye fixations and movements reliably in a real driving context because the techniques required certain restrictions that were incompatible with normal driving—for example, limiting the light available, continuously pointing at or even placing devices near the eye (Miltenburg and Kuiken, 1990). Nowadays, however, there are recording devices that are barely intrusive and can record eye movements just by being placed on the dashboard.

At the same time, given that the driving environment can change suddenly, adequate techniques have to be available to record and analyze what the driver sees along the way. Above all, there have to be appropriate experimental paradigms in place to study the driver's behavior depending on the different factors affecting it. Along these lines, Crundall and Underwood (2001) and Crundall, Underwood and Chapman (1998) have tried to determine what they call the syntax or structure of the visual images occurring when driving a vehicle. In doing so, they try to minimize the difficulties encountered when attempting to relate the eye data with the visual information present when the data were obtained. They consider that in order to find out about the relevant elements and events in a visual traffic scene, they must determine the processing demands required by such a scene. They distinguish between visual and cognitive demands. In principle, the intensity of visual demands should depend on factors such as the density or complexity of the visual elements making up the scene—the geometry of the road or the kind of maneuver carried out—rather than the speed of the vehicle being driven. The intensity of cognitive demands, however, should depend on how relevant the scene is in the context we find ourselves in, and the amount of danger or risk perceived in the scene.

Geoffrey, Scialfa, Caird, and Graw (2001) also analyzed the effects of age and the accumulation of stimuli on the visual search of traffic signs inserted in digitized traffic scenes. They found a greater number of errors in older people and a decline in search efficiency as the number of stimuli in the scene increased. However, older people did not show a disproportionately poor performance compared to young people when faced with a high number of stimuli in the scene.

Underwood, Crundall, and Chapman (2002) also studied how selection comes about in visual search while driving. Their results show that experience plays an important part in detecting dangerous stimuli. They found that experienced drivers respond to dangerous stimuli appearing in the visual periphery with shorter but more frequent eye movements than inexperienced drivers.

Another recent study by Shinoda, Hayhoe, and Shrivastava (2001) looked at drivers' abilities to detect stop signs that are visible for short periods of time in a simulation task. They observed that detection during the task was modulated by the instructions and by the local visual context. Sign visibility requires active searching, and visual search patterns were found to be influenced by previous knowledge of the most likely structure of the traffic environment. All these studies demonstrate the usefulness of recoding eye movement as a modern approach to research perception and attention processes in driving in both simulated driving situations and real ones.

1.6.2 Drivers' Self-Assessments

Self-assessments and memory have also been used to explore how information is acquired from the traffic environment. Self-assessment consists of a verbal explanation given by drivers as to how they collect information, asking them to explain their choices. It is possible to obtain a structure of the general strategy followed by the driver by analyzing the kind of information searched for and the chronological sequence of the search prior to making the decision. Despite the great amount of criticism expressed about this kind of measurement, the data it provides were given greater value in the 1970s and 1980s. The problem is that verbalization involves other mechanisms of a cognitive nature, as well as perceptive ones. Given the level of automation of some of these mechanisms, this approach has serious methodological and practical limitations. In most cases it is only possible to obtain lists of relevant informative elements consciously sought out by the driver.

This category of driving observation tools generally includes questionnaires, inventory lists, scales, interviews, and driving records that are completed according to the information provided verbally by participants. These tools can provide information about the individuals' mental contents related to driving and their driving behavior. They are particularly useful to obtain information about drivers that is difficult to acquire through other techniques, either due to their essentially subjective character (e.g., opinions, feelings, etc.) or because they refer to past experiences (e.g., data on frequency of driving, routes usually taken, or traffic incidents in which the driver may have been involved, etc.) or because any other assessment technique might mean that the driver does not behave in a normal way (e.g., when it comes to recognizing violations). Some examples of inventory lists and questionnaires specifically drawn up to assess such aspects of driving are the Driver Skill Inventory (DSI; Lajunen and Summala, 1995), the Driving Behavior Inventory (DBI; Gulian, Matthews, Glendon, Davies, and Debney, 1989; Glendon, Dorn, Matthews, Gulian, Davies, and Debney, 1993), the Driver Behavior Questionnaire (DBQ; Reason, Manstead, Stradling, Baxter, and Campbell, 1990), and the Spanish *Inventario de Situaciones Ansiógenas en el Tráfico* (Inventory of Situations Producing Anxiety in Traffic [ISAT]; Carbonell, Bañuls, and Miguel-Tobal, 1995).

The main disadvantages of using the drivers' own reports are that the information collected may be biased by their motivations, that drivers may adjust to social desirability or to the evaluator's expectations, and that measurements derived from drivers' reports correspond in many cases to an ordinal scale, which limits the statistical treatment that can be applied to the data.

1.6.3 Memory

Memory is another measurement that is normally used to find out how information is acquired from traffic signs once the driver has passed them (Johansson and Rumar, 1966; Johansson and Blacklund, 1970; Sanderson, 1974; Aberg, 1981; Drory and Shinar, 1982; and Milosevic and Gajic, 1986). The most common procedure is to stop vehicles and ask the drivers to do a recall test on the traffic signs they have seen minutes before. The main criticism of this procedure is the inevitable time lapse between the time the drivers pass the sign and the memory test asking them about it. The possible influence of the drivers' feelings of insecurity and fear at having been stopped suddenly is also criticized. Finally, this kind of experiment loses the naturalness of the driving situation, which means that it is not possible to generalize the results to real contexts where the traffic situation takes place.

Another approach that is also based on the participant's memory is that of studies where the researcher rides in the vehicle with the driver and sporadically asks him or her about some of the traffic signs or other devices they have passed on the road. These studies can determine whether the drivers remember the signs well because the time lapse between passing the sign and asking the question about it is only seconds. Nevertheless, these experiments have to be very well planned because the participants pay more attention than usual to traffic control devices. Another disadvantage is that the number of participants used is necessarily small.

The way drivers acquire information has also been explored by studies in which participants are aware of the aims of the research, and the ease with which stimuli presented to the driver can be perceived is controlled or adapted. The physical characteristics of the stimuli appearing in the driving context are deliberately explored. These studies have analyzed the conspicuity of signs. We can find earlier examples of this concept in Michaud's assertions (1985) when he stated that stimuli with great ease of perception or of being noticed are those that have a high probability of being seen in very short observation times. Also, Cole and Jenkins (1982) defined *conspicuity of a stimulus* as the characteristic of objects that allows them to be seen or noticed easily.

Previous research has shown that various physical parameters are important in determining obviousness, for example, color, amount, size, shape, complexity of context or characteristics of neighboring objects, contrast between the object and the context, eccentricity of the stimulus or angle of the sign and the participant's line of vision, and the frame of the internal structure of the object.

Many studies have explored how different types of traffic situation-related stimuli are perceived and traffic signs are one of the most researched subjects. This insistence can be explained by noting that these devices are used to anticipate information that drivers will be faced with shortly and therefore allow correct maneuvers to be carried out. However, the conclusions reached in this area are by no means satisfactory. The results were applied to driving too soon and, in many cases, the differences between the ease with which signs can be perceived and their intelligibility have not been established.

1.6.4 Sign Recognition

Johansson and Rumar (1966) and Summala and Näätänan (1974) studied traffic signs by having drivers identify signs when traveling in the vehicle as copilots. They recorded a 97% success rate in identifying signs correctly. However, this method is questionable because it is not possible to generalize the results to the normal driving situation.

1.6.5 Traffic Accidents

Given that the causes of traffic accidents are varied and diverse, it is difficult to pin the blame on signs. However, the effectiveness of a sign can be assessed by comparing the number of accidents that happen before and after the sign has been placed is measured (e.g., where accidents often occur). A different example that also used data on the number of accidents is the recent study by Langham et al. (2002) described earlier. It analyzed accident reports involving a collision with a very conspicuous car—a police car—parked on the hard shoulder in order to explore the nature of "looked but failed to see" accidents.

1.6.6 Reaction Time and Other Measurements of Driving Performance

Reaction time (RT) measurements have been widely used in field studies to draw conclusions about drivers' perception of devices in the vehicle or on the road. Their advantage is that there is no time lapse in the experiment. A large number of participants can also be used in each study. Moreover, these measurements play an important role in real driving. However, they also have disadvantages, such as the fact that drivers can react to different stimuli from inside or outside the vehicle in many ways by reducing speed, increasing their activation level, changing their eye movement pattern, and so on. Many of these reactions, with the exception of vehicle speed, are difficult to measure.

Reaction time has also been the dependent variable in many laboratory experiments. Participants are shown drawings, photographs, scenes, or series of drawings and carry out detection, discrimination, and search tasks, among others. The aim of most of these studies has been to analyze the effects of sign details on how easily they are perceived; to compare verbal signs to those using symbols; to explore the effect of quantity and quality of signs on how well they are perceived; to study the possible effect on signs of advertising boards near them; to compare the ease with which alternative signs can be perceived; and to study the effects of different light conditions on sign perception, the simultaneous perception of several stimuli, and the placement of signs within the peripheral area of the field of vision or the effectiveness of horizontal signs, among others. There has also been an analysis of how easily elements inside the vehicle are perceived, such as indicators and needles on the dashboard, new navigation systems, mobile telephones, and indicators placed in the area above the windscreen (windshield).

However, perception as an aim is only the starting point of the driving activity, which begins with incoming information. Even when a secondary task is performed, we cannot talk about simulation because the activity carried out by the participant differs from that of the driver in terms of knowledge of the objectives of the activity. The need to carry out these tasks regarding ease of perception alongside a second task as it happens in real driving is debatable. It is therefore necessary to interpret these results with certain constraints.

Although inherently perceptive aspects of stimuli relevant to the traffic situation have been highlighted—especially regarding traffic signs—the ease with which they can be understood, that is, their intelligibility, has often been studied. Emphasis has been placed on aspects affecting a sign's drawing or design recognition and its clarity, which have an impact on the interpretation of the message. Throughout their study, ease of perception and understanding have not been easy to differentiate. For example, both have been studied measuring reaction time as the dependent variable. These measurements are undoubtedly influenced by one another, which makes it difficult to separate these two factors.

Assuming that the main purpose of traffic signs (especially danger signs) is to induce safe behavior in the driver, obtaining measurements of behavioral responses to traffic signs is an enterprise of great value. This is especially true when measurement of these driver responses is combined with other techniques aimed at discovering the psychological processes that lead to such behavior (questions related to memory, for example). The most important result found by Häkkinen (1965) was that three in every four drivers complied with the message conveyed by the sign and remembered it correctly. More recently, however, Fisher (1992) found considerable incongruence between verbal memory responses and driver behavior. The method used was to hitchhike and ask the drivers about the two previous signs they had passed. He found that memory of the signs was generally poor. Moreover, he concluded that besides studying the capacity of the signs to produce vivid memory prints in drivers, their capacity to sensitize drivers to the danger should be explored as well.

Summala and Hietamäki (1984) assessed changes in speed with different sign conditions and found significant effects of both the "obviousness of the sign" and its content. Specifically, danger signs were more effective when they included a flashing light and danger signs were more effective than indication signs.

All this research shows that motivation factors that lead to sign perception should be taken into account in the design and implementation of traffic signs. In fact, MacDonald and Hoffmann (1991) pointed out that the most important factor influencing sign recognition is the potential action of the sign, potential action being defined as the probability that the driver will have to show a response related to the message supplied by the sign.

1.6.7 Psychophysiological Techniques

Psychophysiological techniques, widely used to study human behavior in general, have also been used to study information acquisition in driving. Human reactions can be measured with neuropsychological systems such as electrooculography, electroencephalography, electromyography, electrocardiography, heart rate record, evoked potentials, body temperature, electrodermal activity, and so on. In other words, practically all the methods used in psychophysiological research have also been used to assess drivers' brain activity, their attention response toward stimuli in the traffic environment, mental effort invested in the driving task, level of anxiety at any particular moment, and the intensity of their emotional reaction to certain stimuli, among other aspects. The drawback of these measurements is that they are easier to take and more reliable in laboratory conditions. In fact, some of these methods—recording evoked potentials, for example—have been dismissed very quickly because of the susceptibility of the devices in such a complex stimulus environment as the traffic environment (De Waard and Brookhuis, 1997). Others, however, have been widely used. Such is the case with electroencephalography, especially in studies on hypervigilance and drowsiness in driving; and electrocardiography, to obtain measurements of the mental effort needed to drive in studies about introducing new technologies in vehicles or the traffic environment and studies on stress in driving. As usual, when it comes to deciding which method, kind of study, or means should be selected, the conclusion that can be drawn from this sketchy display of pros and cons is undoubtedly that the choice should be made according to the nature of the problem to be studied.

1.7 ADAPTATION TO DRIVING: A CHALLENGE OVERCOME OR STILL A CHALLENGE?

Over the last few centuries we human beings have abandoned our ecological environments and have begun to take up artificial activities, such as driving cars in surroundings that are also not natural, such as modern roads. In these cases, our perceptual system has become partially obsolete. Typical examples of the lack of an adequate human adaptation to these environments are the delay or even failure in detection of vehicles coming into the peripheral field of vision during night driving.

As early as 1938, Gibson and Crooks likened driving to any other activity involving locomotion, such as walking or running, although driving requires an instrument (a vehicle) to carry it out. Any type of locomotion implies avoiding obstacles to prevent collision. Locomotion is steered through vision, which takes individuals along a route within their field of vision to avoid obstacles and reach their destination. This is the same for any type of locomotion: for a child beginning to walk, a footballer who starts running, or a driver starting off on a journey, for example.

If we make a comparison between the different ways man has of moving from one place to another, driving or walking, we obtain an example of the importance of relevant information acquisition. Driving and walking differ more in quantitative terms than qualitative ones. In both cases we make similar use of sensory information, mainly visual, in order to go on our way avoiding other living beings and inert objects. However, driving and walking take place in different surroundings. Let us compare, for example, the movement of someone walking in the country with that of a car along a road. The wealth of stimuli and redundancy is much greater in the country. We probably need that redundancy to make correct decisions and the mistakes made in traffic accidents probably lie in providing such relevant information by means of devices or artifacts, such as traffic signs, which provide questionable information (Rumar, 1990).

The driving task has characteristics that make it difficult to perform. Without a doubt, the great speed at which we drive implies that visual stimulation only makes an impression on our retinas for a very short time. The immediate consequence of this is the loss of information redundancy. At the same time, a large amount of driving takes place in conditions of diminished natural light, as is the case of night driving, driving at dusk or dawn, or in adverse weather conditions. Therefore, driving is an unnatural activity to which man is still adapting, and so deserves a more careful study.

One of the determining factors of these differences between walking in the country and driving along a road is therefore the speed we reach while carrying out these activities. If we ran at the same speed at which we drive, we would probably have the same problems acquiring visual information in an adequate way. Speed is the key factor that makes it necessary to put into effect better traffic systems that are visible at great speed. Among alternative improvements emerging in an attempt to reach this objective, we can mention the following: making drivers aware of the speed at which they are driving, increasing the number of traffic signs, repeating and anticipating signs, and emphasizing the contrast, frame, and so forth of signs. All these devices help save time and anticipate and accelerate perception while carrying out a conduct—driving—in which a millisecond can help in avoiding an accident, lessen an accident's effects, or simply lead to making the right maneuver for the current traffic conditions at the right time. The difference in milliseconds is not only a determining factor in the length of one process or another but takes on an immediate application—it becomes just the right and necessary amount of time to carry out other parallel activities (changing gears, turning, etc.), which can be of vital importance.

Another important factor hampering drivers from obtaining adequate information is the lack of light during night driving. Man is a creature who naturally lives during the daytime. The human visual system is not suited to night vision and therefore any activity we carry out at night is necessarily impaired and made worse, especially if we are doing it at high speed. These days many journeys or parts of them are done at night when lighting levels on the road or from the vehicle are only a fraction (1/10,000) of daylight levels. This means that receptors in the retina have to work at levels for which they are not developed and put up with changes in lighting for which they are not adapted, such as the frequent occurrence of being dazzled by other headlights. At the same time, the lack of light at night impedes the depth of vision that is so useful during daytime driving and that allows our visual system to estimate the distance between us and other cars and elements on the road. At night we cannot make adequate use of key elements such as overlapping, degree of texture, parallax movement, aerial perspective, or shadow because the lights on our vehicle only light up a small section of our field of vision, leaving in darkness all the horizon and peripheral vision, which is so important for detecting movement. In these poor lighting conditions, only the information provided by other vehicles' headlights can give us some point of reference to estimate the distance between them and us. We are undoubtedly requiring from our perceptual system too much adaptation in too little time, without knowing if this adaptation we are demanding will one day be possible.

REFERENCES

Aberg, L. (1981). *The Human Factor in Game-Vehicle Accidents: A Study of Drivers' Information Acquisition* (Studia Psychological Upsaliensa, 6). Uppsala, Sweden: University of Uppsala.

Barjonet, P.E., and Tortosa, F. (2000). Transport Psychology in Europe: A Historical and General Overview. In P.E. Barjonet (Ed.), *Traffic Psychology Today* (pp. 13–30). Amsterdam: Elsevier.

Blaauw, G.J. (1982). Driving experience and task demands in simulator and instrumented car: A validation study. *Human Factors, 24,* 473–486.

Brookhuis, K.A., and De Waard, D. (1994). Measuring driving performance by car-following in traffic. *Ergonomics, 37*(3), 427–434.

Brown, I. (1997). Highway hypnosis: Implications for road traffic researchers and practitioners. In A.G. Gale, C.M. Haslegrave, I. Moorhead, and S.P. Taylor (Eds.), *Vision in Vehicles III.* Amsterdam: North-Holland.

Caparrós, A. (1985). Aspectos históricos de la Psicología aplicada a la conducción. In *Primera Reunión Internacional de Psicología del Tráfico y la Seguridad Vial.* Madrid: DGT.

Carbonell, E., Bañuls, R., and Miguel-Tobal, J.J. (1995). El ambiente de tráfico como generador de ansiedad en el conductor: Inventario de situaciones ansiógenas en el tráfico (ISAT). *Anuario de Psicología, 65,* 165–183.

Castro, C., Horberry, H., Tornay, F.J., Martínez, C., and Martos, F.J. (2003). Efectos de facilitación de repetición y semántica en el reconocimiento de señales de indicación y peligro. *Cognitiva, 15(1),* 19–23.

Castro, C., and Martos, F.J. (1995). Effect of the size, number and position of traffic signs mounted on the same post. *Scientific Contributions to General Psychology: Perceiving and Imaging the Space, 13,* 89–100.

Castro, C., and Martos, F.J. (1998). Effect of background complexity in perception of traffic signs: The distracting effect of advertisements in the proximity of the sign. *General Psychology,* Mental Architectures, 143–153.

Cole, B.L., and Jenkins, S.E. (1982). Conspicuity of traffic control devices. Australian Road Research, 12, 221-238. Citado en: Cole, B.L. y Jenkins, S.E. (1984). The effect of variability of background elements on the conspicuity of objects. *Vision Research, 24,* 261–270.

Crundall, D.E., and Underwood, G. (2001). The priming function of road signs. *Transportation Research Part F: Traffic Cognitive Psychology, 16*(4), 459–475.

Crundall, D.E., Underwood, G., and Chapman, P.R. (1998). How much do novice drivers see? The effects of demand on visual search strategies in novice and experienced drivers. In G. Underwood (Ed.), *Eye Guidance in Reading and Scene Perception* (pp. 395–417). Oxford: Elsevier.

De Waard, D., and Brookhuis, K.A. (1991). Assessing driver status: A demonstration experiment on the road. *Accident Analysis and Prevention, 23*(4), 297–307.

De Waard, D., and Brookhuis, K.A. (1997). On the measurement of driver mental workload. In T. Rothengatter and E. Carbonell (Eds.), *Traffic and Transport Psychology: Theory and Application* (pp. 161–171). Amsterdam: Pergamon.

Drakopoulos, A., and Lyles, R.W. (2001). An evaluation of age effects on driver comprehension of flashing traffic signal indications using multivariate multiple response analysis of variance models. *Journal of Safety Research, 32*(1), 85–116.

Drory, A., and Shinar, D. (1982). The effect of roadway environment and fatigue on sign perception. *Journal of Safety Research, 21,* 25–32.

Ells, J.G., and Dewar, R.E. (1979). Rapid comprehension of verbal and symbolic traffic sign messages. *Human Factors, 21*(2), 161–168.

Endsley, M.R. (1995). Towards a theory of situation awareness in dynamic systems. *Human Factors, 37,* 32–64.

Fell, J.C. (1976). A motor vehicle accident causal system: The human element. *Human Factors, 18,* 85–94.

Fisher, J. (1992). Testing the effect of road traffic signs' information value on driver behavior. *Human Factors, 34*(2), 231–237.

Fitts, P.M., and Posner, M.I. (1968). *El rendimiento humano.* Alcoy, Spain: Marfil.

Geoffrey, H., Scialfa, C.T., Caird, J.K., and Graw, T. (2001). Visual search for traffic signs: The effects of clutter, luminance, and ageing. *Human Factors, 43*(2), 194–207.

Gibson, J.J. (1986). *The Ecological Approach to Visual Perception.* Hillsdale, NJ: Lawrence Erlbaum Associates. (Original work published 1979)

Gibson, J.J., and Crooks, L.E. (1938). A theoretical field-analysis of automobile-driving. *The American Journal of Psychology, 51*(3), 453–471.

Glendon, A.I., Dorn, L., Matthews, G., Gulian, E., Davies, D.R., and Debney, L.M. (1993). Reliability of the driving behavior inventory. *Ergonomics, 36,* 727–735.

Gulian, E., Matthews, G., Glendon, A.I., Davies, D.R., and Debney, L.M. (1989). Dimensions of driver stress. *Ergonomics, 32,* 585–602.

Häkkinen, S. (1965). *Perception of highway traffic signs.* Report 1. Helsinki: TAIJA.

Hills, B.L. (1980). Vision, visibility and perception in driving. *Perception, 9,* 183–216.

Horberry, T.J., Halliday, M., Gale, A.G., and Miles, J. (1998). Road signs and markings for railway bridges: Development and evaluation. In A.G. Gale, I.D. Brown, C.M. Haslegrave, and S.P. Taylor (Eds.), *Vision in Vehicles, VI.* Amsterdam: Elsevier.

Jacobs, R.J., Johnston, A.W., and Cole, B.L. (1975). The visibility of alphabetic and symbolic traffic signs. *Australian Road Research, 5*(7), 68–73.

Johansson, G., and Backlund, F. (1970). Drivers and road signs. *Ergonomics, 13,* 749–759.

Johansson, G., and Rumar, K. (1966). Drivers and road signs: A preliminary investigation of the capacity of car drivers to get information from road signs. *Ergonomics, 9,* 57–62.

Lajunen, T., Hakkarainen, P., and Summala, H. (1996). The ergonomics of road signs: Explicit and embedded speed limits. *Ergonomics, 39*(8), 1069–1083.

Lajunen, T., and Summala, H. (1995). Driving experience, personality and skill and safety-motive dimensions in drivers' self-assessments. *Personality and Individual Differences, 19,* 307–318.

Langham, M., Hole, G., Edwards, J., and O'Neil, C. (2002). An analysis of "looking but failed to see" accidents involving parked police cars. *Ergonomics, 45*(3), 167–185.

Luna-Blanco, R., and Ruiz-Soler, M. (2001). Within factors involved in the vertical signaling perception: Holistic processing vs. analytical. *Psicothema, 13*(1), 141–146.

Luoma, J. (1986). *The acquisition of visual information by the driver: Interaction of relevant and irrelevant information.* Report 32. Helsinki: The Central Organization for Traffic Safety.

MacDonald, W.A., and Hoffmann, E.R. (1991). Drivers' awareness of traffic sign information. *Ergonomics, 34*(5), 585–612.

McLane, R.C., and Wierwile, W.W. (1975). The influence of motion and audio cues on driver performance in an automobile simulator. *Human Factors, 17,* 488–501.

Michaud, G. (1985). Los factores perceptivos en la conducción. *Primera Reunión Internacional de Psicología de Tráfico y Seguridad Vial* (pp. 219–243). Madrid: DGT.

Miller, G.A. (1956). The magical number seven, plus or minus two: Some limits on our capacity for processing information. *The Psychological Review, 63, 81–97.*

Milosevic, S., and Gajic, R. (1986). Presentation factors and driver characteristics affecting road-sign registration. *Ergonomics, 29,* 325–335.

Miltenburg, P.G.M., and Kuiken, M.J. (1990). *The Effect of Driving Experience on Visual Search Strategies: Result of a Laboratory Experiment.* Haren, Groningen, Netherlands: Rijksuniversiteit Groningen.

Moore, R.L. (1969). Some human factors affecting the design of vehicles and roads. *Journal of the Institute of Highway Engineers, 16,* 13–22.

Mourant, R., and Rockwell, T. (1970). Mapping eye-movement patterns to the visual scene in driving: An exploratory study. *Human Factors, 12*(1), 81–87.

Mourant, R., and Rockwell, T. (1972). Strategies of visual search by novice and experienced drivers. *Human Factors, 14*(4), 325–335.

Olson, P.L. (1993). Vision and perception. In B. Peacok and W. Karwoski (Eds.), *Automotive Ergonomics.* London: Taylor & Francis.

Reason, J.T., Manstead, A.S.R., Stradling, S.G., Baxter J.S., and Campbell, K.A. (1990). Errors and violations on the roads: A real distinction? *Ergonomics, 33*(10/11), 1315–1332.

Recarte, M.A., Nunes, L.M., and Conchillo, A. (1998). *Atención a señales de tráfico.* Comunicación en el II Congreso Iberoamericano de Psicología (published on the conference CD). Madrid.

Reed, M.P., and Green, P.A. (1999). Comparison of driving performance on-road and in a low-cost simulator using a concurrent telephone dialing task. *Ergonomics, 42*(8), 1015–1037.

Rothengatter, J.A. (1997). Psychological aspects of road user behavior. *Applied Psychology: An International Review, 46*(3), 223–234.

Rumar, K. (1982). The human factor of road safety. *ARRB Proceedings, 11*(1), 63–80.

Rumar, K. (1990). The basic driver error: Late detection. *Ergonomics, 33,* 1281–1289.

Sanderson, J.E. (1974). *Driver recall of roadside signs.* Traffic Research Report, 1. Ministry of Transport. New Zealand: Wellington.

Shanon, C.E., and Weaver, W. (1949). *The Mathematical Theory of Communications.* Urbana, IL: University of Illinois Press.

Shinoda, H., Hayhoe, M., and Shrivastava, A. (2001). What controls attention in natural environments? *Vision Research, 41*(25-26), 3535–3545.

Sivak, M. (1996). The information that drivers use: Is it indeed 90 percent visual? *Perception, 25,* 1081–1089.

Steyvers, F.J.J.M., and De Waard, D. (2000). Road-edge delineation in rural areas: Effects on driving behavior. *Ergonomics, 43*(2), 223–238.

Summala, H. (1998). Forced peripheral vision driving paradigm: Evidence for the hypothesis that car drivers learn to keep in lane with peripheral vision. In A.G. Gale, I.D. Brown, C.M. Haslegrave, and S.P. Taylor (Eds.), *Vision in Vehicles, VI* (pp. 51–60), Amsterdam: Elsevier.

Summala, H., and Hietamäki, J. (1984). Drivers' immediate responses to traffic signs. *Ergonomics, 27,* 205–216.

Summala, H., and Näätänan, R. (1974). Perception of highway signs and motivation. *Journal of Safety Research, 6,* 150–154.

Summala, H., Nieminen, T., and Punto, M. (1996). Maintaining lane position with peripheral vision during in-vehicle tasks. *Human Factors, 38*(3), 442–451.

Tanner, W.P., Jr., and Swets, J.A. (1954). A decision making theory of visual detection. *Psychological Review, 61*, 401–409.

Underwood, G., Crundall, D., and Chapman, P. (2002). Selective searching while driving: The role of experience in hazard detection and general surveillance. *Ergonomics, 45*(1), 1–12.

Whitaker, L.A., and Sommer, R. (1986). Perception of traffic guidance signs conflicting symbolic and direction information. *Ergonomics, 29,* 699–711.

Wogalter, M.S., and Laughery, K.R. (1996). Warning sign and label effectiveness. *Current Directions in Psychological Science, 5*(2), 33–37.

Wogalter, M.S., Kalsher, M.J., Frederick, L.J., Magurno, A.B., and Brewster, B.M. (1998). Hazard level perceptions of warning components and configurations. *International Journal of Cognitive Ergonomics, 2*(1-2), 123–143.

2 Visual Requirements of Vehicular Guidance

Frank Schieber, Ben Schlorholtz,
and Robert McCall

CONTENTS

Reflection 31
2.1 Introduction 32
2.2 Theoretical Framework for Linking Vision and Driving 32
2.2.1 Two Modes of Visual Processing: The Ambient–Focal Dichotomy 33
2.2.2 Properties of Ambient (Dorsal) versus Focal (Ventral) Vision 34
2.2.3 A Two-Level Model of Driver Steering 35
2.2.4 Diagnostic Signature of Ambient versus Focal Mediators of Steering Behavior 36
2.3 Empirical Evidence for Ambient–Focal Mechanisms of Steering 37
2.3.1 Experimental Reductions of Visual Acuity 37
2.3.2 Experimental Reductions of Roadway Luminance 40
2.3.3 Experimental Restrictions of Driver's Field of View 42
2.3.4 Experimental Reductions of Roadway Preview Distance/Time 44
2.4 Heuristic Value of the Ambient–Focal Framework 45
2.5 Conclusions: Visual Requirements of Vehicular Guidance 47
References 48

REFLECTION

We are interested in the very broad, yet deep, questions regarding the minimal visual requirements of driving an automobile. How does one begin such an enormous effort? Like many others, our search starts with a consideration of the evolutionary history of our species. Obviously, it makes no sense to ask how the demands of driving have influenced the evolutionary development of our visual system since the automobile has been in use for only a century. However, when we consider driving as "locomotion via technology" (to paraphrase a 1938 paper by Gibson and Crooks), we instantly arrive at a proposition that links the abilities of today's drivers to the contributions of the thousands of generations that have come before us. We can now easily imagine how our visual systems might have evolved in order to meet the demands of terrestrial navigation and, thus, the driving task itself. Given the context of "driving as terrestrial locomotion" the ambient–focal dichotomization of the visual brain appears to us, and many others, to represent a natural construct upon which to build a visual theory of the driving task. This focus on the ambient–focal dichotomy also provides a powerful mechanism for conceptualizing another major interest of our laboratory; namely, the emerging visual information processing problems of older drivers. Presently, we believe that the proclivity of older drivers to suffer the now infamous "looked but didn't see" type automobile crash to be a manifestation of a diminished efficiency of the ambient visual system's ability to preemptively alert the focal systems as to the occurrence and general location of significant events in the peripheral field of view. Yet, the validity of this proposition—which we have named the *ambient insufficiency hypothesis* of visual aging—has yet to be rigorously tested. This, we hope, shall occupy the pages of book chapters to be published in the not too distant future.

2.1 INTRODUCTION

What are the visual requirements of driving? On one level, this is both an interesting and important question. However, on another level, it is simply too broad of a question to be answered given the current state of our knowledge. Instead, a simpler question will be addressed in this chapter; namely, what are the visual requirements of vehicular guidance. Even this question is not easy to answer given the current state of knowledge. To begin to do so, one needs both a theoretical framework that links observable aspects of steering performance to the rich database of contemporary vision science as well as a family of experimental protocols that can be used to test and refine the theory.

This chapter begins by introducing the reader to the ambient–focal dichotomization of visual functioning and shows how this approach provides a heuristic for contextualizing steering behavior within the domain of neurophysiological and psychophysical vision science. Next, a series of experimental studies that directly and indirectly support the validity of this theoretical framework will be explored. Finally, conclusions regarding current support for the ambient–focal heuristic are summarized together with some speculation regarding future research directions aimed at employing this framework to improve our understanding of visually guided driving behavior.

2.2 THEORETICAL FRAMEWORK FOR LINKING VISION AND DRIVING

Herschel Leibowitz and his colleagues have developed a linking hypothesis that provides a powerful heuristic for conceptualizing vehicle guidance behavior within the rich domain of vision science (see Leibowitz and Owens, 1977; Leibowitz, Owens, and Post, 1982; Owens and Tyrrell, 1999; Andre, Owens, and Harvey, 2002). This heuristic is based upon an anatomical and functional dichotomization of the visual system into two parallel streams of processing, which have been labeled the *ambient* and *focal* subsystems. In order to more fully appreciate the potential utility of this ambient–focal dichotomy for driving research, it is first necessary to consider the origin and characteristics of this functional approach to describing the visual system.

2.2.1 Two Modes of Visual Processing: The Ambient–Focal Dichotomy

The proposition that visual processing proceeds along two parallel streams—one specialized for visual guidance through the environment (ambient system) and the other subsuming the functions of object recognition and identification (focal system)—can be traced back at least as far as the late 1960s. Numerous studies during this period demonstrated that visual functions related to spatial orientation (i.e., those required to answer the question "Where am I?") appeared to be heavily reliant upon subcortical pathways in the brain; while visual functions related to object recognition (i.e., those required to answer the question "What is it?") relied more heavily upon cortical levels of processing. The most direct illustration of this functional and anatomical dissociation of the processing of *what* versus *where* information can be found in the work of Schneider (1967; 1969). He trained hamsters to successfully perform a visual pattern discrimination task in order to achieve a food reward. When the primary visual cortex was surgically damaged in a subgroup of these mammals, Schneider noted that the animals maintained their ability to visually orient within the experimental apparatus despite the fact that they completely lost their ability to perform the visual form discrimination task. In another subgroup of these mammals, he surgically destroyed the subcortical visual pathway involving the superior colliculus while leaving the primary visual cortex intact. These animals lost the ability to visually guide their behavior in the experimental apparatus yet maintained the ability to perform the visual form discrimination task. Schneider had demonstrated a double dissociation between what he termed the *what* and *where* modes of the visual system. Around this same time, other researchers demonstrated a similar anatomical and functional dissociation within the visual systems of fish and amphibians (Ingle, 1967, 1973) as well as cats, monkeys, and humans (Held, 1968, 1970; Trevarthen, 1968). In fact, the terms *ambient* and *focal* (to denote the where and what visual subsystems, respectfully) were originally coined by Trevarthen (1968) who discovered a dissociation between the "vision of space" around the body and the "vision of things" within the environment based upon his work with "split-brain" monkeys.

The classical work, described above, attributed ambient (where) vision to subcortical pathways, while focal (what) vision was thought to be mediated by cortical pathways. However, more recent investigations suggest that a what–where functional dichotomy also exists in two anatomically distinct cortical pathways in the primate. Ungerleider and Mishkin (1982) identified two such pathways and called them the dorsal stream and ventral stream, respectively. Their dorsal stream interconnects the striate (primary) visual cortex, prestriate, and inferior parietal areas, and enables visual location behavior. Their ventral stream, on the other hand, interconnects the striate, prestriate, and inferior temporal areas, and enables the visual identification of objects (see Mishkin, Ungerleider, and Macko, 1983). Norman (2002) has provided a detailed history of the evolution of the ambient–focal construct and its subsequent augmentation by the cortico–centric dorsal–ventral stream dichotomization of visual system along both anatomical and functional lines. He notes one final extension of the dorsal–ventral dichotomy based upon the work of Goodale and Milner (1992). This involves a subtle yet important modification of the allocation of the where function. For Goodale and Milner, the dorsal stream supports the visual control of guidance and motor behavior. As such, the dorsal stream operates using a framework rooted in egocentric coordinates which allow the organism to manipulate the environment and move through it (e.g., grasping and locomotion). The ventral stream remains principally involved with the recognition and identification of objects. Toward this aim, the ventral visual system must process some aspects of spatial information. However, its representation of spatial information employs an allocentric (rather than egocentric) framework. That is, the ventral system carries information about the relative position of objects with respect to one another. Hence, Goodale and Milner support that the ventral system represents space in the relative coordinates needed for the perception of object interrelationships, while the dorsal system represents space in body-centered absolute coordinates needed to support manual interaction with the environment (including locomotion by foot and vehicle).

2.2.2 Properties of Ambient (Dorsal) versus Focal (Ventral) Vision

The previous section reviewed the conceptual evolution of the ambient–focal dichotomization of the visual system into its more contemporary dorsal–ventral stream manifestation. In deference to the pioneering efforts of Leibowitz and his associates, and in order to maintain a consistency in the terminology employed in the surface transportation research literature, this report will continue to use the terms *ambient* and *focal* when referring to the parallel modes of visual representation and processing. Hereafter, any reference to ambient vision will subsume the properties of the dorsal stream, and references to focal vision will subsume the properties of the ventral stream.

In this section, the distinctive functional characteristics of the primate ambient and focal visual streams will be delineated and briefly discussed. These characteristics represent consensus views arising from a large body of physiological, neuropsychological, and psychophysical research and are summarized in Table 2.1. The ambient visual stream receives some input from subcortical areas such as the superior colliculus and the pulvinar region; however, its major source of input comes from magnocellular projections. Since it relies so heavily upon the magnocellular branch of the retino–geniculate–cortical pathway, several special functional characteristics may be attributed to processing within the ambient stream. That is, compared to the focal (ventral) stream, the ambient stream can be thought of as being: capable of resolving high temporal variations (i.e., motion and/or flicker), insensitive to high-spatial-frequency information, especially sensitive to low-contrast/low-spatial-frequency information, and insensitive to color contrast (Merigan and Maunsell, 1993; Fortes and Merchant, 2006). Input to the focal visual stream stems almost exclusively from the primary visual cortex and, unlike the ambient stream, depends heavily upon information from the parvocellular branch of the retino–geniculate–cortical pathway. This dependence upon parvocellular visual input indicates that

TABLE 2.1
Functional Characteristics and Response Properties of the Ambient and Focal Visual Systems

	Ambient System	Focal System
Primary functions	Visual guidance; motor control	Form recognition; identification
LGN (lateral Geniculate Nucleus) source	Magnocellular	Parvocellular
Cortical stream	Dorsal stream	Ventral stream
Field of view	Peripheral (significant rod input)	Central
Spatial resolution	Low	High
Contrast sensitivity	Asymptotic at low (10%) contrast	Requires mid-to-high contrast
Spatial frame of reference	Egocentric (absolute body coordinates)	Allocentric (relative object space)
Temporal resolution	High	Low
Primary control mode[a]	Closed-loop	Open-loop
Memory requirements[b]	Low	Moderate-high

[a] See Donges (1978).
[b] See Norman (2002).

the focal (ventral) stream, compared to the ambient system, may be characterized as being: relatively insensitive to high-temporal-frequency stimulus modulations, insensitive to low-spatial-frequency/low-contrast information, capable of resolving high-spatial-frequency stimuli (i.e., fine spatial detail), capable of fine wavelength (color) discrimination, and limited primarily to information delivered to the macular (central) region of the retina (Livingstone and Hubel, 1988; Milner and Goodale, 1996).

2.2.3 A Two-Level Model of Driver Steering

Building upon previous work aimed at constructing a control theory based model of vehicular guidance (see McRuer, Allen, Weir, and Klein, 1977), Edmund Donges (1978) developed and successfully tested his *two-level model of driver steering*. Donges' model has been highly influential in shaping the way the field of human factors psychology has subsequently conceptualized steering behavior. This model has direct parallels to the ambient–focal theoretical framework. Using Donges' own words, the two-process model can be succinctly described as:

> The steering task can be divided into two levels: (1) the guidance level involving the perception of the instantaneous and future course of the forcing function provided by the forward view of the road, and the response to it in an anticipatory open-loop control mode; and, (2) the stabilization level whereby any occurring deviations from the forcing function are compensated for in a closed-loop control mode. (1978, p. 691)

Expressed in terms of the ambient–focal heuristic, Donges' guidance level process is highly reliant upon the focal mode of processing. That is, foveal vision is used to garner information from the "far" road ahead and the driver uses this information to anticipate and prepare for future alterations in the course of the road. In addition to the need for such anticipatory interaction with the visual environment, the driver also depends upon visual information regarding current (i.e., instantaneous) deviations between the vehicle's actual path and its desired path. This later visual requirement, represented by the stabilization level, requires information from the "near" road ahead and, hence, is primarily dependent upon peripheral vision and the ambient mode of visual processing. Other aspects of the two-process model of driver steering map smoothly onto the ambient–focal dichotomy. For example, the foveal/anticipatory process periodically samples the far road ahead in an open-loop fashion and, hence, must be heavily dependent upon higher-level cognitive resources such as the strategic allocation of attention and memory capacity. These characteristics match those of the focal/ventral processing stream (see Norman, 2002). Thus, Donges' two levels (or processes) underlying visually guided steering behavior can be thought of as consisting of an *ambient/near mechanism* that uses peripheral vision to track and null instantaneous errors in lane position and a *focal/far mechanism* that uses central vision and higher-level visual cognition to anticipate (predict) the changing path ahead and to adequately prepare for such changes.

2.2.4 Diagnostic Signature of Ambient versus Focal Mediators of Steering Behavior

Perhaps the most well-known evidence supporting the existence of separate ambient/near and focal/far visual processes as mediators of visually guided steering behavior can be found in the work of Land and Horwood (1995, 1998). Participants in this simulator-based study were required to drive along a narrow and (extremely) winding virtual roadway while lane position performance was recorded. On experimental trials, the view was restricted to narrow horizontal samples of the road ahead (full horizontal extent with 1° vertical height). The relative position of this narrow sample of the road ahead was varied across trials from 1° to 9° below the horizon. At very low speed (i.e., 28 mph), optimal steering performance was achieved when the available visual information was positioned 7°–8° below the horizon (i.e., the very near road ahead). However, at higher speed (i.e., 44 mph), drivers were unable to achieve criterion (baseline) levels of steering stability when limited to a single narrow cross section of the road ahead (no matter where it was positioned). Instead, normal steering performance was maintained only when drivers were permitted to view a second 1° tall horizontal cross section such that the two visible regions of the road ahead sampled the lower (nearest) and upper (farthest) segments of the simulated road scene. Land and Horwood's partial visual occlusion paradigm demonstrated that both near and far visual information are needed to achieve normal levels of steering performance. This finding is highly consistent with the two modes of vision construct; namely, that parallel ambient/near and focal/far visual mechanisms combine to mediate vehicular guidance performance.

Over the past decade, a number of other investigators have used techniques that are somewhat analogous to Land and Horwood's (1995, 1998) partial occlusion technique to provide more direct assessments of both the existence and the dissociation of ambient versus focal contributions to vehicular guidance. The approach used in all of these studies is based upon the logic of systematically degrading a dimension of information in the visual environment that is thought to support either ambient or focal processes, and then observing which dimensions of driving-related performance change across the experimental manipulation (versus which dimensions of performance remain invariant). For example, by systematically decreasing the relative amount of high-spatial-frequency information available to the driver (by using progressively increasing levels of stimulus blur), one would expect driving behaviors related to focal visual mechanisms to become markedly degraded; while those related to ambient visual mechanisms would be expected to show little or no change across the experimental manipulation. Similarly, if one systematically decreased the peripheral field of view available to the driver, one would expect driving performance measures related to ambient visual mechanisms to become markedly degraded while those mediated by focal mechanisms would be expected—in many cases—to demonstrate little or no decline. Such dissociations between performance indices across theoretically significant categories of visual stimulus manipulation represent *diagnostic signatures* supporting the existence of parallel ambient and focal system mediators of driving-related behaviors. Recent studies providing data that can be used to generate such diagnostic signatures are reviewed in the pages that follow. Taken together, these studies provide considerable support for the working hypothesis that the ambient–focal heuristic represents both a valid and potentially powerful tool for improving our understanding of driving behavior.

2.3 EMPIRICAL EVIDENCE FOR AMBIENT–FOCAL MECHANISMS OF STEERING

2.3.1 Experimental Reductions of Visual Acuity

According to the ambient–focal framework, the relative contributions of focal visual processes to driving performance should be markedly reduced when high-spatial-frequency information in the driving scene is attenuated via poor acuity or optical blur. However, the efficiency of ambient visual processes should remain invariant under optical blur because of their insensitivity to high-spatial-frequency information. Higgins, Wood, and Tait (1998) examined the effects of experimental reductions in central visual acuity upon performance while driving around a 5.1 km closed-course road circuit characterized by complex horizontal geometry as well as a slalom course constructed from closely spaced traffic cones. Twenty-four young drivers (mean age = 23.1 years) drove while wearing modified swimming goggles equipped with binocular convex lenses of varying power. Increasing the power of these lenses resulted in decreasing the central acuity of the participants due to blurring of the retinal image. Lens powers were selected to yield functional acuity levels of 1, 2, 5, and 10 minutes of arc (i.e., 20/20, 20/40, 20/100, and 20/200 equivalent Snellen acuity, respectively). Among the driving performance measures recorded in this study,

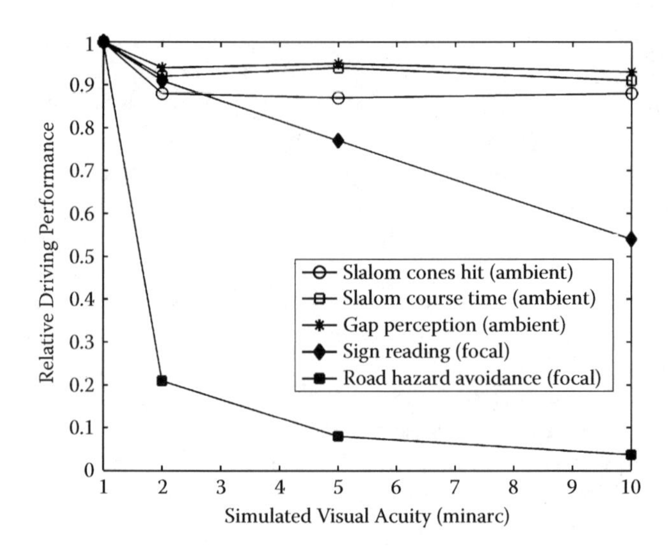

FIGURE 2.1 Relative performance upon five driving subtasks as a function of experimental reductions in central visual acuity. AMBIENT functions are largely unaffected by dramatic reductions in visual acuity, whereas FOCAL functions appear quite sensitive to variations in available visual acuity. (Data source: Higgins, Wood, and Tait, 1998.)

five are considered here because they appear to map well to the presumed ambient–focal dichotomization of visual function. Performance on each of these dependent variables has been normalized to a common relative scale and graphically presented in Figure 2.1 for comparison.

Reference to Figure 2.1 reveals several interesting outcomes regarding the effects of manipulating visual acuity. The first thing to note is that three of the five dependent measures remained virtually invariant as visual acuity was reduced. That is: (1) the number of cones hit while traversing a slalom course delimited by traffic cones (Slalom Cones Hit), (2) the time required to traverse the tight curves of the slalom course (Slalom Course Time), and (3) the ability to judge whether the space between traffic cones was wide enough to permit one's vehicle to pass (Gap Perception) did not significantly decline as simulated visual acuity was reduced from 20/20 (normal vision) to 20/200 (i.e., legally blind in the United States). The fact that these three measures of performance remained essentially invariant across large reductions in the availability of high-spatial-frequency information strongly suggests that they are dependent upon ambient/near visual processes rather than focal/far mechanisms. Yet, two other performance measures demonstrate just the opposite effect. Sign Reading and Road Hazard Avoidance were both found to decline precipitously as visual acuity was degraded. Since these later performance categories unambiguously depend upon the focal mode of visual processing, such effects were clearly anticipated. The dissociation of these two groups of performance functions (labeled AMBIENT and FOCAL in Figure 2.1) across experimental reductions in visual acuity represent the diagnostic signature consistent with the expectations of the ambient–focal heuristic.

That is, evidence that driving behavior is mediated by separate ambient and focal visual streams.

Although the Higgins et al. (1998) investigation can be characterized as having high face validity (i.e., real drivers in a real vehicle), the indices of driving performance collected in this study did not include traditional continuous measures of visually guided steering performance such as variability of lane position or time-to-line crossing. This makes it difficult to integrate their findings with the general scientific literature on visually guided steering behavior. Fortunately, a recent simulator-based study has replicated the effects of the experimental degradation of acuity upon driving performance while extending the results to the domain of continuous measures of steering efficiency. Brooks, Tyrrell, and Frank (2005) used a high-fidelity, fixed-base driving simulator (DriveSafety, Inc.) to investigate the effects of experimentally induced reductions in visual acuity upon various indices of steering performance in a sample of 10 young adults (mean age = 21.2 years). Again, acuity was manipulated through the use of convex lenses that varied in optical power from 0 to 10 diopters (yielding average observed acuities ranging from 1 to 32 minutes of arc (20/20 to 20/647 Snellen acuity, respectively)).

Remarkably, two measures of continuous steering performance remained almost unchanged across this wide range of simulated acuity. The percent time spent entirely within the lane boundaries (mean = 91%; range = 95%–88%) and the standard deviation of lane position (mean = 0.23 m; range = 0.22–0.25 m) remained virtually unchanged while experimental acuity varied from normal levels to well below the criterion for being classified as legally blind. These results reinforce the interpretation that *time-in-lane* and *standard deviations of lane position* are both indices of performance that reflect the level of functioning of the ambient visual system (which relies upon low-spatial-frequency input that is relatively immune to the deleterious effects of blur). These results were also consistent with an earlier report by Owens and Tyrrell (1999) who found that mean lane position error remained unchanged across large reductions in experimental visual acuity. Yet, it is interesting to note that Brooks et al. (2005) also measured two other continuous indices of steering performance that were not robust with respect to the experimental degradation of visual acuity. Both mean *lateral speed* and the *number of lane excursions* (i.e., edge line crossings) demonstrated sizable declines in performance with reductions in experimental visual acuity. This pattern of results suggests that these latter two indices of performance may reflect constraints imposed by focal/far visual processes, while the former indices reflect unconstrained ambient visual processing across the blur manipulation (see Figure 2.2). The classification of lateral speed and lane excursion performance as being mediated by focal/far visual processes is not immediately obvious and cannot be predicted in an *a priori* fashion given the current state of development of the ambient–focal framework. One possible reason for why these two indices became degraded under blur could be as follows: As the level of blur increased, drivers became less able to gather information about approaching curves that would allow them to anticipate large changes in vehicle heading (clearly a task requiring focal/far processes). Entry into such unanticipated curves might thereby be expected to be accompanied by edge line crossings and sudden compensatory increases in lateral speed to restore satisfactory lane position. Such behavior would,

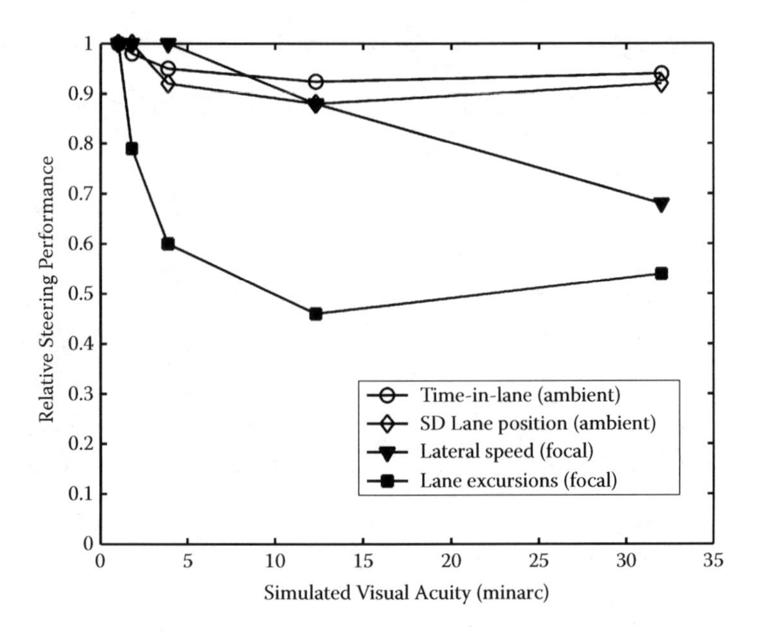

FIGURE 2.2 Relative performance upon four driving subtasks as a function of experimental manipulation of visual acuity levels. Note that two performance indices are invariant across acuity levels (demonstrating the AMBIENT signature), while two other indices decline markedly (demonstrating the FOCAL signature). (Data source: Brooks, Tyrrell, and Frank, 2005.)

in turn, be expected to yield increases in mean lateral speed and the number of lane excursions as the level of visual blur was increased. Indeed, such behavior has been explicitly noted in a related study where visual access to the road ahead was curtailed using an altogether different approach to manipulating preview distance (see COST 331, 1999).

In summary, studies manipulating the level of visual acuity via blur have yielded empirical signatures that strongly support the validity of the ambient–focal framework for understanding vehicular guidance. In addition, these studies have also demonstrated that such signatures can be quite diagnostic in terms of classifying how various dependent measures appear to map onto the ambient–focal dichotomy of visual function in a given experimental scenario.

2.3.2 Experimental Reductions of Roadway Luminance

Although the illumination provided by the sky varies widely between dawn and dusk, the luminance of objects in the daytime driving scene remains well within the eye's photopic range of luminance adaptation (Wyszecki and Stiles, 1982). As a consequence of the eye's ability to quickly adapt to variations in light level within the photopic regime, the amount of light reaching the driver's eyes rarely represents a limiting factor upon vehicular guidance during daytime driving. However, as night begins and environmental light levels fall below the threshold of *civil twilight** numerous visual functions begin to become degraded. Vehicle headlamps and overhead lighting assist the driver at night. Yet, even with the augmentations provided by these artificial sources of light, the luminance of the typical roadway environment is too low to adequately maintain photopic levels of light adaptation (Olson and Aoki, 1989; Eloholma, Ketomäki, and Halonen, 2004). As a result, the visual adaptation level of the typical nighttime driver can be said to be in the *mesopic* range—a poorly understood middle ground of vision in which neither cone (photopic) nor rod (scotopic) visual functions perform optimally.

The vast majority of the scientific literature on human vision has been conducted under either photopic or scotopic adaptation conditions. It is, therefore, often quite difficult to generalize results from classical laboratory studies to the situation facing the driver at night. This makes the ambient–focal heuristic especially useful in the domain of nighttime driving. Accordingly, driving functions thought to be mediated by ambient visual processes—such as certain aspects of vehicular guidance—should remain robust as luminance is reduced from photopic to near-scotopic levels (i.e., across the full range of mesopic vision). Driving processes thought to be mediated by focal visual processes—such as sign and hazard recognition at a distance—should become increasingly degraded at such low luminance levels. Several recent studies of driving performance have yielded results consistent with this view.

Owens and Tyrrell (1999) used a low-resolution, part-task driving simulation environment to investigate the effects of reductions in roadway luminance upon steering behavior. The delineators marking the edges of the simulated roadway were presented at four different luminance levels: 0.003, 0.03, 1, and 30 cd/m^2. The lowest luminance (0.003 cd/m^2) represented vision in the scotopic regime while the highest luminance (30 cd/m^2) was selected to represent the photopic adaptation state. The remaining two levels were selected to simulate low (0.03 cd/m^2) and high (1 cd/m^2) reflectance objects observed at civil twilight (representing the low and high ends of the mesopic luminance regime). Consistent with the predictions of the ambient–focal heuristic, Owens and Tyrrell found that a continuous measure of steering performance (lane position error) was unchanged as luminance conditions were varied from photopic to low mesopic levels. Only in the scotopic condition was steering performance significantly reduced relative to photopic viewing conditions. Similar findings have been reported by Brooks et al. (2005) using a more sophisticated simulation platform and a related measure of steering performance (percent time in lane). That is, both studies revealed that steering performance—as indexed by lane position—was quite robust across the full range of mesopic luminance. Again, this relative invariance of performance across experimental degradation of the visual environment appears to represent the hallmark signature of the ambient/near visual mechanism (see Figure 2.3).

* Since the days of antiquity, astronomers have observed that skylight is sufficient to support most normal tasks until the end of *civil twilight*—the point at which the sun falls more than 6 degrees below the horizon. Typical roadway illumination provided by the sky at this time is approximately 3 lux at northern latitudes in the United States.

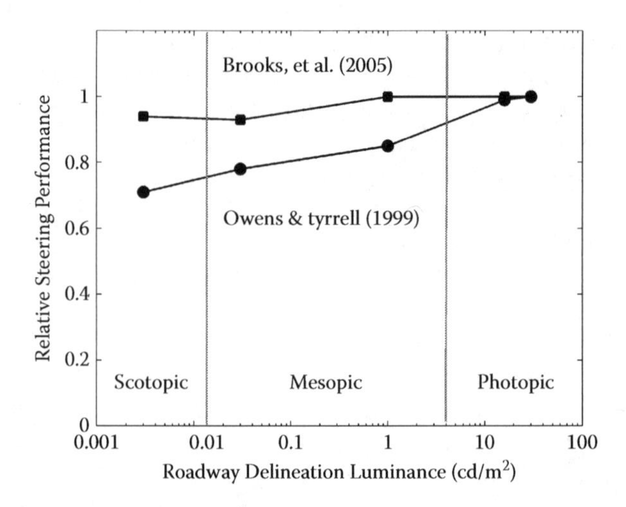

FIGURE 2.3 The robust nature of visually guided steering performance across the scotopic through photopic luminance regimes represents the hallmark signature of a process mediated by the ambient visual system.

2.3.3 Experimental Restrictions of Driver's Field of View

According to the ambient–focal framework, the ambient visual system's contributions to driving performance should become degraded as the field of view is restricted beyond some critical level.

Wood and Troutbeck (1992, 1994) conducted a series of studies which, when combined together, permit a parametric examination of the effects of reductions in the field of view upon various aspects of driving performance. Specially designed goggles were used to restrict the driver's field of view to one of four levels: 20, 40, 90, or 150° (monocular baseline). Performance data were collected while drivers negotiated a slalom course constructed of tightly packed traffic cones and while drivers completed several circuits around a 5.1 km closed course characterized by complex horizontal geometry (see p. 292 of Wood and Troutbeck, 1992, for a map of the Mount Cotton driver training course used in these investigations). Select performance data from both studies have been normalized to foster global comparisons. The effects of reductions in the field of view upon these measures of driving performance are depicted in Figure 2.4.

Two of the performance curves depicted in Figure 2.4 represent indices of vehicular guidance: Lanekeeping and Slalom Cones Hit. Lanekeeping performance, as defined in these studies, is a composite index combining subjective rating scale data and mean lane position data sampled at 45 discrete locations along the test track. The Slalom Cones Hit index was derived from the relative number of traffic cones touched or knocked down while drivers negotiated a slalom course consisting of several very tight curves (Wood and Troutbeck, 1994). Performance on these guidance functions demonstrated little or no decline as available field of view was reduced from 150 to 90 deg. However, remarkable decrements in these indices of vehicular guidance behavior were clearly evident when the field of view was reduced below 40

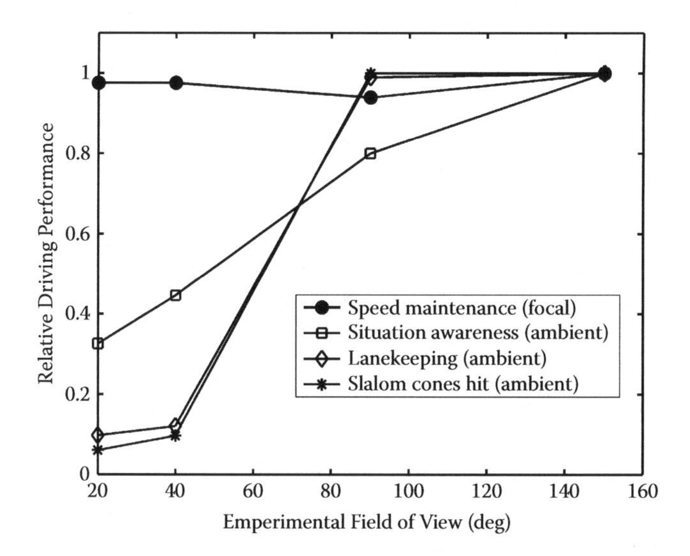

FIGURE 2.4 The effects of systematic restrictions in the field of view upon several measures of driving performance. Note that three of the performance measures demonstrate marked declines as the field of view is diminished. Such changes in performance would be expected for driving skills mediated by the ambient visual system. The surprising invariance of the Speed Maintenance data across changes in the field of view matches the signature of a focal visual process. (Data source: Wood and Troutbeck, 1992, 1994.)

deg. The shapes of these performance functions across experimental variations in the available field of view are consistent with the interpretation that both Lanekeeping and Slalom Cones Hit represented indices of guidance behavior that are mediated by ambient/near visual mechanisms. At least one other performance measure collected by Wood and Troutbeck (1994) also demonstrated a signature that was diagnostic with respect to the ambient visual system. Situation Awareness, representing the relative number of traffic signs and roadside pedestrians that were detected, also declined significantly as the driver's field of view was diminished. Finally, the time required to traverse the closed-course driving circuit (Speed Maintenance) was demonstrated to have been invariant across the experimental manipulation of the driver's field of view. This suggests that visual input into the regulation of driving speed may have involved focal rather than ambient visual mechanisms. This finding is significant because it tends to discount the role of "optic flow" in the far periphery with regard to the regulation of vehicle speed (a function often assigned to the domain of the ambient system; see Riemersma, 1987).

Brooks et al. (2005), whose work was described in some detail earlier, also used their high-fidelity driving simulator to assess the effects of restricted field of view upon vehicular guidance behavior. Young drivers (mean age = 18.5 years) were tested under the following field of view conditions: 1.7, 3.4, 11, 23, and 46 deg. Both monocular and binocular full field (150 deg) baseline conditions were also observed for all participants. The experimental field of view restrictions were implemented by mounting small aluminum cylinders in eyeglass frames and positioning them just anterior to the pupil of the left eye. Data from four measures of vehicle guidance were collected. All four indices of vehicle guidance performance declined significantly as the field of view was restricted. Like the data of Wood and Troutbeck (1992, 1994), the quality of vehicle guidance appeared to decline when the driver's visual field fell below 40 deg. These investigators failed to demonstrate a pattern of dissociation between the performance variables that was consistent with the ambient–focal framework. However, the interpretation of these results is complicated by the fact that their field of view manipulation was limited to monocular viewing from the nondominant eye in most participants. This work needs to be repeated under binocular viewing conditions.

2.3.4 Experimental Reductions of Roadway Preview Distance/Time

In 1999, a comprehensive study of the visual needs of drivers was completed under a European Cooperation in the Field of Scientific and Technical Research initiative (hereafter referred to as the COST 331 study). One of the goals of the COST 331 study was to set minimum guidelines for the retroreflectivity of pavement marking in support of nighttime vehicular guidance. In order to establish this minimum requirement, the investigative team first had to answer the question: How far down the roadway do edge lines need to be visible to support optimal steering behavior? The primary method used to address this question was based upon a unique driving simulation protocol conducted at the Swedish Road and Transport Research Institute (VTI). Twelve young (25–35 years old) and 12 middle-aged (55–65 years old) participants drove along alternating straight and curved segments of a simulated roadway at a fixed speed of 90 km/h while the visibility distance of the road ahead was systematically manipulated across five levels: 20, 30, 45, 67, and 100 m, respectfully. Figure 2.5 depicts the appearance of the simulated roadway at several different preview distances. Results indicated that the *standard deviation of lane position* was elevated (approximately 0.4 m) at the shortest preview distance (20 m). However, asymptotic performance (approximately 0.23 m) was achieved at preview distances between 30 and 45 m. Providing the driver with additional preview distance was not accompanied by improvements in steering performance. Converting this result to a format that can be easily generalized across various driving speeds (i.e., preview time), these results indicate that asymptotic steering performance is achieved once the driver is provided with 2 seconds of preview time.

These findings from the COST 331 study are especially significant when considered within the framework of the ambient–focal dichotomy. Lateral lane position variability reached minimum levels (optimal performance) with just 2 seconds of roadway preview time. Yet, most investigations of the minimum visual requirements for roadway delineation (including COST 331) estimate that much longer preview times are required for safe and efficient operation of a motor vehicle. For example, Zwahlen and Schnell (2000) estimate a minimum preview time requirement of 3.65 seconds, whereas other investigations conclude that a minimum preview time of 5 seconds is necessary (Weir and McRuer, 1968; Godthelp and Riemersma, 1982; CIE, 1992). This pattern of findings suggests that the point at which lateral road position variability performance becomes asymptotic marks the transition point at which

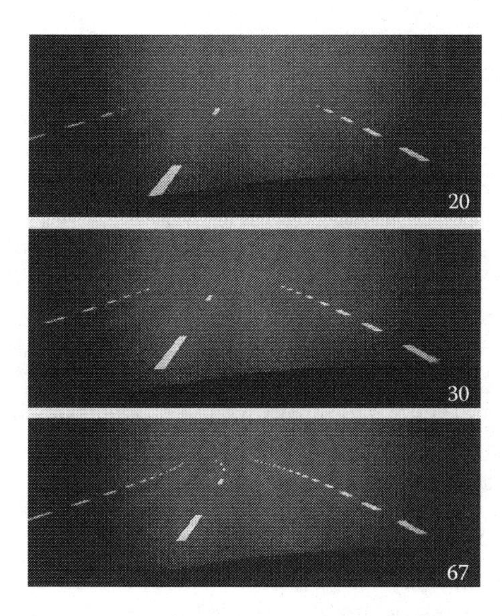

FIGURE 2.5 Examples of three different simulated driving preview distances. Edge lines are visible out to 20, 30, and 67 m, respectively. (Source: COST 331, 1999.)

ambient/near visual processes give way to focal/far visual processes. The ability to delineate the transition point between the near and far visual environments may represent an important advancement for the application of the ambient–focal framework in the service of improving our understanding of visually guided driving behavior.

2.4 HEURISTIC VALUE OF THE AMBIENT–FOCAL FRAMEWORK

The investigations reviewed in the previous section represent a select subset of the available studies on vehicle guidance. They were chosen for inclusion based upon one primary criterion: their ability to link the ambient–focal dichotomization of the visual system to the study of visually guided driving behavior. Two experimental protocols appear to have been particularly effective in demonstrating a dissociation between ambient and focal visual mediators of driving performance. Progressive blurring of the visual scene has been shown to systematically degrade focal mode contributions to driving performance while sparing ambient mode contributions. Progressive reductions in the driver's field of view have been shown to degrade various aspects of driving performance that can be attributed to the ambient mode of visual processing—although perhaps not as unambiguously as the dissociations revealed using the progressive blurring protocol. Additional research is needed to better understand the changes in visual dynamics resulting as a consequence of manipulations of the driver's field of view. Simultaneous records of eye movement behavior during such protocols could contribute much to our understanding of these dynamics.

The functional distinctions between the ambient and focal modes of vision outlined in Table 2.1 suggest several other approaches that could be used to experimentally isolate ambient versus focal contributions to driving-related behavior.

For example, given the strong dependence of ambient vision upon magnocellular input, reductions of display luminance contrast to levels approaching 10% would be expected to maintain the efficiency of ambient mechanisms while dramatically reducing the effectiveness of focal (primarily parvocellular) visual mechanisms. Such a manipulation could be easily instantiated in a driving simulator environment. Yet another approach could be used to attenuate ambient mode contributions in the service of partially isolating focal mode mechanisms of visually guided behavior. That is, a driving simulator could be used to render the visual environment at near-isoluminant conditions (i.e., color contrast with little or no simultaneous luminance contrast). Since the magnocellular inputs making up the ambient stream are "color blind," such an isoluminant stimulus configuration would bypass ambient vision, and, as a consequence, might yield important new insights regarding the differential contributions of focal mechanisms of driving performance.

Another, less speculative approach focuses upon exploring the ambient–focal dichotomy by systematically restricting visual information to the "near" versus "far" domains of vision, respectively. As discussed within the context of the COST 331 (1999) study, this could be accomplished by progressively increasing the preview distance available to the driver until asymptotic levels of lane-keeping behavior were achieved (i.e., minimum standard deviation of lane position). Within the context of the ambient–focal theoretical framework, this point (at which additional preview distance no longer yielded improvements in lane keeping) would mark the end of the "near" range of the visual space subsumed by ambient visual mechanisms. Thus, by restricting available preview to include only the visual world up to this boundary point one could isolate ambient/near visual mechanisms. Similarly, by restricting visual preview to the visual world only beyond this boundary point one could isolate focal/far visual mechanisms. Such manipulations could be accomplished in a driving simulator as well as in a real vehicle on a test track using a very simple visual occlusion technique (see Riemersma, 1987, for an example). It would be most informative to discover which, if any, driving performance indices remained invariant under the "far preview only" condition. Such invariance would be indicative of a performance measure that was sensitive to focal/far visual processes. Myers (2002) has proposed that *time-to-line crossing* (see Van Winsum and Godthelp, 1996) represents a likely candidate for such a diagnostic measure of focal/far visual processing. If experimentally verified, this would allow one to evaluate the ability of a roadway delineation system to support focal/far visual requirements based upon time-to-line crossing data while evaluating its ability to support ambient/near visual requirements based upon simultaneous measures of the standard deviation of road position (relative to their asymptotic levels).

Finally, there is evidence that the ambient–focal theoretical framework may provide a basis for investigating driving behavior well beyond the realm of mere vehicular guidance. For example, this approach holds much promise for improving our understanding of age-related driving problems. Normal adult aging is accompanied by systematic and deleterious changes in the visual system (see Schieber, 1992, 2006). There is a growing body of evidence that ambient visual functions such as low-spatial-frequency motion perception and the useful field of view are especially susceptible to age-related decline (Kline and Schieber, 1981; Owsley et al., 1998; Schieber, 2006). In addition, there is evidence that age-related deterioration of ambient visual information processing may mediate the increased frequency of "looked but didn't see" crashes among older drivers, especially at intersections (Schieber, 1994, 2000). Perhaps the most direct evidence for this *ambient insufficiency hypothesis* of age-related visual difficulties with driving has been presented by Owens and Tyrrell (1999) who reported that steering performance in older adults was not as resistant to systematic reductions in roadway luminance as their young counterparts. Consistent with this interpretation, Wood (2002) has demonstrated that global motion sensitivity and the ability to rapidly detect and localize targets in the peripheral field were strong predictors of age-related decrements in closed-course driving performance. Additional work is needed to more rigorously assess the ambient insufficiency hypothesis. The techniques described hold great potential for more in-depth evaluation of ambient mode processing efficiency among older drivers.

2.5 CONCLUSIONS: VISUAL REQUIREMENTS OF VEHICULAR GUIDANCE

The studies examined to this point have yielded some noteworthy conclusions about the nature of the visual information required to successfully steer an automobile. Vehicular guidance was found to be remarkably robust in the face of great reductions in available high-spatial-frequency information that accompanies experimental degradations in visual acuity via blur. Drivers with a simulated visual acuity level of 20/200 (i.e., legally blind) demonstrated no systematic reductions in their ability to maintain lane position in straight road driving (ambient/near vision) but did appear to demonstrate some deficiencies in terms of preparatory vehicular positioning in anticipation of sharp curves approaching in the distance (focal/far vision). Similarly, the ability to maintain lane position was found to be quite robust across marked reductions in roadway luminance—becoming significantly degraded only when luminance was reduced to levels approaching those provided by mere moonlight (i.e., the scotopic state of light adaptation). A broad range of driving-related skills, including those related to lane-keeping performance, were found to become significantly degraded when the driver's field of view was experimentally reduced to levels smaller than 40°. Remarkably, speed maintenance was shown to be invariant across even the most severe reductions in the field of view—a finding that is difficult to reconcile with optic flow accounts of speed maintenance.

In summary, it can be concluded that successful vehicular guidance is reliably maintained in the absence of high-spatial-frequency information (20/200 acuity), throughout the full range of mesopic roadway luminance levels (1 cd/m^2 and above), and with a minimum forward field of view (approximately 40 degrees). These are the minimum requirements necessary to achieve nominal levels of steering performance. However, it should be obvious that successful driving involves much more than the simple ability to maintain a vehicle's position on the road. Factors such as sign legibility, hazard detection and anticipation, situation awareness, and many additional vision-based behavioral skills are required to support safe and effective driving in a real-world environment. The ambient–focal heuristic, together with its associated family of protocols for systematically manipulating qualitatively distinct categories of visual information, appears to hold significant potential for exploring and better understanding the visual inputs necessary to support these higher-order functions. It is the hope of the authors that this review of the ambient–focal framework will help foster such developments.

REFERENCES

Andre, J., Owens, D.A., and Harvey, L.O., Jr. (Eds.). (2002). *Visual perception: The influence of H.W. Leibowitz*. Washington, DC: American Psychological Association.

Brooks, J.O., Tyrrell, R.A., and Frank, T.A. (2005). The effects of severe visual challenges on steering performance in visually healthy young drivers. *Optometry and Vision Science, 82,* 689–697.

CIE. (1992). Fundamentals of the visual task of night driving (CIE Publication 100). Paris: Commission Internationale d'Eclairage.

COST 331. (1999). *Requirements for horizontal road markings* (European Commission, Directorate General Transport). Luxembourg: Office for Official Publications of the European Communities. http://www.cordis.lu/cost-transport/src/cost-331.htm.

Donges, E. (1978). A two-level model of driver steering behavior. *Human Factors, 20,* 691–707.

Eloholma, M., Ketomäki, J., and Halonen, L. (2004). *Luminances and visibility in road lighting: Conditions, measurements and analysis*. Report 30. Espoo, Finland: Lighting Laboratory, Helsinki University of Technology. http://www.lightinglab.fi/publications/files/report30.pdf.

Fortes, A., and Merchant, H. (2006). Investigating higher order cognitive functions in the dorsal (magnocellular) stream of visual processing. In R. Pinaud, L.A. Tremere, and P. De Weerd (Eds.), *Plasticity of the visual system: From genes to circuits* (pp. 285–306). New York: Springer.

Gibson, J.J., and Crooks, L.E. (1938). A theoretical field-analysis of automobile driving. *American Journal of Psychology, 11,* 453–471.

Godthelp, J., and Riemersma, J.B.J. (1982). Vehicle guidance in road work zones. *Ergonomics, 25,* 909–916.

Goodale, M.A., and Milner, A.D. (1992). Separate visual pathways for perception and action. *Trends in Neuroscience, 15,* 20–25.

Held, R. (1968). Dissociation of visual functions by depravation and rearrangement. *Psychologische Forschung, 31,* 338–348.

Held, R. (1970). Two modes of processing spatially distributed information. In F.O. Schmitt (Ed.), *The neurosciences* (pp. 317–324). New York: Rockefeller University Press.

Higgins, K.E., Wood, J., and Tait, A. (1998). Vision and driving: Selective effect of optical blur on different driving tasks. *Human Factors, 40,* 224–232.

Ingle, D. (1967). Two visual mechanisms underlying behavior in fish. *Psychologische Forschung, 31,* 44–51.

Ingle, D. (1973). Two visual systems in the frog. *Science, 1181,* 1053–1055.

Kline, D.W., and Schieber, F. (1981). Visual aging: A transient-sustained shift? *Perception and Psychophysics, 29,* 181–182.

Land, M., and Horwood, J. (1995). Which parts of the road guide steering? *Nature, 377,* 339–340.

Land, M., and Horwood, J. (1998). How speed affects the way visual information is used in steering. In A.G. Gale, I.D. Brown, C.M. Haslegrave, and S.P. Taylor (Eds.), *Vision in vehicles–VI* (pp. 43–50). Amsterdam: Elsevier.

Leibowitz, H.W., and Owens, D.A. (1977). Nighttime accidents and selective visual degradation. *Science, 197,* 422–423.

Leibowitz, H.W., Owens, D.A., and Post, R.B. (1982). *Nighttime driving and visual degradation* (SAE Technical Paper No. 820414). Warrendale, PA: Society of Automotive Engineers.

Livingstone, M., and Hubel, D. (1988). Segregation of form, color, movement and depth: Anatomy, physiology and perception. *Science*, *240*, 740–749.

McRuer, D.T., Allen, R.W., Weir, D.H., and Klein, R.H. (1977). New results in driver steering control models. *Human Factors*, *19*, 381–397.

Merigan, W.H., and Maunsell, J.H. (1993). How parallel are the primate visual pathways? *Annual Review of Neuroscience*, *16*, 369–402.

Milner, A.D., and Goodale, M.A. (1996). *The visual brain in action*. Oxford: Oxford University Press.

Mishkin, M., Ungerleider, L.G., and Macko, K.A. (1983). Object vision and spatial vision: Two cortical pathways. *Trends in Neuroscience*, *6*, 414–417.

Myers, J. (2002). *The effects of near and far visual occlusion upon a simulated driving task*. Unpublished master's thesis, University of South Dakota, Vermillion, SD. http://www.usd.edu/~schieber/pdf/MyersThesis.pdf

Norman, J. (2002). Two visual systems and two theories of perception: An attempt to reconcile the constructionist and ecological approaches. *Behavioral and Brain Sciences*, *25*, 73–144.

Olson, P.L., and Aoki, T. (1989). *The measurement of dark adaptation level in the presence of glare* (Report No. UMTRI-89-34). Ann Arbor, MI: University of Michigan, Transportation Research Institute.

Owens, D.A., and Tyrrell, R.A. (1999). Effects of luminance blur and age on nighttime visual guidance: A test of the selective degradation hypothesis. *Journal of Experimental Psychology: Applied*, *5*, 115–128.

Owsley, C., Ball, K., McGwin, G., Sloane, M.E., Roenker, D.L., White, M.F., and Overley, T. (1998). Visual processing impairment and risk of motor vehicle crash among older adults. *Journal of the American Medical Association*, *279*, 1083–1088.

Riemersma, J.B.J. (1987). *Visual cues in straight road driving*. Soest, Netherlands: Drukkerij Neo Print.

Schieber, F. (1992). Aging and the senses. In J.E. Birren, R.B. Sloan, and G. Cohen (Eds.), *Handbook of mental health and aging* (pp. 251–306). New York: Academic Press,.

Schieber, F. (1994). *Recent developments in vision, aging and driving: 1988–1994* (Report No. UMTRI-94-26). Ann Arbor, MI: Transportation Research Institute, University of Michigan.

Schieber, F. (2000). What do driving accident patterns reveal about age-related changes in visual information processing? In K.W. Schaie and M. Pietrucha (Eds.), *Mobility and transportation in the elderly* (pp. 207–211). New York: Springer.

Schieber, F. (2006). Vision and aging. In J.E. Birren and K.W. Schaie (Eds.), *Handbook of the psychology of aging* (6th ed., pp. 129–161). Amsterdam: Elsevier.

Schneider, G.A. (1967). Contrasting visuomotor functions of tectum and cortex in Golden Hamster. *Psychologische Forschung*, *31*, 52–62.

Schneider, G.A. (1969). Two visual systems. *Science*, *163*, 895–902.

Trevarthen, C.P. (1968). Two mechanisms of vision in primates. *Psychologische Forschung*, *31*, 299–337.

Ungerleider, L.G., and Mishkin, M. (1982). Two cortical visual systems. In D.J. Ingle, M.A. Goodale, and R.J.W. Mansfield (Eds.), *Analysis of visual behavior* (pp. 549–586). Cambridge, MA: MIT Press.

Van Winsum, W., and Godthelp, H. (1996). Speed choice and steering behavior in curve driving. *Human Factors*, *38*, 434–441.

Weir, D.H., and McRuer, D.T. (1968). A theory for driver steering control of motor vehicles. *Highway Research Record*, *247*, 7–28.

Wood, J.M. (2002). Age and visual impairment decrease driving performance as measured on a closed-road circuit. *Human Factors*, *44*, 482–494.

Wood, J.M., and Troutbeck, R. (1992). Effect of restriction of the binocular visual field on driving performance. *Ophthalmic and Physiological Optics*, *12,* 291–298.

Wood, J.M., and Troutbeck, R. (1994). Effect of visual impairment on driving. *Human Factors*, *36,* 476–487.

Wyszecki, G., and Stiles, W.S. (1982). *Color science: Concepts and methods, quantitative data and formulae* (2nd ed.). New York: John Wiley and Sons.

Zwahlen, H.T., and Schnell, T. (2000). Minimum in-service retroreflectivity of pavement markings. *Transportation Research Record*, *1715,* 60–70.

3 Estimations in Driving

Justin F. Morgan and Peter A. Hancock

CONTENTS

Reflection 51
3.1 Starting from the Real World 51
3.2 Making Estimations 52
3.3 The Likelihood of Collision 54
3.4 A Broad View of Driver Estimation Performance Capacities 55
3.5 Gap Acceptance 57
3.6 Movement Perception 58
3.7 Conclusions 59
References 60

REFLECTION

Our chapter is primarily concerned with the perceptual estimations that an individual is asked to make in the process of their journey. The capacities through which humans succeed (and sometimes fail) in these tasks have been around approximately 4 million years, although vehicle-type prosthetics have only been around for some thousands of years. In fact, fundamentally necessary psychomotor coordination is multiple millions of years older than human beings themselves, but since chariot driving demands these common skills, we can say that power-assisted transportation is not that "new" either. The presence of the vehicle dictates the change in kinematic parameters now involved in the whole process, and it is a testament to human adaptive capabilities that most drivers are very successful even in these changed and constrained circumstances that evolution would be challenged to have foreseen. However, we know to our detriment that such capacitates are not perfect and so our search for improved transport efficiency and driving safety continues.

3.1 STARTING FROM THE REAL WORLD

On a sunny day in a leafy suburb of a major city, a driver is about to begin his everyday commute to work. Characterized generally as the transit between origin and destination, research traditionally begins when the person settles himself behind the wheel and starts the vehicle. However, because people are constantly in motion, this is not necessarily so. By the time he reaches his garage, even before he has placed himself behind the wheel, he has already faced and succeeded in solving any number of transportation demands. He has assessed and navigated around obstacles within his house. He has coordinated strategic and tactical movement decisions and enacted a hierarchy of goals in which movement has played the vital resolutional role. Transitioning to a vehicle now means that his method of locomotion has changed from upright bipedal gait to a four-wheeled power assist. However, the fundamental task itself has not changed. Instead the task is now directly augmented by a technical prosthetic (see Hancock, 1997).

If we are to follow a traditional route, let us then consider the journey of this driver and the estimates he is required to make between the respective parking places of his vehicle. First, he has to back out of the garage; this can prove to be a tricky maneuver given the clearances available in modern suburban garages. It requires the driver to make fine-grained estimates of mere inches, and so this process is usually conducted in a tentative manner at very low velocities. Also, there may be objects, children, grandparents, or other obstructions behind the vehicle. Thus, before the driver has left his overnight parking spot safely he has already performed numerous feats of estimation that required rapid shifts of attention and continual fine-grained control.

This illustrates the underlying principle that driving is a deceptively simple activity. Upon immediate inspection, it appears to require continuous visual sampling, periodic control, and sustained attention. However, in actuality, it is a task that can often be performed satisfactorily with only intermittent visual sampling and periodic inputs on behalf of the driver. These inputs are triggered largely by the need to respond to the demands of the roadway itself and the transient and ephemeral obstructions that can appear on it. Thus, to be successful, drivers must engage in a sampling of the environment for critical control cues (Senders, Kristofferson, Levison, Dietrich, and Ward, 1966). Most often, these sample points are logical and predictable extrapolations of anticipatable and well-learned circumstances. Indeed, without these consistencies driving would be an exceptionally arduous task. However, occasionally such demands are sudden, urgent, unpredictable, and a source of very strong threat. Each of these regular and exceptional event sequences require drivers to engage in a spectrum of estimations as to what is currently happening in their driving environment and what is liable to happen in the immediate future.

3.2 MAKING ESTIMATIONS

Estimation can be considered as the process upon which decisions and conclusions about any individual's surrounding environment are based. Estimates can be explicit responses to obligatory demands or composed of a more general implicit evaluation of current conditions. These estimations can include estimates in which a decision is primarily based on a single stimulus channel or property (e.g., Is the back of the vehicle moving out of the garage?), as well as estimates involving the comparative evaluation of multiple stimuli (e.g., Is the car moving at a safe speed compared to other traffic on this residential street?).

However, this distinction between single and multiple stimulus channel estimates should not be over-emphasized in the search to understand driver decision making. For example, although as our example driver turns the corner from his neighborhood onto the major connecting street he will make estimates such as determining whether there are any pedestrians in the crosswalk (which rely primarily on a single stimulus property: whether or not the crosswalk is clear), the majority of his estimates in that same situation will be founded on relative observations, such as determining the closing rate of any oncoming traffic as he attempts to cross the intersection (an estimation process which involves the balancing of multiple rates of closures, tracking, and vehicle control tasks). Thus, both estimates along single and multiple stimulus channels most frequently occur simultaneously. Indeed, as with many perceptual processes, the strongest and most accurate estimates are most often derived when both estimation processes are enacted and the driver is able to make decisions based on the integration of such information.

Drivers are generally able to perform these perceptual estimation tasks very successfully. This is clear from the relative level of safety on the roadways of the world. For example, in 2005 the U.S. National Highway Traffic Safety Administration's (NHTSA) Fatality Analysis Reporting System (FARS, 2007) listed accident rate estimates for the entire country. The death rate was approximately 1.47 per 100 million vehicle miles traveled, which equated to a total number of 39,189 fatal accidents for that year. Although the toll that these deaths take on both society and the individuals involved is tremendous, in terms of the expense and loss of human life this frequency is amazingly low considering the sheer number of drivers on the road at any one time and the number of potential opportunities for collision. The relative infrequency of accidents is a testament to overall driver performance in the process of vehicle control (also see Hancock and deRidder, 2003).

The primary issue when examining driver estimation is the importance of each respective process. Humans, by nature, are conservative with respect to their expenditure of both physical and cognitive resources. Thus, the level to which the driver expends the cognitive energy necessary to explicitly analyze each scenario, compare alternatives, draw from previous experience, and apply that information becomes of great importance. This conservation of cognitive effort implies varying levels of attention devoted to the process of estimation, contingent upon the perceived demands of the moment. We propose that in respect of this resource conservation issue, that over 90% of all driver-performed estimations are thus implicit and demand little or no use of active conscious attention.

Related to this concept of demand-based allocation of attentional resources in driver decision making is the idea of ambient and focal vision in driving (Leibowitz and Post, 1982). This concept proposes that two separate visual perceptual mechanisms exist. The first is the ambient mode, which concentrates on factors such as spatial orientation and locomotion. The second, the focal mode, focuses on object recognition and identification. In terms of driver performance, the ambient visual system is more concerned with the concept of vehicle guidance, using information gained from optic flow to inform driver estimations of vehicle positioning. Meanwhile, the focal system concentrates on the detection and identification of objects of importance in the environment such as other vehicles and roadway threats, among others. The interaction between the two visual systems, and their ability to complement one another and to provide mutual reinforcement for items of interest in the environment, allows for an overall reduction in response time. In addition, the two disparate systems serve as a physiological metaphor for the differences in driver decision making: one system is dedicated to positive control, while a separate but equal system serves to inform driver decision making regarding potential sources of hazard in the environment.

Let us return to our example of a driver traveling his daily commute. As this driver begins to travel around a sweeping curve his peripheral vision detects movement to the left of his vehicle. This detected movement begins a physiologically based and psychologically adapted process of focusing both the driver's eyes and attention on the potential hazard. However, as soon as selective attention is applied and the object identified, the source of information is either given further attention or is dismissed from consciousness. For example, if our driver spots a train on a perpendicular trajectory, this would be regarded as an event with great importance. Conversely, if the driver looks to his left and sees an empty field with a tree in the distance, the scenario is unlikely to be the subject of any more detailed cognitive evaluation.

This is partially a function of the human cognitive system attempting to operate at a high degree of efficiency. This accords directly with the notion of satisficing advanced by Simon (1969) in which individuals perform at a level well enough to avoid collision but not at their maximal level of performance, which may well exhaust the driver's cognitive capacities and present subsequent risk of vigilance decrement type failures of detection (Hancock and Scallen, 1999).

3.3 THE LIKELIHOOD OF COLLISION

This resulting, satisficed process (and to some extent the potential cost of the infraction) may be viewed as a function of the speed and the distance of the fixated object of interest, in terms of time-to-contact and in relation to the driver. Looming, or the change in spatio-temporal orientation between the driver-observer and an object in the environment as the visual angle subtended on the driver's retina, only becomes useful as a visual cue to the driver at closer distances and shorter durations of time. These are both situations in which tau (or the inverse of the rate of optic expansion) increases at a more rapid pace (Schiff and Detweiler, 1979). At long distances and times, changes in the visual angle an object present relatively very little useful information regarding the estimates a driver must make. Between the extremely close looming space and the thresholds of visual perception of distance and time is what we here term a *field of useful expansion*, where the driver may detect, perceive, and react to other objects in the driving environment (see Figure 3.1).

This field of useful expansion is produced as a function of time and distance, and the amount of satisficed cognitive (mental) effort spent processing any object falling within the field is a direct function of the driver's assessment of the threat posed by the object. Thus, the difference in driver mental effort between a tree and a locomotive is a function of the differing mental demands that each object imposes on our commuting driver. The cost to the driver in terms of both equipment damage and human injury and death is much different for the two different objects.

Beyond the question of singular estimates lie the frequently more complex problems of the real world. When drivers are processing information in active (dynamic) situations, there is little use for the conception of a simple thing labeled a stimulus.

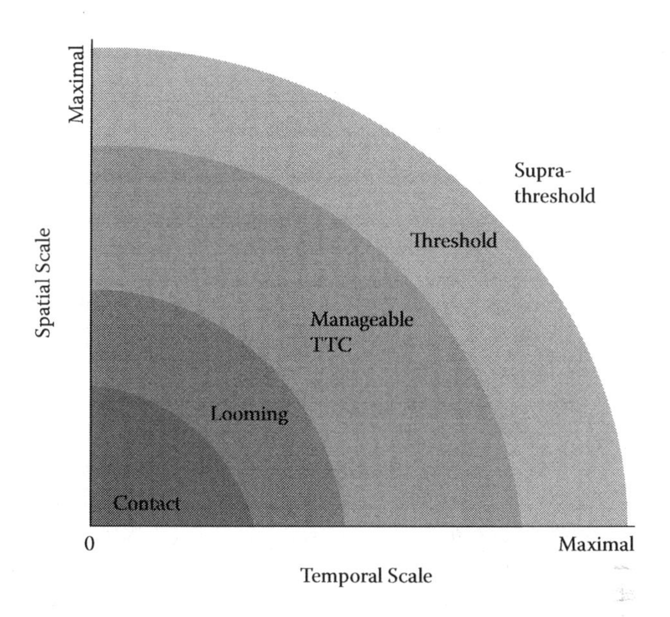

FIGURE 3.1 Cartesian coordinate structure of spatio-temporal regions of human (driver) navigation. The supra-threshold range denotes coordinates that exceed immediate perception. The sequential regions that proceed toward contact describe behavioral ranges that limit potential response. This form of coordinate system accords with previous conceptions of action–space (see Hancock and deRidder, 2003).

Rather, it is what Gibson (1950) so eloquently termed a "flowing array of stimulus energy." This flowing array is not any one individual stimulus, but stimuli in relation to other stimuli and environmental distracters, as well as in relation to the driver's own motion through the environment. Just as a common criticism of laboratory studies is the objection to static observers in front of stimuli of questionable ecological validity, we must view what the driver is perceiving and processing as an active, continuous stream of information, instead of somehow discrete quanta. This distinction may also be termed as the difference between absolute and relative estimates.

3.4 A BROAD VIEW OF DRIVER ESTIMATION PERFORMANCE CAPACITIES

As our driver continues along the main thoroughfare, other drivers slow in front of him or approach him from behind. This means the driver must perform an estimation process where he constantly monitors and adjusts for his own position relative to the environment. The majority of such estimations focus on estimates of time, speed, and distance. Time estimations are commonly manifested in estimates of time-to-contact (TTC) between the driver and another vehicle or object (Hancock and Manser, 1997). In the information processing paradigm, speed estimations are commonly viewed as relative estimates between the driver's own vehicle and either the roadway or other vehicles. Finally, estimations of distance are most often examined (in an experimental paradigm) by relative comparisons between the driver and another object in the environment. Each of these categories of estimations is manifested as our driver brakes for traffic slowed in front of the vehicle; the driver must perform constant, rapid, and highly accurate estimates of space and time (here in terms of TTC based on the closing speed of oncoming traffic and distance to such vehicles).

These estimates have each been thoroughly examined in the literature and we summarize the most pertinent findings herein. One important concept for the reader to keep in mind when considering driver estimations across space and time is that these two concepts (that is, space and time), and the derivation of velocity from them, are as closely interrelated in human navigation as they are in the physical world. The formulae for time, speed (velocity), and distance, with each expressed in terms of the others, are listed in the following expressions:

$$T = \frac{D}{S},\ S = \frac{D}{T},\ D = S \times T$$

This mathematical relationship between time and space is the foundation (whether known to the driver or not) of the information supporting tau-based estimates. Tau (τ) or the inverse of the rate of optic expansion is a measure of TTC using the size an image presents on the retina. Derived by Lee (1974, 1976) from Gibson's earlier work in ecological perception of visual flow (Gibson, 1966), tau describes the process by which moving objects are perceived, thus indicating the closure of spatial gaps between the driver and the intended path of motion. This is a commonly used metric in explaining driver reaction to closing or receding objects when examined as tau-dot ($\dot{\tau}$), which is tau's first temporal derivative allowing for accurate estimates of time to contact (Groeger and Brady, 1999). Early research (Lee, 1976) demonstrated the usefulness of τ as an explanatory variable by showing how a constant rate of τ at or above –0.5 would result in a safe braking maneuver with regard to a lead vehicle. This figure of $\tau \geq -0.5$ has become a standard for examining braking in lead vehicle scenarios (Lee, 1976). The distance threshold for τ may be expressed as

$$D = \left(\frac{WV}{0.003}\right)^2$$

where D is the distance in feet, W is the width of the target vehicle, and V is the closing rate in feet per second (Rock, Harris, and Yates, 2006). Although closure on a target may be detected from great distances, this formula illustrates the relative closeness required of the target for information from optic expansion to become useful to the driver.

However, τ does not provide a complete picture of driver braking. Other important factors in this continuous control task include the *perception* of the object's speed (Andersen, Cisneros, Atchley, and Saidpour, 1999), the ground plane (Fajen and David, 2003), and individual monocular and binocular depth cues (van der Kamp, Savelsbergh, and Smeets, 1997; DeLucia, Kaiser, Bush, Meyer, and Sweet, 2003). Since Lee's original work, alternative derivations of tau-controlled braking have emerged (see Flach, Smith, Stanard, and Dittman, 2004). One such visually guided braking hypothesis is the ideal deceleration strategy (Fajen, 2005), which proposes a process of braking controlled by keeping deceleration in a safe-zone below its maximum. One conclusion that may be drawn from this hypothesis is that ideal braking is less than maximum braking, with the specification of maximum braking changing based on the scenario's demands. Both of these conclusions reinforce the idea of driving as a satisficed process.

3.5 GAP ACCEPTANCE

Our driver, as the daily commute progresses, is entirely likely to encounter one of the most challenging and dangerous types of normal roadway environment: the left turn. The left turn scenario provides a situation in which the previously mentioned estimation processes must interact with another estimation process in which humans have mixed performance: gap acceptance. Similar to visually controlled braking, gap acceptance is a demonstration of estimations of motion and position in action. Humans tend to perform well at estimating safe gaps in intersections, with a few exceptions. An intersection gap acceptance study by Caird and Hancock (1994) demonstrated the largest problem in gap acceptance performance: the left turn. Their findings demonstrated that a variety of factors such as sex, vehicle types, and sight lines all have significant contributions to errors in estimation of time to contact. This is a direct precursor to and causal factor of accidents at intersections.

In an overall study of gap acceptance on minor–major roadway intersections controlled by stop signs, Fitzpatrick (1991) determined the average gaps accepted by drivers. The results of this study, for both low and higher volume intersections, are summarized in Table 3.1. A similar, later study by Harwood, Mason, and Brydia (2000) reported similar findings, with critical gaps of 6.5 and 8.2 s for passenger cars in right- and left-hand turns, respectively. Their findings for commercial vehicles were correspondingly longer, with tractor-trailer vehicles having critical gaps of 11.3 and 12.2 s for right- and left-hand turns, respectively. In addition to this data, they also collected adjustments for various environmental conditions. Multilane roads led to increases of 0.5 s (for cars) and 0.7 s (for commercial trucks) in critical gap acceptance. Deceleration of oncoming traffic led to reduction in critical gaps of 0.5 s across all vehicle types. This led the authors to recommend sight distances allowing 7.5 s for passenger cars along such roads.

TABLE 3.1
Gap Acceptance for Stop Controlled Intersections

	50th Percentile	85th Percentile	
		Moderate/High Volume	Low Volume
Cars	6.5	8.25	10.5
Trucks	8.5	10.0	15.0

Source: Fitzpatrick, 1991.
Note: All times listed in seconds.

One important consideration is the velocity at which vehicles are traveling. With every mile per hour (mph) in velocity equaling 1.467 ft traveled per second, drivers can quickly outrun sight distances. At the 50 mph speed limit, which is common on U.S. highways, just over 73 ft are passing every second. Providing the driver with additional margins of error (or accommodation) in sight distances can have a positive influence reducing the number and severity of accidents.

3.6 MOVEMENT PERCEPTION

Continuing along the roadway, our driver must constantly be aware of self-motion, which is here provided in terms of the vehicle. This is a process that requires an estimation of self-motion within the environment, and is a well studied topic in driving (Brown, 1931; Denton, 1980; Ryan and Zanker, 2001). Perception of one's own movement appears to be driven by a relatively small number of factors such as size of the objects within the driving environment, the distance that individual objects lie from the driver's current position, and the driver's useful field of view (UFOV; Roenker, Cissell, Ball, Wadley, and Edwards, 2003). However, research in specific areas of driver movement perception reveals some interesting subtleties in the human perceptual process.

Earlier studies have established some fairly stable and commonly accepted boundaries for the detection of movement in other vehicles. A study conducted by Mortimer, Hoffman, Poskocil, Jorgeson, Olson, and Moore (1974) found a threshold for angular velocity of 3.5×10^{-3} radians/sec. This perception occurred in relatively brief exposures. However, as velocity variations became less reliable (e.g., oscillations in velocity became of a lower frequency), drivers seemed to focus more on perceived headway. This is an important distinction, as drivers seem to be looking at motion in terms of the ability to safely traverse the roadway (see Gibson and Crooks, 1938) instead of examining for individual variations in the movement of other vehicles.

One possible method of alleviating gap acceptance and left-turn gap acceptance crashes is by reducing the speed of oncoming traffic, thereby increasing the time available for the turning driver to make accurate estimates of the potential hazards within the driving environment. Passive speed control measures provide an interesting modifier of driver perception of movement. These devices attempt to modify the drivers' perception of their own speed to effect change (either positive, negative, or to influence consistency) in driving speed. Several forms of passive speed control have been studied, most in the form of pavement markings. Among those types of pavement markings studied are longitudinal markings (Vey and Ferreri, 1968; Lum, 1984; Davidse, Van Driel, and Goldenbeld, 2004), transverse markings (Denton, 1980; Meyer, 2001), chevron-patterned markings (Griffin and Reinhardt, 1995; Helliar-Symons, Webster, and Skinner, 1995; Drakopoulos and Vergou, 2003), and tunnel markings (Manser and Hancock, 2007). Most studies demonstrate a great reduction in vehicle speed when such markings are installed. In the rare instances in which no main effect was found in speed modification due to the markings, positive guidance of the vehicle was still improved. This demonstrates the power of visual guidance in the perception of motion, and the installation of such passive speed

FIGURE 3.2 A passive speed control measure in the form of transversal lines. The distance (frequency) between the transversal lines, drawn on the roadway, decreases when approaching the roundabout. According to Denton's idea the driver will feel the illusion of driving at a higher speed when approaching the roundabout. (Adapted from Hills, 1980.)

control measures in problematic intersections with common gap acceptance accidents should be considered (see Figure 3.2).

One set of critical questions remains: Do humans estimate time, speed, distance, or some combination of the three? Or indeed, as derived quantitative metrics do perceptual capacities precede any such mathematical formulizations? These questions, with present methodologies, must be approached indirectly and leaves us with an unsatisfactory and incomplete picture. As researchers begin to approach driver decision making and estimation questions with more sophisticated techniques such as in-car "black box" naturalistic recording devices and traffic monitoring solutions, a more realistic and reliable picture of normal driving will emerge. The increase in technological abilities allowing traffic researchers to look at driver behavior in this manner, especially with regards to the combined subjective, performance, and physiological measurement of driver performance, should lead to a more comprehensive view of this problem.

3.7 CONCLUSIONS

Driving is a process that has evolved from our interactions with other humans in a social context. One may view an individual driving as a tripartite system consisting of control, navigation, and communication. Within, and central to, this system is our nature as social animals as all three components require the interaction with both socially controlled and individual driver-interaction scenarios.

It has often been stated that the process of driving is 90% visually controlled (Sivak, 1996). If we are to accept that driving is indeed 90% visual, we suggest that 90% or more of the estimations that drivers make are implicit in nature and therefore never reach the driver's conscious attention. Just as Leibowitz and Post (1982) suggest, by using ambient vision one may easily navigate and direct a car in a relatively empty environment. However, the complex social interactions that are imposed by modern driving necessitate the intense processing of information via estimations that in turn require focal vision. Focal vision may be seen as the counterpart of the social role of driving, where one must control a vehicle in a highly complex environment.

An example of the impact of societal forces on driver decision making and estimations may be observed in Drachten, Netherlands. In Drachten (and later in Christianfield, Denmark) many road signs, which have traditionally been used to influence driver behavior, have been removed, leaving only the social interaction between drivers and, interestingly, reducing the number of accidents (McNichol, 2004). In this situation the social rules are produced by the local interaction of the two (or more) drivers instead of imposed by forces that are remote and distal to this momentary, intimate interaction. Here, focal interactions and the resultant estimations are imposed by this immediate social, interpersonal interaction. This social dynamic between drivers causes the estimation process to shift from being primarily relative to mostly absolute in a rapid manner. A driver traveling into city limits will perform estimations that serve mainly to allow guidance of the vehicle along a path in the environment. However, once in the city environment the driver must identify other vehicles and determine their interactions, all while performing the primary tasks of driving and navigating safely. Removing the third-party regulation imposed by traffic control road signs transitions the entire process to the drivers, who at this point appear to be more accurate in their abilities of estimation than regulatory bodies in their attempts to order.

Driving is a dynamic process in which the driver must make both absolute and relative estimates and estimations. These estimations are made in relation to a multitude of objects in the driver's environment: the driver's own field of travel, the possibility of intruding objects, and the roadway surface. These estimations may be viewed as a satisficed process, which individual estimations may be suboptimal in order to produce an overall safe and relatively high performing system. The fact that so few accidents occur on the roadways of the world demonstrates this satisficed system is in itself a remarkable feat considering that we are asking humans to perform at a consistent and high level of driving precision with very little provided in means of second chances. Our hypothetical driver, nearing the end of his commute and finishing the drive with many of the same estimations required at the very beginning of the drive, demonstrates the amazing way that humans are able to detect, analyze, and move their automobiles through the roadway environment without (too frequent) intrusions into other vehicles' fields of safe travel.

REFERENCES

Anderson, G.J., Cisneros, J., Atchley, P., and Saidpour, A. (1999). Speed, size, and edge-rate information for the detection of collision events. *Journal of Experimental Psychology: Human Perception and Performance, 25*, 256–269.

Brown, J.F. (1931). The visual perception of velocity. *Psychologische Forschung, 14*, 199–232.

Caird, J.K., and Hancock, P.A. (1994). The perception of arrival time for different oncoming vehicles at an intersection. *Ecological Psychology, 6*(2), 83–109.

Davidse, R., Van Driel, C., and Goldenbeld, C. (2004). *The effect of altered road markings on speed and lateral position.* Leidschendam, Netherlands: SWOV, Institute for Road Safety Research.

DeLucia, P.R., Kaiser, M.K., Bush, J.M., Meyer, L.E., and Sweet, B.T. (2003). Information integration in judgments of time to contact. *Quarterly Journal of Experimental Psychology: Human Experimental Psychology, 56*(A), 1165–1189.

Denton, G.G. (1980). The influence of visual pattern on perceived speed. *Perception, 9,* 393–402.

Drakopoulos, A., and Vergou, G. (2003, July). *Evaluation of the converging chevron pavement marking pattern at one Wisconsin location.* Washington, DC: AAA Foundation for Traffic Safety.

Fajen, B.R. (2005). Calibration, information, and control strategies for braking to avoid a collision. *Journal of Experimental Psychology: Human Perception and Performance, 31,* 480–501.

Fajen, B., and David, R. (2003). Speed information and the visual control of braking to avoid a collision. *Journal of Vision, 3,* 555a.

Fitzpatrick, K. (1991). Gaps accepted at stop-controlled intersections. *Transportation Research Record, 1303,* 103–112.

Flach, J.M., Smith, M.R.H., Stanard, T., and Dittman, S.M. (2004). Collisions: Getting them under control. In H. Hecht and G.J.P. Savelsbergh (Eds.), *Time-to-contact* (pp. 67–91). Amsterdam: Elsevier.

Gibson, J.J. (1950). *The perception of the visual world.* Oxford: Houghton Mifflin.

Gibson, J.J. (1966). *The senses considered as perceptual systems.* Boston: Houghton Mifflin.

Gibson, J.J., and Crooks, L.E. (1938). A theoretical field-analysis of automobile-driving. *American Journal of Psychology, 51,* 453–471.

Griffin, L., and Reinhardt, R. (1995, August). *A review of two innovative pavement marking patterns that have been developed to reduce traffic speeds and crashes.* Washington, DC: AAA Foundation for Traffic Safety.

Groeger, J.A., and Brady, S.J. (1999). Tau dot or not? Visual information and control of car following. In A.G. Gale (Ed.), *Vision in vehicles VII.* Amsterdam: Elsevier.

Hancock, P.A. (1997). *Essays on the future of human–machine systems.* Eden Prairie, MN: Banta.

Hancock, P.A., and deRidder, S.N. (2003). Behavioral accident avoidance science: Understanding response in collision incipient conditions. *Ergonomics, 46*(12), 1111–1135.

Hancock, P.A., and Manser, M.P. (1997). Time-to-contact: More than tau alone. *Ecological Psychology, 9*(4), 265–297.

Hancock, P.A., and Scallen, S.F. (1999). The driving question. *Transportation Human Factors, 1,* 47–55.

Harwood, D.W., Mason, J.M., Jr., and Brydia, R.E. (2000). Sight distance for stop-controlled intersections based on gap acceptance. *Transportation Research Record, 1701,* 32–41.

Helliar-Symons, R., Webster, P., and Skinner, A. The M1 chevron trial. *Traffic Engineering and Control, 36,* 563–567.

Hills, B.L. (1980). Vision, visibility, and perception in driving. *Perception, 9*(2), 183–216.

Lee, D.N. (1974). Visual information during locomotion. In R.B. MacLeod and H.L. Pick, Jr. (Eds.), *Perception: Essays in honor of James J. Gibson* (pp. 250–267). Ithaca, NY: Cornell University Press.

Lee, D.N. (1976). A theory of visual control of braking based on information about time to collision. *Perception, 5,* 437–459.

Leibowitz, H.W., and Post, R.B. (1982). The two modes of processing concept and some implications. In J. Beck (Ed.), *Organization and representation in perception* (pp. 343–363). Hillsdale, NJ: Lawrence Erlbaum.

Lum, H. (1984). The use of road markings to narrow lanes for controlling speed in residential areas. *ITE Journal, 54*(6), 50–53.

Manser, M.P., and Hancock, P.A. (2007). The influence of perceptual speed regulation on speed perception, choice, and control: Tunnel wall characteristics and influences. *Accident Analysis and Prevention, 39*, 69–78.

McNichol, T. (2004). Roads gone wild. Retrieved June 20, 2007, from http://www.wired.com/wired/archive/12.12/traffic.html.

Meyer, E. (2001). A new look at optical speed bars. *ITE Journal, 71*(11), 44–48.

Mortimer, R.G., Hoffman, E.R., Poskocil, A., Jorgeson, C.M., Olson, P.L., and Moore, C.D. (1974, November). *Studies of automobile and truck rear lighting and signaling systems* (Report No. UM-HSRI-HF-74-25). Ann Arbor, MI: Highway Safety Research Institute.

National Highway Traffic Safety Administration (NHTSA). (2007). Fatality Analysis Reporting System (FARS). Database retrieved from ftp://ftp.nhtsa.dot.gov/FARS.

Rock, P., Harris, M.G., and Yates, T. (2006). A test of the tau-dot hypothesis of braking control in the real world. *Journal of Experimental Psychology: Human Perception and Performance, 32*, 1479–1484.

Roenker, D.L., Cissell, G.M., Ball, K.K., Wadley, V.G., and Edwards, J.D. (2003). Speed-of-processing and driving simulator training result in improved driving performance. *Human Factors, 45*, 218–233.

Ryan, J., and Zanker, J.M. (2001). What determines the perceived speed of dots moving within apertures? *Experimental Brain Research, 141*, 79–87.

Schiff, W., and Detweiler, M.L. (1979). Information used in judging impending collision. *Perception, 8*(6), 647–658.

Senders, J.W., Kristofferson, A.B., Levison, W., Dietrich, C.W., and Ward, J.L. (1966). *An investigation of automobile driver information processing* (Report No. 1335). Washington, DC: U.S. Dept of Commerce, Bureau of Public Roads.

Simon, H.A. (1969). *The sciences of the artificial.* Cambridge, MA: MIT Press.

Sivak, M. (1996). The information that drivers use: Is it indeed 90% visual? *Perception, 25*(9), 1081–1089.

van der Kamp, J., Savelsbergh, G., and Smeets, J. (1997). Multiple information sources in interceptive timing. *Human Movement Science, 16*, 787–821.

Vey, A.H., and Ferreri, M.G. (1968). The effects of lane width on traffic operation. *Traffic Engineering, 38*(11), 22–27.

4 A Two-Dimensional Framework for Understanding the Role of Attentional Selection in Driving

Lana M. Trick and James T. Enns

CONTENTS

Reflection 63
4.1 Introduction 64
4.2 Dimensions of Attentional Selection in Driving 64
4.3 Four Modes of Attentional Selection 65
4.3.1 Reflexive (Automatic-Exogenous) Selection 67
4.3.2 Habitual (Automatic-Endogenous) Selection 68
4.3.3 Exploratory (Controlled-Exogenous) Selection 69
4.3.4 Deliberate (Controlled-Endogenous) Selection 70
4.4 Conclusions 71
References 72

REFLECTION

This chapter was written in reaction to the fragmentation that we perceived in both the basic research on attention and the applied research on driving. Within the basic research, there are a series of micro-theories explaining performance in specific experimental tasks, but there is no overarching theory of attention. Similarly, within the driving research, there are three largely independent traditions: the experimental research, which investigates the effects of various situational variables on driving performance using driving simulators or closed-circuit test courses; the individual differences research, which investigates the attributes of the collision-prone driver, often using psychological tests; and the automation research, which investigates the impact of devices designed to provide drivers with information or take over various aspects of the driving task. Our framework was designed to unify the basic and applied research, and integrate different streams within the driving literature. The advantage to this common framework is that it helps organize the research, revealing situations where researchers from very different traditions may be doing related research, and conversely, situations where researchers from the same tradition may be producing conflicting results because they are actually investigating different mechanisms of attentional selection. Although it was impossible to list all types of driving research, this chapter was designed to give a sense of the breadth of the field and the diverse ways that attentional selection can affect driving performance.

4.1 INTRODUCTION

Lapses in selective attention, either through inattention or distraction, cause many crashes (e.g., Neale, Dingus, Klauer, Sudweeks, and Goodman, 2005). Although there is consensus that attention has an important role to play in explaining driver performance, the research is fragmented. The goal of this chapter is to provide a framework to unify the literature, drawing together diverging threads to reveal underlying commonalities. In this chapter we summarize fundamental features of the framework, highlighting findings that exemplify key principles. The chapter is divided into three sections: the first introduces the two global dimensions that serve as the basis for the framework; the second outlines the framework, summarizing the four modes of selection; and the third touches on practical implications for crash prevention.

4.2 DIMENSIONS OF ATTENTIONAL SELECTION IN DRIVING

Selective attention is thought to be necessary because there are too many things in the environment to perceive and respond to at once. Specifically, individuals fail to select the appropriate information from the stimulus environment (i.e., they look but fail to see) or fail to select the appropriate response at the appropriate time (i.e., they know what to do but fail to do it). However, at present there is no general theory of selective attention. Instead there are micro-theories for specific tasks, tasks such as orienting, visual search, filtering, multiple-action monitoring (dual task), and multiple-object tracking. It is our view that differential performance on these tasks reflects the presence of two underlying dimensions of attentional selection (Trick, Enns, Mills, and Vavrik, 2004).

One concerns the issue of awareness and involves the distinction first made by Shiffrin and Schneider (1977) between automatic and controlled processes. Specifically, there are two ways that a selection process might work. First, stimuli and responses might be selected without awareness. *Selection without awareness* has been called preattentive, inattentional, subconscious, unconscious, and unintentional by different authors, but regardless, this type of selection is *automatic*. Automatic selection is rapid, effortless, and unconscious, and is difficult to stop or modify once initiated. These processes are triggered by the presence of certain stimuli in the environment, and they run to completion without interfering with other processes. Second, stimuli and responses might be selected deliberately with awareness. *Selection with awareness* (variously called attentive, conscious, or intentional) involves *controlled processing*, which is to say that selection is effortful and slow, but it can be started, stopped, or modified at will, a feature that makes this type of processing flexible and intelligent. Controlled processes can cause changes in long-term memory through learning, and with adequate practice some types of controlled process may even become automatic. The fundamental problem with controlled processing is that it is difficult to carry out several controlled processes at once. Though the distinction between automatic and controlled processing is often discussed as if it were a strict dichotomy, we believe that it is probably more useful to consider it a continuum. Some processes are more automatic than others in the sense that they are initiated more quickly, require less effort, are more likely to be evoked unintentionally in a given situation, and are thus more difficult to bring under deliberate control.

The second dimension in the framework concerns the origin of the process, whether it is innately specified and thus common to all (exogenous) or engendered by a person's specific goals and thus idiosyncratic (endogenous). *Exogenous selection* occurs as a result of the way humans are built: the nervous system is structured to respond preferentially so that there is an innate continuum of salience, with some stimuli and responses more likely to receive exogenous selection than others. In contrast, *endogenous selection* results from what people know about an environment and what they want to achieve, and it is thus idiosyncratic and situation-specific. People actively search the environment for information relevant to their specific goals or intentions and perform these tasks in ways that are consistent with their own expectations and previous learning. These expectancies may act as a form of "perceptual set" causing people to look for specific objects at certain locations.

Considering attention in this way explains findings within the driving literature that otherwise might not be explained. As well, it clarifies a confusion that exists within both the basic and applied literatures: the conflation of the distinction between automatic and controlled processes with the distinction between exogenously triggered and endogenously initiated processes. We believe this confusion has caused some issues to be neglected, particularly those relating to how innately determined (exogenous) factors that affect human behavior in general might influence attentional selection when driving, as will be discussed next.

4.3 FOUR MODES OF ATTENTIONAL SELECTION

The combination of automatic and controlled processing with exogenous and endogenous selection produces the framework shown in Figure 4.1. In this framework, two forms of selection involve automatic processes. We call these reflex and habit. *Reflexes* of selection are automatic processes that are innately specified and triggered by the presence of certain stimuli in the environment. These processes initiate effortless, unconscious, and obligatory responses that occur even when inappropriate. Reflexes are not learned and consequently they cannot be unlearned. At best, when a reflexive process is counterproductive, the response can be reduced in intensity or "undone" after the fact, but in most cases this requires controlled-endogenous processing.

Habits are processes that come into existence when the operations necessary to fulfill a certain goal are carried out so often in a certain stimulus context that the processes become automatic and are carried out as soon as the person is in that context.

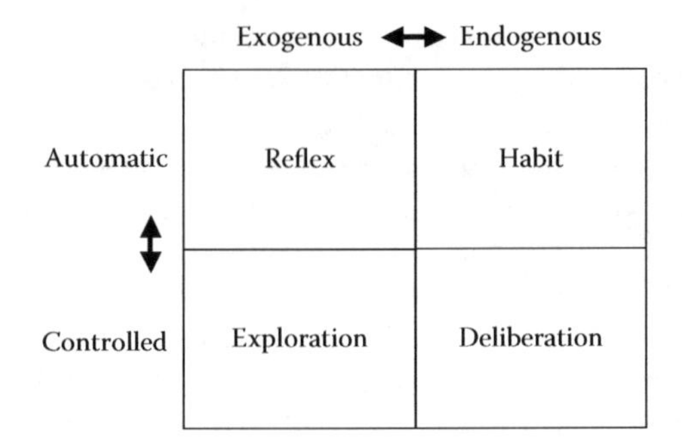

FIGURE 4.1 Four modes of attentional selection.

Although habits are the basis of skilled behavior, they can be problematic. If a habit is strongly associated with a specific situation, once an individual is in that situation it will require effort and planning (controlled-endogenous processing) to avoid acting in accordance with the habit. There are similarities between reflexive and habitual selection, but they differ in two important ways. First, though both are "triggered" by the presence of certain stimuli, the triggers for reflexes are innately set, whereas triggers for habits are learned, and this makes habits idiosyncratic and reflexes common to all. Second, a developmental timetable determines when reflexes emerge, but they are stable once acquired. Habits can be formed at any time and can also be replaced or fade at any time due to lack of practice or new learning. In the continuum of automaticity, where some processes are more automatic than others, reflexes retain their position near the extreme end of the automaticity continuum, whereas habits change their level of automaticity based on the frequency and recency of practice.

The other two modes of performance are controlled (exploration and deliberation). *Exploration* is the default mode for controlled processing—a type of selection that is carried out in absence of a specific task goal. Instead, exploration involves a generic goal—one common to all human beings in any environment—to maximize the acquisition of new information. Confronted with a new environment, humans find exploration rewarding in itself. Cognitive neuroscientists have recently linked this behavioral tendency of humans to the neurochemical finding that areas in the brain that are used to process visual information are arranged in a hierarchy, with natural opiates (chemicals linked to the experience of pleasure) present in only small amounts in the areas used for the simple registration of information, but present in increasingly larger amounts in areas used in the comprehension and interpretation of images (Biederman and Vessel, 2006). Thus, there is a growing understanding that the human tendency to be both "infovores" and to be "pleasure seeking" are inextricably linked in the way we are wired. We argue that in the absence of specific goals, these innate preferences set the default for what is attended when humans explore new environments for which they have no specific expectations, that is, environments lacking the stimulus triggers necessary to evoke specific reflexes or habits. The generic goal to acquire new information can be overridden without undue effort or planning once an individual undertakes a specific goal. To date, there is little research on exploratory selection.

In contrast, there are many studies of deliberate selection. *Deliberate selection* involves the execution of a chosen attention-demanding process at the expense of other processes. This type of processing involves conscious goals that reflect an individual's specific knowledge, plans, and strategies for a certain situation, and these goals determine what is selected. Deliberate processing is flexible and responsive to new information because it is conscious and internally directed. With this type of processing there is hope of changing behavior rapidly (within seconds) in response to symbolic information, such as an oral command or written message. Processes that involve deliberate selection are necessary whenever the situation is difficult or novel, and when unruly habits or reflexes must be brought under control. However, deliberate selection is noticeably effortful. Moreover, because controlled processes interfere with one another, processes that involve deliberation preclude general exploration and impede other deliberate processes.

In the following sections we will work through the four modes of selection, applying the framework to studies chosen to represent a range of topics in the driving literature.

4.3.1 Reflexive (Automatic-Exogenous) Selection

Certain stimuli initiate effortless, unconscious, obligatory responses that occur even when counterproductive. This type of selection is not learned and may even be present in the very young. Generally, bringing reflexive selection under control requires deliberate (controlled-endogenous) processes though in some cases it may be possible to learn to compensate using a habitual response if a deliberate compensation process is practiced often enough. Regardless, it is important to note that the reflexive response is always there. It must be brought under control by other processes if it is to be avoided.

There is little research in this area, and at present some of the clearest demonstrations involve visual illusions. These are cases where certain stimulus configurations are selected and processed to yield a percept at odds with reality. Processing is clearly automatic (because it occurs effortlessly, even when counterproductive and inaccurate), and it is exogenous (prompted by natural reactions to environmental stimuli). Hills (1980) described an accident-inducing "perceptual trap" created when two unconnected roads appear to be coextensive from the driver's perspective due to automatic perceptual grouping processes (grouping by good continuation). Similarly, there is an illusion that causes vehicles to appear to be moving more slowly than they actually are when viewed through fog (Krekelberg, van Wezel, and Albright, 2006). Illusions can also be used to encourage safe driving, as shown by Shinar, Rockwell, and Malecki (1980) when they induced drivers to slow down on a dangerous curve by using transverse road markings that produced an illusion of speed.

Sudden luminance onsets can trigger automatic eye movements and the reassignment of the attentional focus (Theeuwes, Kramer, Hahn, and Irwin, 1998), and this can be understood as an example of reflexive selection. It occurs whether or not the onsets are predictive of future events (e.g., stimulus cues in covert orienting tasks) and is even evident in young children (Plude, Enns, and Brodeur, 1994). Moreover, overcoming this tendency requires deliberate (endogenous-controlled) processing—either advanced planning (deliberately directing attention to another location) or compensation after the fact (Theeuwes et al., 1998). This finding has relevance given the use of flashing lights on emergency vehicles and a recent suggestion that they also be used in brake lights (Berg, Berglund, Strang, and Baum, 2007). Certain types of auditory and tactile stimuli can be selected reflexively as well, and there are a number of researchers exploring the use of auditory and vibrotactile warnings to prevent collisions (e.g., Abdelsalam, Desroches, Famewo, Nonnecke, and Varden, 2006; Desroches, Varden, Nonnecke, and Trick, 2007; Ho, Reed, and Spence, 2006).

There are also situations where reflexive selection guides response selection. For example, there is literature on reflexive processing of faces, and it is interesting to note that when drivers encounter cyclists, they reflexively direct their gaze towards cyclists' faces when trying to infer their intentions—a practice that can lead to problems (Walker and Brosnan, 2007). There are also cases where one action causes the reflexive selection of another, as shown by an effect often observed by driving instructors. Novice drivers tend to turn the steering wheel in the same direction as they move their eyes, for example, steering left when looking left. This can cause crashes and drivers have to be trained not to do it. Controlling this reflexive tendency seems to require attentional resources (deliberate selection). Readinger, Chatziastros, Cunningham, Bulthoff, and Cutting (2002) showed that experienced drivers also tend to steer in the direction of their gaze when they are required to perform a secondary task. Because this tendency to steer in the direction of gaze can even be seen in young children learning to drive tricycles, this may represent an example of a reflexive association between responses.

4.3.2 Habitual (Automatic-Endogenous) Selection

When a goal is enacted repeatedly, carrying it out can become habitual and unconscious, and the processes associated with it may become effortless. When this occurs, it becomes possible to carry out those operations while performing another task with little interference. In fact, these habits are a large part of what is meant by driving skills. Although habits are often thought of as overlearned actions, we propose that there can be habits of stimulus selection as well, and these govern what type of information is selected, and where, when, and how drivers scan the driving environment.

Novice drivers are extraordinarily prone to accidents and it is commonly believed that this is partly because they lack automaticized behaviors that form the basis of driving skill. This is supported by many studies that show experienced drivers have less difficulty than inexperienced drivers when carrying out secondary tasks while driving (e.g., Shinar, Meir, and Ben-Shoham, 1998). These problems are especially notable in teen drivers, who are disproportionately at risk when there are passengers in the vehicle, or when using in-vehicle devices such as cellular phones (Chen, Baker, Braver, and Gouhua, 2000; Neyens and Boyle, 2007). Experienced drivers do not require as much controlled processing to carry out the basic operations necessary for driving, and as a result, they multitask with less deterioration to their driving performance.

When drivers learn to drive, one of the things that develop is the ability to sense hazards quickly and efficiently. Drivers develop habits of sensory selection that enable them to automatically encode the safety-related aspects of a driving scene. As a result, there is less change blindness (an inability to notice safety-related change in an image) in experienced drivers (e.g., Famewo, Trick, and Nonnecke, 2006). There is also evidence that with experience, road signs begin to cause automatic priming, not only for the appearance of the same sign but for the related road scene (Crundall and Underwood, 2001). In experienced drivers, this priming causes the same sign to be perceived more quickly and accurately when presented a second time, but it also causes faster recognition of the associated hazard on the road.

Although habits of selection can be helpful, they can also put drivers at risk. The most obvious examples involve cases where drivers import their well-developed habits into new situations that require different behaviors, as occurs when they drive in rented cars (Al-Balbissi, 2001) or foreign countries (Summala, 1998). Habits of stimulus selection form a perceptual set, which yields an advantage in familiar situations but a disadvantage in atypical situations, temporarily "blinding" drivers to things that run counter to their expectations—things in plain view. This can even occur when the object in question is as brightly colored as a police car. Langham, Hole, Edwards, and O'Neil (2002) found that experienced drivers were faster to detect the presence of a parked police car than inexperienced drivers when the police car was in the standard position (angle parked) but they were slower than inexperienced drivers when the car was in a nonstandard (straight on) position (see also Most and Astur, 2007).

The driving environment is forgiving, insofar as collisions are rare, and dangerous practices typically do not produce adverse consequences on a day-to-day basis. As a result, bad habits can develop over time, reversing the positive effects of driver training. Duncan, Williams, and Brown (1991) noted that drivers who had received their license in the last year checked their mirrors more often than more experienced drivers and were also more likely to leave an adequate distance when following another car. Drivers may develop the habit of exceeding the speed limit when they drive. Conforming to speed restrictions seems to require deliberate (controlled-endogenous) processing because secondary tasks interfere (Recarte and Nunes, 2002).

4.3.3 Exploratory (Controlled-Exogenous) Selection

Exploratory selection governs where attention goes when there is no specific task goal. Driving researchers were among the first to notice the importance of exploratory selection. Hills (1980) observed that experienced drivers, when not fully occupied with the driving task, look away from the relevant driving-related information and explore roadside advertising, low-flying planes, trees, etc. It may be impossible to prevent exploratory selection (Hughes and Cole, 1986). If driving does not require drivers' full attention, they devote their attention elsewhere.

What determines what things attract attention when an individual is exploring the environment without a specific goal? Although there has been relatively little research on this topic, Hughes and Cole (1986) noted that the "sensory conspicuity" of an object, its tendency to attract attention even when it is not deliberately sought, is determined in part by its size, eccentricity, and contrast with the background. Crundall, van Loon, and Underwood (2006) found that street-level advertisements were more likely to divert eye movements than raised advertisements. The study of exploratory selection is relevant for the development of effective safety-related signs.

However, sometimes the stimuli that attract attention are inside the vehicle. Exploratory selection is especially likely to occur in conditions of novelty, and this may explain why drivers spend disproportionate amounts of time looking at in-vehicle devices when they are new (e.g., Dingus, Hulse, Mollenhauer, Fleischman, McGehee, and Manakkal, 1997). These devices are often engineered with colorful, high-contrast displays to maximize legibility, and as a result they may compete successfully with extravehicular stimuli for exploratory selection. This may be particularly problematic if the display for the in-vehicle device is superimposed on the outside world, as occurs with "heads-up" displays. Tufano (1997) argues that the salience of heads-up displays interferes with a driver's ability to see what is going on in the outside world, particularly when the outside events are unexpected.

4.3.4 Deliberate (Controlled-Endogenous) Selection

Deliberate selection is the most flexible and intelligent of all, and when driving, these processes are necessary in a variety of situations: (a) when conditions are challenging (low visibility, heavy traffic, unexpected events, unfamiliar environments); (b) when individuals perform unfamiliar activities or combinations of activities (dual tasks) that require an action plan to be constructed on-line using moment-to-moment feedback from the environment; (c) when individuals are acting strategically, and not simply reacting to events in the immediate environment; (d) when individuals react to symbolic information that must be interpreted to be acted upon; and (e) when maladaptive habits and reflexes must be monitored and overcome.

There are a large number of studies relevant to deliberation, including those looking at the influence of secondary tasks such as using cellular phones (e.g., Horrey and Wickens, 2006) or using automated systems designed to provide drivers with information (e.g., Burnett, Summerskill, and Porter, 2004; Hampton and Langham, 2005; Stanton and Young, 1998). It is clear that when the resources for deliberate selection are focused on a task, it can produce prolonged inattentional blindness for things in plain view—as occurred when a road crew inadvertently paved over the carcass of a deer (see Most, Scholl, Clifford, and Simons, 2005).

There are also a variety of studies that examine factors that reduce the resources available for controlled (deliberate) processing, such as alcohol and fatigue (e.g., Mascord, Walls, and Starmes, 1995). Alcohol depresses neural activity, and thus has significant effects on a wide range of sensory, cognitive, and motor processes. However, controlled (deliberate) selection processes are especially vulnerable because they are the most demanding and time-consuming. Similarly, even before drivers fall asleep, fatigue affects deliberate selection as drivers begin to withdraw resources necessary for deliberate (effortful) selection. It has also been noted that automated systems that take over various aspects of the driving task may put drivers at risk because they encourage drivers to withdraw resources used for deliberate selection, thus leaving drivers "out of the loop" when the systems fail (Stanton and Young, 1998).

Factors related to deliberate selection also predict individual differences in crash risk, particularly among senior drivers. Several components of deliberate selection are affected by age (Plude et al. 1994): attention switching becomes more difficult (as shown by orienting tasks), attentional search is slowed, and dual-task interference is exaggerated. In addition, with age there are exaggerated reductions in "useful field of view" with increases in primary task complexity and number of distractors. There are also age-related reductions in both sensory and motor function in older adults. Nonetheless, in general, it is factors related to deliberate selection that best predict accident risk in older drivers (e.g., Ball and Owsley, 1991; Lundberg, Hakamies-Blomqvist, Almkuist, and Johansson, 1998).

Deliberate selection involves executive functions associated with the frontal lobes, and there have studies that look at the impact of frontal lobe damage on driving performance (e.g., Wikman et al., 2004). Similarly, one of the dominant accounts of attention deficit hyperactivity disorder (ADHD) suggests that individuals with this disorder have deficits in frontal lobe function that cause them to be disproportionately at risk for accidents of all sorts, but most notably, accidents when driving (e.g., Barkley, 1997).

4.4 CONCLUSIONS

This way of understanding attentional selection has ramifications for policymakers and driving-safety professionals interested in reducing the number of traffic accidents. It is widely agreed that driver inattention is the cause of many vehicle crashes, and, generally speaking, there are two approaches that might be used to prevent such incidents: either modify the driving environment (change road and vehicle design) or modify the drivers so that they change their behavior (induce them to change their personal goals, expectations, knowledge, and behavioral repertoire). The framework proposed here provides a straightforward way of identifying problems best remedied with environmental interventions and those that require behavioral interventions. The origin of the process (exogenous vs. endogenous) determines what type of intervention will work best. Problems originating from exogenous selection are most effectively dealt with using environmental solutions; it is better to work with the nervous system than against it. Problems relating to endogenous selection will be more amenable to behavioral solutions.

There are also implications for determining the type of behavioral intervention that will be most effective. Automatic (habitual) and controlled (deliberate) processes respond best to different types of intervention. Behavioral interventions that involve aversive or pleasant stimuli in the immediate driving environment will be most effective with habits. If the problematic behavior is the result of deliberate selection then there is hope for interventions that involve attitude change, education, and long-term penalties and incentives (providing the driver is not suffering from some form of frontal lobe dysfunction). Maladaptive driving behaviors are often maintained by both habit and deliberation, and consequently two-level interventions may be necessary.

REFERENCES

Abdelsalam, W., Desroches, P., Famewo, J., Nonnecke, B., and Varden, A. (2006, June 11–14). *Effectiveness of in-vehicle moose warning modalities: An experimental study.* Proceedings CMRSC-SVI of the Canadian Association of Road Safety Professionals, Winnipeg, Manitoba.

Al-Balbissi, A. (2001). Unique accident trend of rental cars. *Journal of Transportation Engineering, 127*, 175–177.

Ball, K., and Owsley, C. (1991). Identifying correlates of accident involvement for the older driver. *Human Factors, 33*, 583–595.

Barkley, R. (1997) Behavioral inhibition, sustained attention, and executive functions: Constructing a unified theory of ADHD. *Psychological Bulletin, 121,* 65–94.

Berg, W.P., Berglund, E.D., Strang, A.J., and Baum, M.J. (2007). Attention-capturing properties of high frequency luminance flicker: Implications for brake light conspicuity. *Transportation Research Part F, 10*, 22–32.

Biederman, I., and Vessel, E.A. (2006). Perceptual pleasure and the brain. *American Scientist, 94*, 247–253.

Burnett, G.E., Summerskill, S.J., and Porter, J.M. (2004). On-the-move destination entry for vehicle navigation systems: Unsafe by any means? *Behavior and Information Technology*, *23*, 265–272.

Chen, L., Baker, S., Braver, E., and Gouhua, L. (2000). Carrying passengers as a risk factor for crashes fatal to 16–17-year-old drivers. *Journal of the American Medical Association, 283*, 1578–1617.

Crundall, D., and Underwood, G. (2001). The priming function of road signs. *Transportation Research Part F, 4*, 187–200.

Crundall, D., van Loon, E., and Underwood, G. (2006). Attraction and distraction of attention with roadside advertisements. *Accident Analysis and Prevention, 38,* 671–677.

Desroches, P., Varden, A., Nonnecke, B., and Trick, L. (2007, April 16–19). *Validating driving simulators: Do simulators and circuit tracks yield similar results for moose-related research?* (SAE technical paper 07B-507 Booth B3a). SAE World Congress and Exhibition, Detroit, MI.

Dingus, T., Hulse, M., Mollenhauer, M., Fleischman, R., McGehee, D., and Manakkal, N. (1997). Effects of age, system experience, and navigation technique on driving with an advanced traveler information system. *Human Factors, 39*, 177–199.

Duncan, J., Williams, P., and Brown, I. (1991). Components of driving skill: Experiences does not mean expertise. *Ergonomics, 34,* 919–937.

Famewo, J., Trick, L.M., and Nonnecke, B. (2006, July 27–30). *How does driving experience affect allocation of attention in complex signaled intersections?* Paper presented at 11th International Conference on Vision in Vehicles, Dublin, Ireland.

Hampton, P., and Langham, M. (2005). A contextual study of police car telematics: The future of in-car information systems. *Ergonomics, 2,* 109–118.

Hills, B. (1980). Vision, visibility, and perception in driving. *Perception, 9*, 183–216.

Ho, C., Reed, N., and Spence, C. (2006). Assessing the effectiveness of "intuitive" vibrotactile warning signals in preventing front-to-rear-end collisions in a driving simulator. *Accident Analysis and Prevention, 38*, 988–996.

Horrey, W.J., and Wickens, C.D. (2006). Examining the impact of cell phone conversations on driving using meta-analytic techniques. *Human Factors*, *48*(1), 195–205.

Hughes, P., and Cole, B. (1986). What attracts attention when driving? *Ergonomics, 29*, 377–391.

Krekelberg, B., van Wezel, R., and Albright, T. (2006). Interactions between speed and contrast tuning in the middle temporal area: Implications for the neural code for speed. *Journal of Neuroscience, 26*(35), 8988–8998.

Langham, M., Hole, G., Edwards, J., and O'Neil, C. (2002). An analysis of "looked but failed to see" accidents involving parked police vehicles. *Ergonomics, 45*(3), 167–185.

Lundberg, C., Hakamies-Blomqvist, L., Almkuist, O., and Johansson, K. (1998). Impairments of some cognitive functions are common in crash-involved older drivers. *Accident Analysis and Prevention, 30*(3), 371–377.

Mascord, D., Walls, J., and Starmes, G. (1995). Fatigue and alcohol: Interactive effects on human performance in driving-related tasks. In L. Hartley (Ed.), *Fatigue and driving: Driver impairment, driver fatigue, and driving simulation* (pp. 189–205). London: Taylor & Francis.

Most, S.B., and Astur, R.S. (2007). Feature-based attentional set as a cause of traffic accidents. *Visual Cognition, 15*(2), 125–132.

Most, S.B., Scholl, B.J., Clifford, E.R., and Simons, D.J. (2005). What you see is what you set: Sustained inattentional blindness and the capture of awareness. *Psychological Review, 112*(1), 217–242.

Neale, V.L., Dingus, T.A., Klauer, G.S., Sudweeks, J., and Goodman, M. (2005). *An overview of the 100-car naturalistic study and findings* (DOT HS Publication 05-0400). Washington, DC: National Highway Traffic Safety Administration.

Neyens, D.M., and Boyle, L.N. (2007). The effect of distractions on the crash types of teenage drivers. *Accident Analysis and Prevention, 39*, 206–212.

Plude, D., Enns, J.T., and Brodeur, D.A. (1994). The development of selective attention: A lifespan overview. *Acta Psychologica, 86*, 227–272.

Readinger, W., Chatziastros, A., Cunningham, D., Bulthoff, H., and Cutting, J. (2002). Gaze-eccentricity effects on road position and steering. *Journal of Experimental Psychology: Applied, 8*(4), 247–258.

Recarte, M.A., and Nunes, L.M. (2002). Mental load and loss of control over speed in real driving. Towards a theory of attentional speed control. *Transportation Research: Part F, 5*, 111–122.

Shiffrin, R., and Schneider, W. (1977). Controlled and automatic human information processing: II. Perceptual learning, automatic attending, and a general theory, *Psychological Review, 84*, 127–190.

Shinar, D., Meir, M., and Ben-Shoham, I. (1998). How automatic is manual gear shifting? *Human Factors, 40*(4), 647–654.

Shinar, D., Rockwell, T., and Malecki, J. (1980). Effects of changes in driver perception on rural curve negotiation. *Ergonomics, 23*, 263–275.

Stanton, N., and Young, M. (1998). Vehicle automation and driving performance, *Ergonomics, 41*, 1014–1028.

Summala, H. (1998). American drivers in Europe: Different signing policy may cause safety problems at uncontrolled intersections. *Accident Analysis and Prevention, 30*(2), 285–289.

Theeuwes, J., Kramer, A., Hahn, S., and Irwin, D. (1998). Our eyes do not always go where we want them to go: Capture of the eyes by new objects. *Psychological Science, 9*, 379–385.

Trick, L.M., Enns, J., Mills, J., and Vavrik, J. (2004). Paying attention behind the wheel: A framework for studying the role of selective attention in driving. *Theoretical Issues in Ergonomic Science*, *5*(5), 385–424.

Tufano, D. (1997). Automatic HUD's: The overlooked safety issue. *Human Factors, 39*, 303–311.

Walker, I., and Brosnan, M.B. (2007). Drivers' gaze fixations during judgments about a bicyclist's intentions. *Transportation Research Part F, 10*, 90–98.

Wikman, A., Haikonen, S, Summala, H., Kalska, H., Hietanen, M., and Vilkki, J. (2004). Time-sharing strategies in driving after various cerebral lesions. *Brain Injury, 18*(5), 419–423.

5 Driver Distractions

Miguel A. Recarte and Luis M. Nunes

CONTENTS

Reflection 75
5.1 Introduction 76
5.2 Attention, Distraction, and Driving 76
5.2.1 Distraction, Attentional Efficiency, and Errors 76
5.2.2 Distraction and Accidents 77
5.2.3 Distraction, Consciousness, and Attribution of Responsibility 77
5.2.4 Information Processing Errors and Types of Distraction 78
5.3 Visual Attention in the Driving Context 79
5.3.1 Characterization of the Driver's Scenario 80
5.3.2 Visual Parameters, Visual Search, and Attention 80
5.4 Environment and Distraction 81
5.4.1 Distraction Due to Systems Designed to Attract Attention 81
5.4.2 Distraction Due to Systems Imposing High Visual or Cognitive Demands 82
5.4.3 Distraction Due to Systems Affecting Drivers' Expectations 82
5.5 Thinking and Driving 83
5.5.1 Coping with Everyday Driving: Cognitive Effort and Internal Distraction 83
5.5.2 Measuring Mental Load 83
5.5.3 Mental Load, Vision, Performance, and Inattentional Blindness 84
5.5.4 Mental Load, Speed Perception, and Speed Control 85
5.6 Conclusions 86
Acknowledgments 87
References 87

REFLECTION

The role of our chapter on attention/distraction in this book on visual and cognitive performance in driving is to point out the importance of attention for adequate perception and its insertion in the higher cognitive processes involved in driving. Attention is a necessary condition to achieve a complete perception. Here, attention is discussed from its dysfunctional side, usually called distraction. Our particular contribution stresses the effects of purely cognitive load as opposed to those involving visual load. Along these lines, the chapter condenses more than a decade of experimental research activity on visual search and attention in driving. The main focus of our research has been to stress the relevance of cognitive load to explain changes in visual search, cognitive processing, and action. A great effort was made to implement the appropriate technology (the Argos instrumented vehicle), allowing realistic on-road experiments, including several hours per driver of continuous and accurate quantitative data on gaze, pupil, and other behavioral parameters. The results provided a solid empirical basis for the discussion on relevant issues regarding the effects of cognitive activity unrelated to driving on traffic safety and drivers' efficiency.

5.1 INTRODUCTION

The traffic environment is an example of a paradigm in which survival relies on attention, particularly on visual attention. Distraction, as a common explanatory concept for traffic accidents, suggests, as the opposite of attention, any kind of attentional inefficiency: a dysfunction in information processing leading to increased risk and human error. From this wide perspective, distraction invokes a multiplicity of psychological processes. This chapter attempts to analyze some of the relevant attentional processes that explain error and risk in the driving context, by stressing the contribution of the authors to the field of cognitive load effects on visual processing.

5.2 ATTENTION, DISTRACTION, AND DRIVING

5.2.1 Distraction, Attentional Efficiency, and Errors

Distraction is attention to irrelevant stimuli or actions, and this implies a definition of what is relevant or irrelevant for a given goal. As what defines attention is the fitness for purpose of the assigned resource, a discussion on distraction involves discussing how we should pay attention, since information acquisition to action performance processes: how information should be acquired, how it should be processed, and how an action should be performed. In practice, when looking for distraction and human error, we rely more often on theoretical assumptions about what should not have been done. For example, an erroneous decision may be explained by lack of awareness, maybe because relevant information was disregarded although it was available. But driving without conscious attention may also be successful as in the case of automatic driving, more frequent under undemanding and monotonous conditions when no relevant targets needing attention appear. In this case, attending to irrelevant objects or performing irrelevant actions may seem harmless because no observable driving errors occur. Although environmental cues may trigger the shift from automatic to consciously controlled mode in an appropriate manner, this may not occur when needed if the driver's attention is focussed elsewhere: a critical target is left unattended or processed late and an accident occurs. Being involved in something else is a common definition of distraction. By considering the wider scope of attentional processes, distraction can be extended to other kinds of attentional inefficiency. First, failures or omissions derived from erroneous or nonexistent anticipatory hypotheses (due to inexperience, lack of knowledge) about what can be expected to happen and how to detect the appropriate clues. A bad mental model leads to a poor attentional strategy. Second, failures involving the intensive component of attention (partially coincident with the not so well-defined construct of arousal). It seems clear that those factors contributing to reduced activation states (fatigue, sleepiness, drugs) cause attention processes to run less efficiently.

5.2.2 Distraction and Accidents

The weight of distraction on accident statistics produces a variety of estimates depending on the criteria used to attribute distraction. Most estimations fall within the range of 25% to 50%. A more striking result arises from the "100 car study" (National Highway Traffic Safety Administration [NHTSA], 2006). Based on wide-ranging records of driver behavior in natural situations, this study reports that driver inattention contributed to 78% of the recorded crashes and 65% of the near-crashes. If such general estimations are useful to stress the importance of attention as a safety issue, more detailed estimates, focusing on a limited range of situations, are of more interest for applied purposes. For example, Neyens and Boyle (2007) analyze the relationships among three types of accidents (angular collision, rear-end, fixed object) and four types of distractions (cognitive, cell phone, in-vehicle, passenger-related) among young drivers. Self-reported descriptions by drivers involved in accidents also confirm the high predominance of attentional inefficiency. The argument "looked but failed to see" has became a colloquial expression, maybe because its meaning is easy to understand just by recalling past experiences of crashes or traffic conflicts. In addition, other self-reported descriptions explicitly citing "I was distracted" or "I had my mind elsewhere" (Rumar, 1990) suggest that distraction failures are common events in our lives as road users, although many of them do not result in accidents.

5.2.3 Distraction, Consciousness, and Attribution of Responsibility

In the traffic context, we tend to believe that distraction is mostly involuntary, assuming that every driver wants to arrive successfully at a destination and that no one wants to crash. On the other hand, drivers are expected to do their best to stay alert. William James (1890) distinguished voluntary and involuntary attention. Accepting such a distinction, a distraction failure can also be seen as an involuntary outcome and it is possible that some drivers would rather report an involuntary distraction than accept responsibility for a deliberate risky maneuver. On the other hand, attention is to a large extent voluntary in normal perception, action planning, and performance, and traffic laws presume the driver is responsible for the consequences of his behavior.

However, increasing knowledge of environmental factors affecting attentional performance and distraction also has legal implications that may alter the balance of individual versus public responsibility. New regulations to prevent distraction affecting in-vehicle information and communication systems (IVIS) or other information systems located on the road, such as variable message signs, place increasing responsibility on car manufacturers, road designers, and public authorities. Thus, some distraction accidents could be attributed to "bad design" or "malfunction of information systems." In addition, regulations and enforcement policies directly addressing specific driver distraction behavior (do not use your cell phone, do not access the

FIGURE 5.1 This cartoon by Forges, a popular Spanish humorist, shows a mind driving test as an analogy of a preventive alcohol test. The policeman checks if the driver's thoughts allow him to drive safely. Reproduced with permission from *El Pais*, March 2000.

menu of your navigation system while driving, etc.) may also produce adverse effects by leading drivers to presume that nonprohibited activities can be considered safe because they are legal. An example of this is if you are using a hands-free phone (allowed by law); you may feel that you are allowed to access all available functions, for example, navigating manually through the "hands-free phone" menu to look for a phone number entry. That is one risk of legislation focusing on devices rather than on psychological processes (Nunes and Recarte, 2002). Of course, the optimization of psychological processes to improve attentional performance is difficult to fit into classical enforcement strategies.

5.2.4 Information Processing Errors and Types of Distraction

Trick, Enns, Mills, and Vavrik (2004) suggest four types of attentional phenomenon resulting from combining the exogenous/endogenous dimension with the automatic/controlled one. These four selective processing modes may help to describe attentional processes in a variety of traffic situations and also explain distractions (Crundall, Van Loon, and Underwood, 2006). We can consider distraction in relation to successive stages of visual information processing. In order to be aware of the visual scenario, drivers have to start by keeping their eyes on the road most of the time; here, distraction may lead to missing relevant incoming information. Then the information must be evaluated; here, distraction may lead to poor perception and poor understanding. Finally, decisions have to be taken on the available information; here, distraction causes indecision, wrong decisions, or poor performance. To consider how different factors can cause distraction, we suggest, for practical purposes, the following distraction categories: visual, cognitive, activation, and anticipation (knowledge/expectations related).

Visual distraction: Caused by tasks involving visual demands (looking at an advertisement hoarding, searching for a phone number on the cell phone menu, or checking a temperature display), which give rise to a direct conflict at the level of visual input.

Cognitive distraction: Caused by tasks involving cognitive processing, that do not explicitly require looking (listening to the radio, searching by touch for a button to open a window, or planning activities for your journey), or caused by the cognitive effort derived from a visual input.

Activation distraction: Attentional dysfunction attributable to the energy aspect of attention (low activation level, drowsiness, tiredness) or to altered states (i.e., related to alcohol or other substances), affecting the availability of attentional resources.

Anticipation distraction: Related to learning and expertise (missing relevant information due to a lack of search skills or selecting a wrong response due to lack of training).

It is likely that an inexperienced driver will have a higher probability of errors when attempting to perform additional tasks and less chance of overcoming the negative effects of fatigue. Assuming that such categorization is not exclusive and not seeking to deny the importance of the last two types of distraction, we will focus in this chapter on the first two types, assuming as a starting point, a healthy, well-trained driver who is neither intoxicated nor sleep-deprived.

The most obvious distraction is looking away from the driving scene. Gazing at objects whose line of sight is far away from relevant locations has a potential risk that increases depending on the time a driver spends looking away from the traffic scene. The critical time spent looking away depends greatly on the traffic situation: half a second while following a car at a close distance on a winding road may be more critical than 2 seconds while driving on a straight, wide, and empty motorway. Nevertheless, distraction times over 2 seconds are considered unacceptable as general criteria for driving (NHTSA, 2006). Of course, you can be distracted even while keeping your eyes on the road. As a driver must prioritize where to search for relevant information, a bad choice of where to look is inefficient; successful visual scanning depends on expertise, expectations, and so forth. In addition, even while keeping your eyes on the road, cognitive activity can be a source of distraction, that is, current thoughts unrelated to driving or associated with the driving context and irrelevant at that precise moment. In the case of high cognitive load, this type of distraction may cause dramatic impairment, including preventing the further processing of a relevant visual input coming from a spatially well-oriented ocular fixation due to lack of attention. Missing the brake lights of the car in front or just being unable to elicit the braking response while being involved in a complex thought are examples of looking without really seeing.

5.3 VISUAL ATTENTION IN THE DRIVING CONTEXT

Assuming general agreement on the predominance of visual information for driving, it seems logical that the analysis of visual search can provide relevant information on drivers' attentional issues. Even accepting the distinction between attention and gaze, in the dynamic road environment, attentional changes can be inferred from visual search (Moray, 1993).

5.3.1 Characterization of the Driver's Scenario

The visual driving scenario displays a global optical flow dependent on our own trajectory and speed, reflecting different textures and contrasts depending on the road surface and its surroundings, whereas local flow features reflect other moving road users. Longitudinal road markings allow automatic processing of lane keeping (Land and Lee, 1994), and other standardized features such as vehicle size, indicator and/or brake lights of other vehicles, their braking and accelerating capacities, along with the traffic rules, contribute to depicting a more or less predictable world.

5.3.2 Visual Parameters, Visual Search, and Attention

A useful description of visual behavior with regard to attention requires meaningful measures. With no attempt at being exhaustive, we present several examples of measures that can be used to characterize ordinary driving and to address attentional changes that occur with cognitive load. Studying visual behavior in both ordinary driving and while performing different cognitive tasks, Recarte and Nunes (2000, 2002, 2003) considered, in several of their road experiments, the following measures:

1. *Spatial distribution (SD) of ocular fixations,* including their mean position, their spatial variability, and their concentration in particular areas, such as mirrors or instruments, allows us to identify general patterns of sensitivity to increased mental load. In our studies, spatial gaze variability (expressed as the product of 1 horizontal SD × 1 Vertical SD) was significantly reduced as a function of mental load. Similar results were found by Harbluk, Noy, Trbovich, and Eizenman (2007), who interpreted this effect as "a concentration of the visual inspection area due to the cognitive demands of the task." (For alternative interpretations see Underwood, Chapman, Bowden, and Crundall, 2002). Nunes and Recarte (2005) also obtained spatial concentration effects due to increased speed. However, the results showed that the effect cannot be due to a hypothetical increased load attributable to speed because unlike cognitive load, speed did not affect the visual discrimination task. Besides, in neither cognitive load nor speed was gaze concentration associated with tunnel vision (visual performance impairment gradient, higher in the periphery and lower in the center of the visual field in a discrimination task). Thus, spatial gaze concentration is a predictable effect in certain well-known circumstances, but its interpretation is complex and may possibly have different meanings in different situations.
2. *Fixation duration* has been interpreted, in the context of reading, as the amount of information processed in a fixation. Accepting this empirical law as a means to quantify the information extracted from individual targets, fixation duration is sensitive to the demands imposed by the complexity of

the traffic scenario; a greater need to pick up information results in shorter fixations. Nunes and Recarte (2005) found evidence of this effect even with moderate traffic densities. With regard to increased cognitive load, different effects occur depending on qualitative aspects of cognitive load. Whereas cognitive processing involving verbal codes produces shorter fixations, mental imagery contents produce the opposite effect—a pattern of long fixations mixed with those of normal duration. In our interpretation, this pattern of long fixations, which we call "eye-freezing," could indicate a distraction mode consisting in looking without seeing. When dealing with the mental image, the eye could temporarily become blind to external visual input. In the case of image rotation, included in some of the mental tasks studied, a possible explanation could be the following: As mental rotations are incompatible with saccades (Irwin and Brockmole, 2000), the long-fixations pattern could be a simple result of the saccadic inhibition subsequent to mental rotation tasks.

3. *Pupil dilation* proved to be a reliable indicator of mental effort, being sensitive to cognitive load or traffic complexity. With appropriate experimental control, it is possible to eliminate the effect of changing light conditions in the natural environment.
4. *Blinks,* depending on the type of load, visual or cognitive, display different patterns. Complex traffic scenarios produce blink inhibition, but the addition of a cognitive loading task produces an increase in the blink rate. Such an increase could be interpreted as a double inhibition process; assuming that blink inhibition due to visual load requires resources, the performance of an additional cognitive task could interfere with the resources dedicated to blink inhibition.

5.4 ENVIRONMENT AND DISTRACTION

Accepting that environmental design can contribute to prevent distraction, increasing the saliency of what should be paid attention to (i.e., enhancing conspicuousness), or removing irrelevant competing information (suppressing potential distractions), are means to prevent problems of detectability. But in order to address action planning and performance errors, further processing stages have to be considered. If reducing the demands of a task is a way of addressing workload problems, a monotonous environment with too few demands may also cause inattention. Setting limits to the amount or complexity of incoming information or designing systems able to manage complex information makes possible to reduce errors. But do these solutions achieve their goals? As Chapters 8, 9, and 10 deal directly with these systems, we will just comment here on a few points regarding our approach to the distraction problem.

5.4.1 Distraction Due to Systems Designed to Attract Attention

A driver is alerted by the flashing lights of a police car approaching from behind. While watching the police car and tracking its trajectory through the rear-view mirror, in order to plan a maneuver to give way, he or she hits a pedestrian who suddenly crossed the street. Many accidents can occur because our attention, captured by a previous event, does not let us process something else that happens unexpectedly and simultaneously.

In terms of distraction, because the flashing lights of the police car are a warning system specifically designed to attract visual attention, one could not say that the driver of our story was "distracted" in the same sense as if he or she was looking at the screen of a navigation system or manipulating a cell phone. But if we think in terms of efficiency, we can conclude that, in this case, the attentional capture of the warning lights of the police car caused a distraction. As environmental distraction is normally seen as involving secondary activities unrelated to traffic, one tends to presume that a driver is "on the task" (not distracted) if he or she is attending traffic-related information and particularly if it is displayed by means of conspicuous, specific visual information devices like electronic panels or conventional signs. The point we wish to stress is that in order to evaluate attentional efficiency (and environmental distraction), besides the distinction between traffic-related and traffic-unrelated information, more refined criteria are needed to evaluate the relative relevancy of traffic targets to a specific traffic situation. General visual search evaluation criteria such as "eyes on the road" or generic conspicuousness enhancement solutions may lead to misunderstandings or attentional errors due to saliency conflicts.

5.4.2 Distraction Due to Systems Imposing High Visual or Cognitive Demands

Meanwhile, despite the lack of highly refined models of what we should attend to and how in each situation, we can still agree on some general criteria to detect negative signs of inappropriate visual search: If our eyes are off the road for too long, it is impossible to know what is going on there. Consequently, in-vehicle information and communication systems (IVIS) and any type of traffic sign on the road would be inappropriate for driving if they imposed a high visual demand due to the amount or complexity of the visual information displayed. In practical terms, for the purpose of evaluation, the analysis of visual search may reveal inadequate visual scanning attributable to the excessive visual demands of an information system. But beyond the visual demands imposed by incoming information, it is important to stress the problem of the cognitive load involved in interaction with the system, including subsequent actions. Discarding unacceptably risky conditions and identifying impairment factors is a basic necessity, but is insufficient. A more complex methodological approach is needed. Besides visual search changes, a careful selection of meaningful driving performance measures is critical in order to provide a reliable evaluation in terms of a positive definition of efficiency.

5.4.3 Distraction Due to Systems Affecting Drivers' Expectations

Besides the possible problems of overloading, other attentional dysfunctions may occur while interacting with the more advanced driver assistance systems (ADAS) affecting drivers' expectations of how their vehicle will interact with the environment, either by providing warnings requiring immediate decisions from drivers or by automatically taking appropriate actions of vehicle control. ADAS suggest a means to increase traffic efficiency by making the driving task less demanding and more resistant to human lack of expertise or failures in vigilance or perception. The probability of human failure is hypothetically reduced by transferring part of the task from the driver's decision repertoire to the system's repertoire and driving on a shared basis (co-driving). Thus the driver must trust the system and have a mental model of when and how the system will act and, at the same time, stay alert to monitor the system's behavior in order to respond if the system fails to do what is appropriate (or expected). Such significant changes in drivers' expectations may cause a particular kind of inattentional inefficiency leading to behavioral adaptation. Hoedemaeker and Brookhuis (1998), Hoedemaeker and Kopf (2001), and Rudin-Brown and Parker (2004) provide evidence of some of these relevant attentional problems.

5.5 THINKING AND DRIVING

5.5.1 Coping with Everyday Driving: Cognitive Effort and Internal Distraction

Except in heavily complex traffic conditions, thinking about our personal concerns while driving is the paradigm of most current driving activity. As previously pointed out, activities with high cognitive load affect visual search and such changes in visual processing may lead to perceptual impairment and increased risk. At the same time, there are many undemanding tasks that do not hinder performance. However, different cognitive activities have different effects. Consequently, both quantitative and qualitative aspects of cognitive load have to be considered.

5.5.2 Measuring Mental Load

Significant efforts have been made to find measures of mental workload. We can find examples of the usefulness of such a concept for addressing information-processing errors in the field of air traffic control operators (Wickens, 1992) and drivers' performance (De Waard and Brookhuis, 1997). O'Donnell and Eggemeier (1986) consider three categories of load measure: subjective self-reported measures, performance measures, and physiological measures. In a laboratory study, Recarte et al. (unpublished results) used pupil dilation, the NASA TLX workload index, and blink rate as dependent measures to evaluate the cognitive load of cognitive tasks (used in previous driving experiments) when performed alone (single task) and also when combined with a continuous visual detection task (dual task). They found that consistency between the three workload measures is clear in a single task but becomes critical in the dual-task condition. While NASA TLX reflects the subjective impression that two tasks carry more load that one single task, independently of performance, the pupil reflects the amount of brain activity with cognitive versus visual load (activation vs. inhibition respectively). For applied purposes, this level of complexity poses an additional problem: If risk has to be predicted from workload, then care has to be

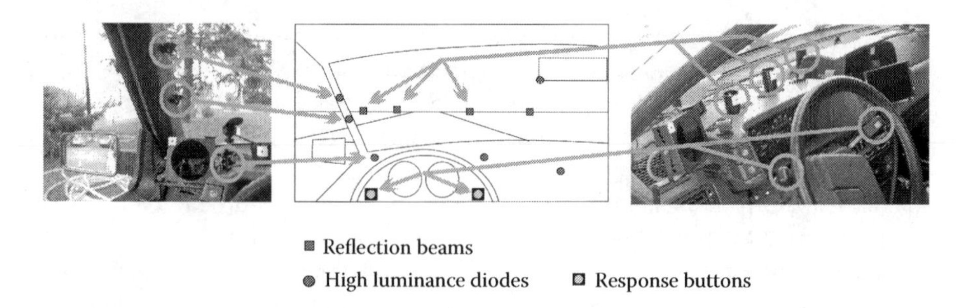

FIGURE 5.2 Experimental setting of visual stimuli.

taken when comparing heterogeneous and complex tasks such as those occurring in the driving context.

5.5.3 Mental Load, Vision, Performance, and Inattentional Blindness

High cognitive load also affects visual detection, discrimination, and response selection capacities (Recarte and Nunes, 2003). Figure 5.2 represents the experimental setting of visual stimuli and Figure 5.3 represents the main trends in visual search and performance effects due to cognitive load.

This perceptual impairment is a representative example of "looking but not seeing," and thus it can be seen as a kind of inattentional blindness, in a very broad sense,

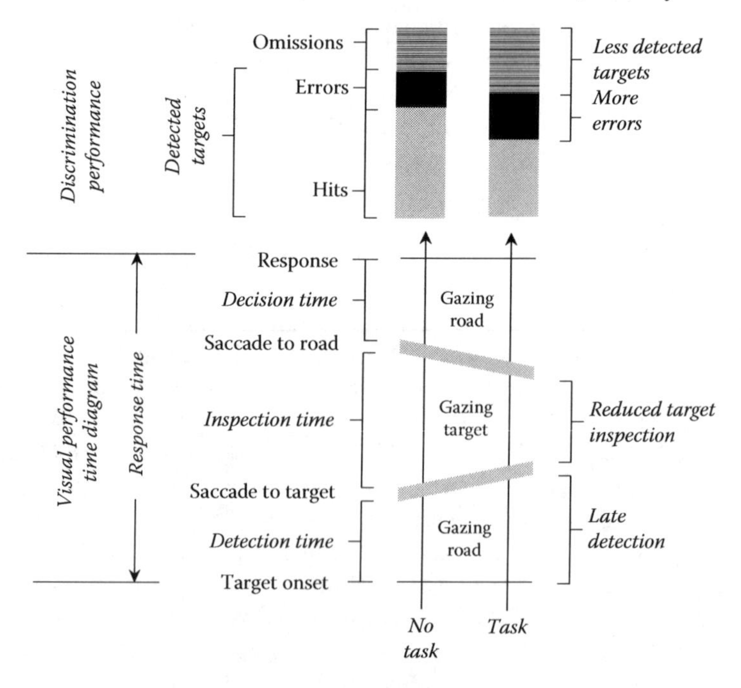

FIGURE 5.3 Task effects on visual behavior (bottom) and discrimination performance (top).

although it does not fit into the strict inattentional blindness paradigm (Mack and Rock, 1998). In our visual discrimination experiments, the undetected stimuli were not considered irrelevant (or we would not have interpreted errors as distractions), nor were they unexpected, as drivers were alert, instructed, and trained to respond to them. Moreover, Most, Scholl, Clifford, and Simons (2005), demonstrated the phenomenon using a continuous stimuli presentation and concluded that "the most influential factor affecting noticing is a person's own attentional goals." This study provides clearer evidence of mechanisms that facilitate perception than of perceptual impairment. The importance of awareness with regard to improving visual search and anticipatory skills as a means to reduce the probability of these distraction errors is undeniable (Underwood et al., 2002). However, that kind of distraction is beyond the scope of this chapter. Lavie (2006), systematically varying the perceptual load of the relevant target, concluded: "Awareness reports in both the inattentional blindness and change blindness paradigms were found to depend on the extent to which an attended primary task loads attention." The importance attributed to attentional load is in agreement with our results on undetected stimuli. Besides the above-mentioned effects of high cognitive load on visual processing, including the previously cited differences relating to the type of content (spatial vs. verbal), no signs of impairment were observed when listening to messages with no relevant content for action or when having an irrelevant talk with a passenger or using a hands-free cell phone. However, care should be taken with regard to the generalization of speech-based interactions involving actions, for example, in the case of some in-vehicle systems. If merely listening to an incoming message does not necessarily distract, its further processing may lead to a high load situation, cognitive or visual.

5.5.4 Mental Load, Speed Perception, and Speed Control

Traditionally, speed has been a relevant issue in safety discussions and certain assumptions in particular have been made regarding attention and perception. In Section 5.2 we made a reference to the experimental evidence against the interpretation of speed-induced spatial gaze concentration as a tunnel vision effect. However, besides this, the most common assumption is that increased speed implies higher visual load. In an experiment on highway driving explicitly designed to evaluate the effect of speed and traffic complexity, using a visual detection and discrimination task with free speed choice, Nunes and Recarte (2005) found no evidence of any of the impairment effects observed with cognitive load and the observed changes in the observed visual search were interpreted as an adaptive resource reallocation effect. Traffic complexity, independent of speed, negatively affected the visual discrimination task and produced pupil dilation, as did cognitive load.

Another aspect of speed relevant for attentional issues is when a driver has to adapt to mandatory speed limits. Recarte and Nunes (2002) demonstrated that explicit speed restrictions impose not only an additional visual load (more glances at the speedometer are needed) but also cognitive load. Drivers, having the tendency to drive at their preferred speed, must keep the speed restrictions in mind to keep their speed under control. If a cognitive task is performed, they tend to drive at the preferred speed, presumably because the preferred speed is controlled with minimal resource consumption.

To summarize, in this section we have shown how cognitive distraction with no visual load and with high or moderate cognitive load affects visual search patterns, visual information processing, and driving performance, as in the case of speed control required by explicit speed restrictions. Given voluntary acceptance of a speed restriction, compliance is compromised due to the interference of increased cognitive load.

In our opinion, the attentional dimension of speed is far from being well understood. The tendency to treat the speeding problem as a whole limits the scope of the psychological approach to the speed control problem, reducing it to a mere rule-compliance issue, unconnected with considerations of the dialectics between traffic conditions and human performance. Considerations about what speed is appropriate are seen as an engineering or merely legal issue. The modern concept of dynamic speed limits attempts to go beyond the traditional static and generic considerations by providing more precise models of what speed is appropriate in which circumstances. To achieve this, a psychological approach to speed control behavior is needed. Dynamic speed regulation should be imposed in a speed management framework sensitive to human performance under specific traffic conditions. Appropriate speed should involve something more than setting upper limits. It should contribute to optimizing human performance by setting a model of longitudinal control in which appropriate speed is the desirable speed for a given situation, in which attentional variables are crucial, along with traffic flow, headway optimization, road and vehicle physical constraints (including visibility constraints), and meteorological and time of day conditions. Intelligent speed adaptation should reduce crash probability by lowering or raising speed, depending on the particular ongoing conditions. Instead of involving an additional input, dynamic traffic rules on speed would be a reflection of this model. Rule compliance and risk perception discussions would benefit from the more detailed contribution of attention and performance issues in speed control.

5.6 CONCLUSIONS

The analysis of visual behavior is a powerful way to understand and quantify attentional processes while driving. Besides providing objective evidence of not looking, visual search data demonstrate the effects of cognitive load on visual processing, offering a valuable approach to address the "looked but failed to see" problem. But if increased workload can cause distraction, reducing workload may not always be successful as a general means to improve safety. Besides educational or enforcement measures addressing individual responsibility for inattention, environmental measures focusing on public responsibility need to be founded on a thorough knowledge of the processes underlying distraction errors. There is a need for more detailed models addressing visual search and driving performance in well-defined and specific traffic situations. This is the challenge of applied visual search studies seeking measures to increase attentional efficiency.

ACKNOWLEDGMENTS

The authors thank the support of the Dirección General de Tráfico for the funding of the experimental work within its Argos Research Program, which made this chapter possible.

REFERENCES

Crundall, D., Van Loon, E., and Underwood, G. (2006). Attraction and distraction of attention with roadside advertisements. *Accident Analysis and Prevention, 38*(4), 671–677.

De Waard, D., and Brookhuis, K.A. (1997). On the measurement of driver mental workload. In J.A. Rothengatter and E. Carbonell (Eds.), *Traffic and transport psychology* (pp. 161–173). Amsterdam: Elsevier.

Harbluk, J.L., Noy, Y.I., Trbovich, P.L., and Eizenman, M. (2007). An on-road assessment of cognitive distraction: Impacts on drivers' visual behaviour and braking performance. *Accident Analysis and Prevention, 39*(2), 372–379.

Hoedemaeker, M., and Brookhuis, K.A. (1998). Behavioural adaptation to driving with an adaptive cruise control (ACC). *Transportation Research Record Part F, 1*(2), 95–106.

Hoedemaeker, M., and Kopf, M. (2001, August 19–22). *Visual sampling behaviour when driving with adaptive cruise control.* Proceedings of the Ninth International Conference on Vision in Vehicles, Brisbane, Australia.

Irwin, D.E., and Brockmole, J.R. (2000). Mental rotation is suppressed during saccadic eye movements. *Psychonomic Bulletin and Review, 7*(4), 654–661.

James, W. (1890). *The principles of psychology.* New York: Holt.

Land, M.F., and Lee, D.N. (1994). Where we look when we steer. *Nature, 369*(6483), 742–744.

Lavie, N. (2006). The role of perceptual load in visual awareness. *Brain Research, 1080*, 91–100.

Mack, A., and Rock, I. (1998). *Inattentional blindness.* Cambridge, MA: MIT Press.

Moray, N. (1993). Designing for attention. In A. Baddeley and L. Weiskrantz (Eds.), *Attention: Selection, awareness, and control* (pp. 111–134). Oxford, England: Clarendon Press.

Most, S.B., Scholl, B.J., Clifford, E.R., and Simons, D.J. (2005). What you see is what you set: Sustained inattentional blindness and the capture of awareness. *Psychological Review, 112*(1), 217–242.

National Highway Traffic Safety Administration (NHTSA). (2006). *The impact of driver inattention on near-crash/crash risk.* Available at: http://wwwnrd.nhtsa.dot.gov/departments/nrd-13/810594/images/810594.pdf.

Neyens, D.M., and Boyle, L.N. (2007). The effect of distractions on the crash types of teenage drivers. *Accident Analysis and Prevention, 39*(1), 206–212.

Nunes, L.M., and Recarte, M.A. (2002). Adaptation to increased workload and risk of distraction: Focussing prevention on devices or on psychological processes. *Dirección General de Trafico: Proceedings of the IX World PRI Congress.* Madrid.

Nunes, L.M., and Recarte, M.A. (2005). Speed, traffic complexity, and visual performance: A study on open road. In G. Underwood (Ed.), *Traffic and Transport Psychology: Theory and Application* (pp. 339–354). Amsterdam: Elsevier.

O'Donnell, R.D., and Eggemeier, F.T. (1986). Workload assessment methodology. In K.R. Boff, L. Kaufman, and J.P. Thomas (Eds.), *Handbook of perception and human performance. Volume II: Cognitive processes and performance* (pp. 42/1–42/49). New York: Wiley.

Recarte, M.A., and Nunes, L.M. (2000). Effects of verbal and spatial-imagery task on eye fixations while driving. *Journal of Experimental Psychology: Applied, 6*, 31–43.

Recarte, M.A., and Nunes, L.M. (2002). Mental load and loss of control over speed in real driving. Towards a theory of attentional speed control. *Transportation Research Part F 5,* 111–122.

Recarte, M.A., and Nunes, L.M. (2003). Mental workload while driving: Effects on visual search, discrimination and decision making. *Journal of Experimental Psychology: Applied, 9,* 119–137.

Rudin-Brown, C.M., and Parker, H.A. (2004). Behavioural adaptation to adaptive cruise control (ACC): Implications for preventive strategies. *Transportation Research Part F, 7,* 59–76.

Rumar, K. (1990). The basic driver error: Late detection. *Ergonomics, 33*, 1281–1290.

Trick, L.M., Enns, J.T., Mills, J., and Vavrik, J. (2004). Paying attention behind the wheel: A framework for studying the role of attention in driving. *Theoretical Issues in Ergonomic Sciences, 5*(5), 385–424.

Underwood, G., Chapman, P., Bowden, K., and Crundall, D. (2002). Visual search while driving: Skill and awareness during inspection of the scene. *Transportation Research Part F: Traffic Psychology and Behaviour, 5*(2), 87–97.

Wickens, C.D. (1992). *Engineering psychology and human performance.* New York: HarperCollins.

6 Experience and Visual Attention in Driving

Geoffrey Underwood,
Peter Chapman, and David Crundall

CONTENTS

Reflection 89
6.1 Introduction: The Relationship between Attention, Distraction, and Accident Liability 89
6.2 Processing Demands in Visual Information Acquisition: Hazards and Scanning 93
6.3 Visual Search by Experienced and Novice Drivers 98
6.4 Driving Experience and the Visual Field 102
6.5 Training and Visual Search in Driving 108
6.6 Conclusion: Proposed Components of Novice Driver Training 112
6.6.1 Predicting the Behavior of Other Road Users 112
6.6.2 Developing a Mental Model of the Situation 113
6.6.3 Dividing and Focusing Attention 113
6.6.4 Hazard Management 113
References 113

REFLECTION

Newly qualified drivers are at disproportionate risk of involvement in a crash. As they gain experience, their road accident liability decreases, and this can be attributed in part to changes in the distribution of attention. As well as knowing better where to look, they are also less distracted by events that are unrelated to the task of driving. This chapter analyzes the role that processing demands in visual information acquisition play on anticipating the behavior of other road users and developing a mental model of the driving situation that will help with this process of anticipation.

6.1 INTRODUCTION: THE RELATIONSHIP BETWEEN ATTENTION, DISTRACTION, AND ACCIDENT LIABILITY

The newly qualified teenage driver has been estimated as being anything up to 10 times as likely to be involved in a nonfatal accident relative to a middle-aged driver,

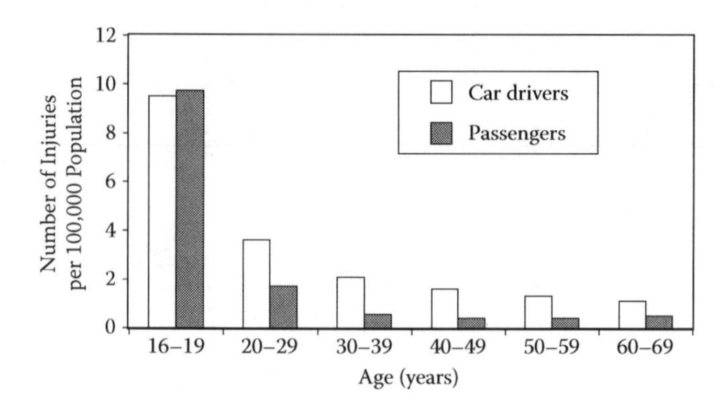

FIGURE 6.1 Fatal and serious injuries and their relationship with age, for drivers and their passengers (UK data for 2005). Serious injuries are defined here as those requiring medical treatment in hospital.

and Figure 6.1 illustrates this relationship between age and liability. For accidents resulting in a fatality or a serious injury, teenager drivers in the United Kingdom are more than twice as likely to be involved as drivers 10 years older. The change in liability with experience is summarized by Maycock, Lockwood, and Lester's (1991) estimate of a decrease of 30% in this liability during the first year of driving. A more recent study by McCartt, Shabanova, and Leaf (2003) has confirmed the exceptionally high accident rates of newly qualified drivers during the first month of holding a license. Telephone interviews were conducted with 911 high school students who were learning to drive, with 7 interviews over a 3-year period. The interviewers asked about accidents and traffic violations, and found that the risk of a crash in the first month after gaining a license (1.43 crashes per 100,000 km driven) was twice that of the second month (0.68 crashes per 100,000 km) and continued to decline over the remainder of the first year. Violation citations followed the same pattern, with twice as many citations (66% were for speeding) than during any other month on the first year of licensed driving.

The dramatic trend shown in Figure 6.1 is the exceptionally high fatality/injury rate that is experienced by teenage drivers, but this is matched by the rate for their passengers, and this requires comment. The passengers of teenage drivers are also likely to be teenagers, as a consequence of this age group being gregarious and tending to use their cars for social activities to a greater extent than older drivers. But it is not simply that the high casualty rate for passengers is a reflection of young drivers having a high crash rate, because the very presence of passengers elevates the crash risk. Williams (2003) summarized the patterns of risk for teenagers, again acknowledging the high crash risk for newly qualified drivers, especially in the period immediately after gaining their license. This survey identified the high-risk situations for drivers generally, and for teenagers particularly. For example, for young drivers, driving at night is approximately three times as risky as driving during daytime, on the basis of fatalities per distance driven, whereas for older drivers there is only a marginal increase in risk at night. However, the pattern is complicated by the finding that there is little difference between day-risk and night-risk if we take all crashes into account—overall crash rates are similar but when teenagers do have crashes they are more likely to involve a fatality. Two important factors are associated with teenage crashes—alcohol use and passenger presence, two factors that stem from the social use of the car. Teenagers tend to drink and drive less than older drivers, but when they do their crash risk is elevated more than that of older drivers. Having passengers in the car also elevates the crash risk of teenagers. The data suggest that older drivers, if anything, have a reduced risk if they have passengers, but for teenagers, the more passengers being carried, the greater the crash risk. When three or more passengers are present, the crash risk for teenage drivers is approximately four times as high as when the driver is alone. There is a case here for restricting the carrying of passengers by newly qualified drivers.

Why does the presence of passengers influence the crash risk of teenagers? One possibility is that passengers distract drivers who are not fully skilled in vehicle control and traffic negotiation, and it is this factor of driver distraction that will occupy much of the remainder of this part of the discussion.

Distraction does not elevate crash liability uniformly for all types of collisions, and Neyens and Boyle's (2007) study of teenage driver crashes used a large national database of crash descriptions to map the type of distraction against the type of collision. When distracted at a road junction by passengers or by being lost in thought, teenagers were more likely to be involved in a rear-end or side-impact crash than they were to hit a stationary object. An in-car distraction, such as adjusting the car's controls or eating, increased the likelihood of collision with a stationary object, and use of a phone increased the likelihood of a rear-end collision. An understanding of the type of distraction to which novice drivers are vulnerable is an important step in determining the remedy.

As with all activities, experience performing a task brings a level of skill that reduces the incidence of errors and that is accompanied by smooth perceptual–motor control. It is not surprising that novice drivers have poorer car control than experienced drivers, with curve negotiation, inconsistent steering, slow gear changes, and slow acceleration out of a bend distinguishing between drivers in a study reported by Duncan, Williams, and Brown (1991). When skilled drivers execute a change to the cars controls—braking, changing gear, or steering, for example—they can initiate the change and then attend to the traffic. They may even perform these actions without attention, and it is probably attention to the feedback from the system that becomes unnecessary (see Underwood and Everatt, 1996). Novices may need to attend to the consequences of their initiated action, however, to take account of the feedback from the car to ascertain that the intended outcome has occurred, for example, a change of vehicle speed, change of engine speed, or position of vehicle. Skilled performers do not need to attend to this feedback to the same extent, and their attention can be directed elsewhere without consequence. The novice needs to attend to this vehicle feedback more consistently, and so distractions are more likely to have a consequence for performance—the chattering of a group of passengers is more likely to distract a teenage driver from the task of maintaining the progress of the car and from the task of monitoring the progress of other road users.

The role of attention in teenage driving has been highlighted in a study of crash types reported by McKnight and McKnight (2003). Descriptions of more than 2000 crashes in police reports were coded to identify common factors in collisions involving teenage drivers, with factors such as car control and speed adjustment, observation of traffic signals, attention, and searching the roadway among those coded. The crashes of older teenagers (18–19 years) with some driving experience were compared with younger, less experienced drivers (16–17 years), and the contributory factors compared. The three most frequently cited deficiencies that were associated with large numbers of crashes were searching the roadway (43.6%), attention (23% of crashes), and speed adjustment (20.8%). Although the experienced and inexperienced teenage groups had generally similar profiles of contributory factors, there were some differences. Notably, the younger drivers had a reliably greater proportion of accidents associated with a failure to search the roadway prior to turning, in addition to failing to monitor the car ahead, driving too fast, and failing to adjust to wet roads. The less experienced drivers had crashes resulting from a failure to search the roadway effectively and from a failure to adjust their speed. A similar pattern emerged from an earlier study that also used police reports (Lestina and Miller, 1994), in which failure to search the roadway was a factor in 39% of teenage crashes. In contrast, roadway searching was a factor in only 10% of the crashes of 35- to 54-year-old drivers. Police reports have limitations as objective data, of course, having been written by many different officers (sometimes from different police districts), and being compiled some time after the crash on the basis of subjective recollections. Failing to perform an action may leave no memory record, and the contribution of this omission can only be inferred on the basis of what a driver has failed to report after the event. The dominance of roadway search failures in separate studies is striking, nevertheless, and points to an attentional failure in inexperienced drivers. This failure to direct attention to the most appropriate part of the visual environment will be exacerbated by demands for the driver to attend elsewhere, of course, with interactions between driver and passenger and between driver and in-car technology being potent here. In these cases a source of communication may require a reply, and with it the attention of the driver, whether this comes from a conversation with a passenger or from a phone call.

Distraction is an important factor in crashes and it appears to be particularly important in novice driver crashes. We can infer the association between inexperience and distraction from the statistics showing increased liability with increasing numbers of passengers, but stronger data come from observations of drivers at the time of the crash. Klauer, Dingus, Neale, Sudweeks, and Ramsey (2006) fitted 100 cars with video cameras pointing at the roadway and at the driver, as well as other instruments to record headway and acceleration. Recording progressed for 1 year, with more than 3 million kilometers of driving being recorded. During this time there were 82 actual crashes and 761 incidents that constituted near-crashes requiring severe evasive action. A further 8295 events required the driver to take evasive action short of becoming a conflict. The video recordings for the 60 s prior to the incident and the 30-s period following the incident were analyzed, with particular interest in what the drivers were doing immediately prior to each of these critical events. In addition, other sequences of video were analyzed to provide baseline measures from noncritical periods that were without crashes, conflicts, or other incidents.

Some form of inattention featured in almost four fifths of the crashes and in more than three fifths of near-crashes during the 3-s interval prior to the event. The length of a glance away from the road ahead was used to determine the direction of attention. If the glance exceeded 2 s, then this was coded as an inattention event. The two major categories of inattention to the road ahead at noncritical times were those involving a secondary task unrelated to driving, such as using a phone or eating, and driving-related inattention, such as checking the rear-view mirror or the car's instruments. These baseline measures recorded under normal, uneventful driving indicated that secondary tasks were engaging the drivers on 54% of occasions and driving-related inattention was observed on 44% of occasions. The two other types of inattention-related behavior that were recorded were driver drowsiness (e.g., eye closure, yawning) and nonspecific glances away from the roadway (e.g., toward a pedestrian or building to one side), but the baseline observations found these behaviors to be relatively rare. The analysis of the critical incidents found that engaging in a secondary task, and the incidence of drowsiness, increased the likelihood of crash or near-crash involvement. For example, reading while driving, applying makeup, and dialing a handheld device (phone) increased the likelihood of event involvement threefold, and driving while drowsy increased crash/near-crash liability sixfold.

Klauer et al.'s (2006) 100-car study shows that failing to attend to the road ahead has serious consequences for a driver's crash liability, and it further shows that some drivers are more likely to show inattention than others. The drivers were assigned to one of two groups on the basis of the number of inattention-related incidents that they were involved in over the course of the year of observations. The high-involvement group, who can be characterized as being less attentive, was then compared with the low-involvement, or attentive, group. The inattentive drivers tended to be both younger and less experienced than their attentive counterparts, and when crash/near-crash incidents are plotted against age groups the correspondence with the pattern in Figure 6.1 is remarkable. Drivers in the 16–20 age band had more than twice the number of inattention-related crashes and near-crashes than any of the other age groups.

From these studies the profile of a teenage novice driver emerges that describes someone who is distracted, violation prone, and accident prone, and who does not search the roadway effectively. A major cause of inattention stems from passengers, who also have a high incidence of crash involvement by being in cars driven by teenagers. The engagement of teenagers with mobile technology—phones, PDAs, and iPods—also provides a major source of distraction. Having established that inexperienced drivers have a high crash risk that is in part attributable to the inappropriate allocation of attention to the visual world, we now consider the visual demands of driving to identify what it is that should gain a driver's attention.

6.2 PROCESSING DEMANDS IN VISUAL INFORMATION ACQUISITION: HAZARDS AND SCANNING

One sort of demanding situation that is particularly important for drivers is the handling of hazardous situations. Other road users who perform unexpected actions are the major source of hazards to be negotiated—the car ahead brakes suddenly, a pedestrian steps into the road, or an oncoming vehicle steers across your pathway to make a turn, for example. Search strategies that may be adequate in most everyday driving are likely to be tested by the occasions in which sudden hazards occur. The idea that attentional strategies may change in stressful situations has a long history in psychology. Easterbrook (1959) is often cited as proposing the general idea that arousal causes a narrowing in the range of cues attended to by an organism. Applied cognitive psychologists have extended this idea to explain a phenomenon in eyewitness testimony known as "weapon focus" (Kramer, Buckhout, and Eugenio, 1990) whereby a witness to a crime may look at the weapon, but fail to remember the face of an assailant. Loftus, Loftus, and Messo (1987) had participants watch scenarios where a shop customer holds either a gun or a check. They found that viewers fixated significantly longer and more often on the gun than they did on the check. In a similar study Christianson, Loftus, Hoffman, and Loftus (1991) observed more frequent and extended fixations on central information in stressful conditions. If such attention focusing occurs in driving situations we might expect that in hazardous situations participants would focus on information directly ahead of them and fail to attend to more peripheral sources of information. Chapman and Underwood (1999) tested this hypothesis by having drivers watch hazard perception videos while their eye movements were recorded. Figure 6.2 shows a comparison of the locations fixated when hazards were present or developing (as indexed by button presses during the hazard perception test) and with the locations fixated when there was no hazard present. As can be seen, there are small differences in the overall pattern of locations fixated, but there is no clear overall tendency for a focus on central locations at the expense of peripheral ones. As Chapman and Groeger (2004) argued, memory tests

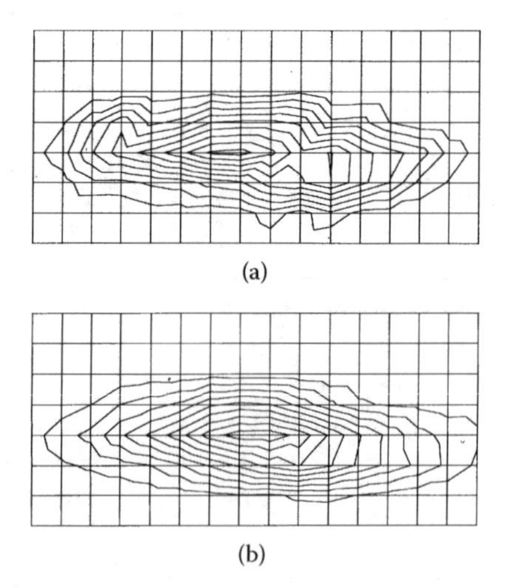

FIGURE 6.2 Scanning the roadway ahead. Fixation density plots for the central 15° by 7° of visual angle while watching hazard perception videos, as a function of the occurrence of a hazard. Plot (a) shows the spread of fixations during periods where hazards were developing or present, and plot (b) shows the spread of fixations in matched environments without hazards. (Adapted from Chapman and Underwood, 1999.)

in such situations are consistent with the idea that what drivers focus on in dangerous situations is not information that is spatially central, but information that is central to the driving task.

As it turns out, although there is little evidence for attention focusing in the spatial sense from records of eye movements, there is plenty of evidence that some form of attention focusing does exist when hazards are present. Chapman and Underwood (1998) made systematic comparisons of a wide range of eye movement variables for safe and dangerous windows while participants watched hazard perception videos. Although the locations viewed did not differ, fixation durations on hazards increased, the mean saccade amplitude decreased, and the overall spread of search (as measured by both vertical and horizontal variance in fixation locations) decreased. The point here is that drivers do appear to focus on hazards. However, the location of hazards is not always spatially central to the driving scene. To observe attention focusing in hazardous situations, it is necessary to define eye movement measures relative to individual hazard locations. Underwood, Phelps, Wright, van Loon, and Galpin (2005) did just this, defining each fixation relative to the appearance of an individual hazard. As shown in Figure 6.3, there is a dramatic increase in fixation duration at the time an individual driver detects a hazard. Although this increase in fixation duration can be detected by comparisons across broadly safe and dangerous situations (e.g., Chapman and Underwood, 1998), the data from Underwood et al. (2005) strongly suggest that the actual duration of attention focusing is likely to be limited to the single fixation at the time a hazard is detected. Interestingly, the effect of the attentional capture by a hazard does not change with the age of an experienced driver. Novice drivers hold their attention on hazards longer than more experienced drivers (Chapman and Underwood, 1998), however, and this is a potential problem in

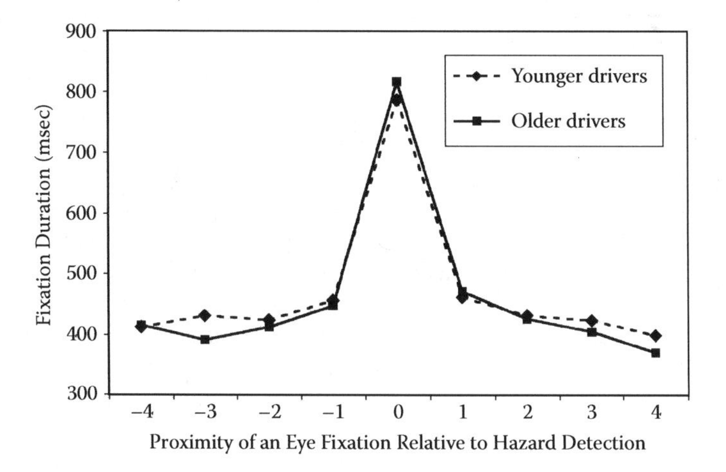

FIGURE 6.3 Attentional capture by hazards. An eye fixation is prolonged at the point where a hazard is detected. Prior to hazard detection, and after detection, the fixation durations of young experienced drivers (30–45 years old) and older drivers (60–75 years old) are half the magnitude of those recorded at hazard detection. (Adapted from Underwood, Phelps, Wright, van Loon, and Galpin, 2005.)

those cases where one hazard causes another. For example, if a cyclist rides out into the path of a line of traffic, the primary hazard (the bicycle) will capture attention, and if another road user responds to the hazard by braking sharply then they become a secondary hazard that also requires attention. If attention is locked onto the primary hazard, the secondary hazard may not receive the attention that it requires. Chapman and Underwood (1998) demonstrated that novices are likely to have their attention captured by a primary hazard for longer than experienced drivers.

While these findings of increased fixation durations on hazardous stimuli are immensely important in understanding driver responses to abrupt and well-defined hazards, how do the results relate to complex driving environments that do not contain neatly packaged hazards?

If we consider the increase in fixation durations on hazards to be due to a corresponding increase in processing demands, then it is easy to look for other driving stimuli and tasks that also raise demands and compare their effects upon visual search. For instance, there are greater visual demands made of a driver's attention when driving around a curve compared to driving on a straight road (e.g., Shinar, McDowell, and Rockwell, 1977). More specific information (e.g., the tangent point; Land and Lee, 1994) needs to be accessed and fed forward into motor control allowing smooth navigation. Increases in traffic density (Rahimi, Briggs, and Thom, 1990), and the proximity to other vehicles (Hella, Laya, and Neboit, 1996), are other sources of demand, while the chosen speed of the vehicle and particular maneuvers that one undertakes have also been noted to influence eye movements and fixation durations (Cohen, 1981; Muira, 1979). Though all these events and stimuli can be considered to raise the level of processing demands, their actual influence upon fixation durations and eye movements is opposite to the attentional capture noted with abrupt hazard onsets. Instead, fixation durations become shorter, and the spread of search may increase as the driver tries to take in more information to cope with an increasingly complex situation.

Thus we are presented with two separate influences of visual demand upon eye movements and visual attention, though both can be reconciled within Easterbrooks' (1959) cue utilization hypothesis. As arousal increases (e.g., increased speed, more traffic), drivers try to rise to the challenge and increase their sampling rates. However, certain events that pose immediate danger to the driver or other road users may increase arousal levels beyond what the driver is comfortable with, leading to a reduction in the amount of cue utilization, with drivers becoming visually locked in to exceptionally dangerous and sudden events.

One related avenue of research pursued by Crundall, Shenton, and Underwood (2004) asked whether the increased fixation durations noted with abrupt hazards could be found with more enduring "hazards" or whether the prolonged nature of such stimuli desensitize drivers to their hazardous nature. In this instance, a lead vehicle was classed as the prolonged hazard. Participants were required to drive through the simulated streets of London either following a lead vehicle or obeying verbal instructions that dictated their direction through the city. Eye movements and behavioral measures (give-way violations, curb impacts, etc.) were recorded.

The results demonstrated increased fixation durations on the car ahead during a following task compared to control trials where there was a lead vehicle but participants had no requirement to fixate it for journey information. Similarly the spread of horizontal search decreased suggesting attentional capture, and the total proportion of time spent looking at the car ahead doubled.

When following verbal instructions drivers increased their horizontal search when pedestrians were present. When following a lead vehicle, however, the drivers failed to increase their search either suggesting that they decided that the car ahead was more important than checking whether pedestrians were about to step into the road, or that they were less aware of the pedestrians due to their increased focus on the lead vehicle. Car following also led to an increase in give-way violations and the severity of give-way collisions and curb impacts.

We consider these findings especially relevant to police-car drivers who are at the other extreme of the expertise continuum. Crundall, Chapman, Phelps, and Underwood (2003) suggested that the prolonged hazard associated with chasing or following a suspect vehicle may capture attention, reducing visual search for other vulnerable road users. At the same time the benefits of police driver training and experience could outweigh such attentional focusing, perhaps through the use of compensatory strategies that prioritize other vital peripheral stimuli that could have an impact on the safety of the drive.

Crundall et al. (2003) invited police and normal drivers into the laboratory to watch video clips of police pursuits, rapid response drives, and control drives. These videos were filmed primarily from dashboard cameras mounted in active police cars and involved real events. Eye movements were recorded while participants watched the clips on a wide screen and rated them for hazardousness. As with previous studies of experience on the spread of visual search, the police drivers were found to have a wider search strategy. They also spent more time inspecting peripheral hazards such as parked vehicles and side roads (Crundall, Chapman, France, Underwood, and Phelps, 2005). Despite this, all participants (including the police) showed evidence of attentional capture on the fleeing vehicle (with longer fixation durations on the car ahead and a relative reduction in fixation of peripheral sources of hazard). The benefits of police training and experience were definitely apparent in this study, yet even their high level of skill could not completely compensate for the focusing effect when compared to control clips.

Though these studies merely define the influences of experience upon eye movements in hazardous and complex environments, the results can be used to suggest ways to decrease accident liability on the road. Although training intervention studies are fraught with difficulties (see Section 6.5), one way to reduce accidents is to manage the environment according to the skills that drivers possess. For instance, a study by Crundall, van Loon, and Underwood (2006) demonstrated the benefits of placing roadside advertisements on poles that were 3 meters above ground level rather than at ground (similar-sized advertisements were placed in bus shelters). As hazardous visual search is mainly contained with a horizontal window (Figure 6.2), it makes sense to declutter this area of the visual scene to prevent inadvertent stimuli from capturing attention. When most legible, the advertisements on the poles were outside this horizontal search window. The rationale behind this is that if experienced drivers have spare attentional capacity they will be able to look at these advertisements. If the drive is especially hazardous, however, they will maintain search within the horizontal window, avoiding the advertisements. Bus-shelter advertisements may be fixated, however, when drivers are searching for hazards (especially pedestrians who congregate in bus shelters). Such fixations may detract from a visual search intended to identify hazards, and thereby increase accident liability. In addition, Crundall et al. (2006) found that bus-shelter advertisements that did capture attention during a hazard-oriented visual search also produced weaker memory traces compared to advertisements on poles that were fixated when the driver felt comfortable to do so. All of the participants in that particular study were experienced drivers, but one can easily imagine the problems that ground-level advertisements might create for inexperienced drivers who already suffer from increased liability to attentional capture.

6.3 VISUAL SEARCH BY EXPERIENCED AND NOVICE DRIVERS

In an innovative and justifiably well-known study, Mourant and Rockwell (1972) recorded the gaze patterns of novice drivers, and this method has now become a standard way of tracking the visual attention of drivers of varying ability and under varying driving conditions. Novices were compared to experienced drivers, and were found to look less far ahead (perhaps checking on the position of the car relative to nearby road markings), to look farther away from the center of the road (perhaps judging the position of the car relative to the edge of the road), and to look less often in the driving mirror. One of the features of the Mourant and Rockwell study is the sample. They tested only six novices and four experienced drivers, and the 16- to 17-year-old novices had no formal training and almost no previous experience (three had no previous experience, and the other three had less than 15 minutes of driving experience). These novices should be expected to focus on perceptual–motor coordination and on keeping the car in the roadway. These drivers are not representative of the novices that feature in the accident statistics, and more informative comparisons come from studies of experienced drivers and novices observed within a few months of passing their driving test.

Crundall and Underwood (1998) recorded the eye movements of novices within 3 months of their gaining a driving license (they were recruited at the driving test centers when they passed the practical on-road driving test), for comparison with older drivers who had an average of 9 years of experience. This study was part of a large project funded by the UK Department for Transport in which novice and experienced drivers were compared on a number of driving and driving-related tasks, including driving an instrumented car (video, acceleration, braking, headway, gear selection, etc.) on a range of rural and urban roads, and watching and responding to video clips showing hazardous driving scenes. The aim of the project was to identify behaviors that discriminated between the two groups of drivers, with a goal to design a training package that would accelerate the progress of the novices through the period in which they were most vulnerable. The training package is described in Section 6.5.

The Crundall and Underwood (1998) study took the eye movement recordings of a sample of 16 novices (not all drivers in the main project delivered usable recordings) on rural and suburban roads and on dual carriageways. The most interesting results concerned the extent of visual scanning, as indicated by the variance of fixation locations. High variance represents fixations that are widely spread, and are found

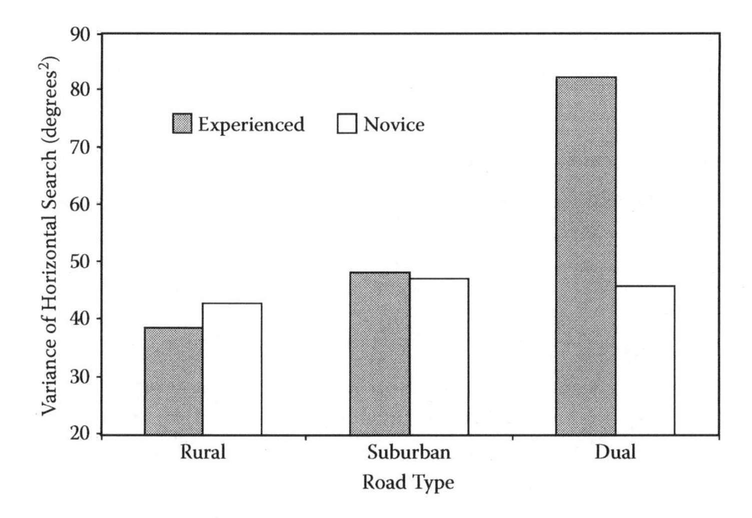

FIGURE 6.4A The extent of scanning as indicated by the variance of fixation locations in the horizontal plane for experienced and novice drivers. (Adapted from Crundall and Underwood, 1998.) The variance of fixations is an indication of the extent of visual search, with greater variance indicating greater scanning.

when the drivers look around them, while low variance represents a driver looking in one direction for an extended period of time. Variance was recorded separately in the horizontal and vertical planes, and the results are shown in Figure 6.4A and Figure 6.4B. The search along the horizontal plane (Figure 6.4A) indicates a difference between novice and experienced drivers only on the dual carriageway. This

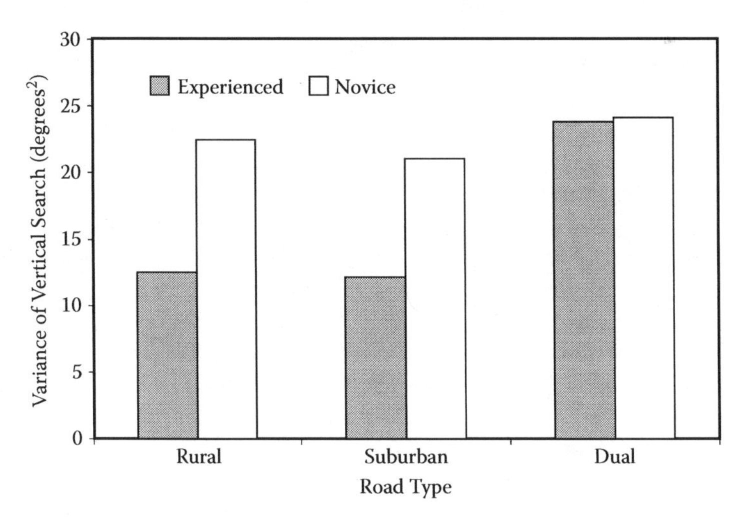

FIGURE 6.4B The extent of scanning as indicated by the variance of fixation locations in the vertical plane for experienced and novice drivers. (Adapted from Crundall and Underwood, 1998.) The variance of fixations is an indication of the extent of visual search, with greater variance indicating greater scanning.

urban motorway is a road that varies between two and three lanes and has sliproads joining and leaving from both left and right. Traffic entering from one side frequently interweaves with existing traffic in order to make an exit on the other side. As well as maintaining a speed consistent with existing traffic and changing traffic lane in order to make the correct exit, drivers must be aware of other vehicles that intend to change lane themselves.

The high variance of horizontal search shown by the experienced drivers on this dual carriageway is an indication of the need to monitor concurrent vehicles and to identify and maneuver into gaps in the traffic. The road demands extensive scanning of the traffic lanes to both left and right, but the novices scanned from left to right on this road no more extensively than they did on a quiet rural or suburban road. The stereotyped scanning behavior of the novices is also shown in the variance of vertical fixations—the inspection of the roadway near to or further from the car (Figure 6.4B). The novices showed a similar amount of scanning from near to far, while the experienced drivers tended to reduce the amount of vertical scanning on quieter roads.

The surprising result from this study is that the experienced drivers increased their scanning in the horizontal plane at times when they were negotiating with other traffic, but the novices continued to scan in the same way that they did on quiet roads. The novices did not adapt to the changing traffic conditions. Underwood, Chapman, Brocklehurst, Underwood, and Crundall (2003) also took the object of fixation into account—what the driver was inspecting on a moment-to-moment basis—rather than the overall distribution of fixations. The point is made just as dramatically. This analysis plotted fixations as they moved from object to object in the visual scene, and determined eye movement transitions that occurred more often than would be expected by chance alone. Novice drivers made more of these regular, stereotyped eye movements, often moving their eyes from one part of the scene to the road directly ahead, while experienced drivers appeared to be more flexible, and changed their fixation patterns according to current roadway conditions. This was especially apparent on the dual carriageway, where there were very few regular transitions in the eye movements of experienced drivers, who looked around them according to the behavior of the other traffic.

A similar result came from a study of mirror use as part of the same project. Underwood, Crundall, and Chapman (2002) found that novices did not use their mirrors selectively, while experienced drivers relied on the external door mirror when changing lanes and when needing to collect information about traffic in the adjacent lane. There was no overall difference in mirror use between groups in this study, a result that is inconsistent with a report by Duncan et al. (1991) who found that novices looked at their mirrors more often than experienced drivers. We attribute the inconsistency to driver selection. Underwood, Crundall, and Chapman's novices were 18 years old on average, and had passed their driving tests quite easily, while Duncan et al. tested age-matched novices who averaged 35 years and who would be expected to have taken a number of attempts to pass their test. When Underwood, Crundall, and Chapman tested older novices as part of the overall project they found high variance and unusual driving behavior according to many of the measures taken, sometimes characterized by caution and lack of confidence and sometimes by lack of control. The old novices also had a higher crash rate than other novices. The inconsistency between these studies of mirror use is likely to arise from differences between the characteristics of the novices—older novices do not behave in the same way as young novices.

The failure of the novices to adapt their scanning behavior to the changing traffic conditions in the Crundall and Underwood (1998) study prompted us to ask whether this was because they were unable to process more information than they were already doing, or whether they were unaware of the changed risks on the dual carriageway. The increase in a driver's workload has the effect of decreasing the extent of their scanning (Recarte and Nunes, 2003), and perhaps the novices were overloaded with the task of keeping the car in the correct lane at an appropriate speed. An alternative hypothesis is that they did not know of the dangers associated with interweaving traffic and of the need to understand the intentions of concurrent traffic. One hypothesis emphasizes the limited perceptual–motor skill of the novices, and the other emphasizes their understanding of the traffic environment. The driver's knowledge of the driving task can be described with the use of Endsley's (1995) situation awareness model of dynamic task control, in which the driver's mental model of a situation can represent what is happening, how that situation arose, and how it will develop in the immediate future. Gugerty (1997); Horswill and McKenna (2004); Kass, Cole, and Stanny (2007); Underwood, Crundall, and Chapman (2007); and Underwood (2007) have applied this model to the driver's control task. The interesting possibility here is that we can record the progression of skill acquisition by reference to the development of the driver's mental model of what is currently happening, how that situation arose, and what might happen next. The novice may have an incomplete mental model that does not facilitate the anticipation of what other drivers may do next, and what the consequences of these actions will be. If this is the case, then we would expect that novices would not adapt their scanning behavior to current traffic conditions, and this is consistent with our analyses of fixation behavior by newly qualified drivers.

The three levels of situation awareness may be associated with the developing skill of the novice driver, with all three levels of understanding being observed more frequently in more experienced drivers. The restricted scanning of novices will result in the formation of an impoverished mental model of the current situation because the model will not represent the complete set of events being created by other road users. Incomplete inspection of the roadway is associated with increased accident liability and may be a product of high workload or of a failure to understand this aspect of driving that requires a mental model of what other drivers are doing and what they intend to do next. Using the situation awareness framework, Kass et al. (2007) compared the performance of unlicensed novices and experienced drivers on a simulator, to look at the effects of phone use as well as situation awareness. Both groups of drivers showed a performance decrement when using a hands-free phone in the driving simulator, and the novice drivers were found to have less situational awareness, as assessed by interrupting the driving task to ask questions about the scene. Drivers were asked about other road users and about the current speed limit, for example, and experienced drivers answered more of these questions than did the novices. Interestingly, the adverse effect of using a phone did not have a differential effect on the situation awareness of the two groups of drivers. Both novices and experienced drivers answered around a third fewer situation questions when distracted by using the phone. If novices are overloaded by perceptual–motor demands of car control then the additional task of holding a phone conversation might be expected to have a differentially damaging effect upon their situation awareness. The absence of an interaction suggests that a novice's failure to develop a mental model of the situation is not dependent upon residual processing capacity. This raises the possibility that novices do not build rich mental models of the driving situation because they do not see this as an important component of the driving task.

Are novices unaware of the need to collect visual information from the driving scene around them, or are they incapable because they are overloaded by the perceptual–motor demands of car control? Underwood, Chapman, Bowden, and Crundall (2002) addressed this question by removing the demands of car control from a task in which different roadways were seen. Novice and experienced drivers watched video recordings of the same roads used in the Crundall and Underwood (1998) experiment in which in-car scanning was observed. The laboratory task was to watch the video and press a response button if anything happened that would prompt the driver to take evasive action, and eye movements were again recorded. If the reduced scanning of novices is a product of perceptual–motor workload, then eliminating the task of car control would eliminate the workload, and so the scanning of novices should be similar to that of the experienced drivers. On the other hand, if reduced scanning is a product of the failure to understand the need for a mental model of the situation, then there should be a difference between the scanning of the novices and the experienced drivers. The results from this study are shown in Figure 6.5, where it is clear that the novices looked around the video screen less than the experienced drivers, and this is consistent with the idea that these novices had not recognized the need to build a rich situation model of the driving scene. When the task of controlling the vehicle was removed, novices still had reduced scanning relative to experienced drivers. This suggests that their restricted scanning is not entirely determined by mental workload.

6.4 DRIVING EXPERIENCE AND THE VISUAL FIELD

As we have noted in the previous section, there is much evidence to suggest that more experienced drivers have a better understanding of where to look when they drive. The research discussed so far has been primarily concerned with overt attention, that is, where the eyes are pointing (Sereno, 1992). However, since the formation of the spotlight theory of attention in the early 1970s (Eriksen and Eriksen, 1974), and the later variants of the spotlight, such as the zoom lens (e.g., Eriksen and Murphy, 1987) and the gradient models (e.g., LaBerge, 1995), there has been an acknowledgment that we can take in valuable information from extrafoveal regions of the visual world (i.e., beyond the item that is being inspected or foveated). The basic spotlight model suggests that attention resembles a beam of light that moves in an analog fashion (however, see Sperling and Weischelgartner, 1995) across the visual scene. Whatever it highlights is thus available for processing, or may even be processed whether you want to or not (Eriksen and Eriksen's initial studies identified

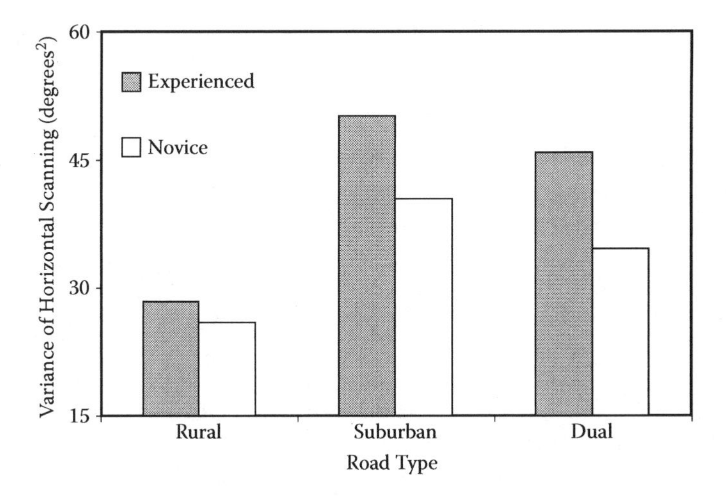

FIGURE 6.5 Scanning while watching video recordings. The variance of horizontal fixations is shown here in a laboratory study in which novice and experienced drivers watch recordings taken from a car traveling along the same roads as used in the Crundall and Underwood (1998) study. See Figure 6.4A for comparison. The values for the dual carriageway represent both a quiet section of road with no junctions and a more difficult section in which sliproads join from the left and the right, to make comparison with our earlier study. (Adapted from Underwood, Chapman, Bowden, and Crundall, 2002.)

unwanted interference from irrelevant letters flanking a target letter). Although this spotlight can move around independently of the eyes, it is often considered that the two are tightly linked, with the focal point of the eyes lying at the center of the spotlight (for a discussion of the linkage between the eye and mind, see Underwood and Everatt, 1992). The zoom lens and gradient models added flexibility to the spotlight model, allowing the spotlight to change in size and to have a variable strength across its diameter. Changes in the size of spotlight can occur for many reasons including fatigue, the influence of alcohol, anxiety, and processing demands. In this latter case, the greater processing demands that one is faced with at the point of fixation, the more resources are then devoted to it. A reduction in the diameter of the spotlight thus increases the resources at the point of fixation, in the same way that focusing a torch into a narrow beam reduces the area over which the light falls but increases the intensity. The impact of these models is the acknowledgment that we might not notice important and salient stimuli if they fall too far into the peripheral visual field, outside our spread of attention, and, more important, that the effective area of the visual field covered by attention changes constantly. Whereas a strict interpretation of the spotlight model might suggest that anything that falls within its boundary is automatically processed, in driving we are more concerned with the area within which a salient event might be noticed (e.g., the errant cyclist emerging suddenly from a side street). This boundary, often termed the functional field of view, is larger than the spotlight though it is proportionally linked in size.

If we assume that the increased fixation durations of inexperienced drivers (see Section 6.2) reflect increased processing demands, then we are faced with the possibility that such novices will also reduce their peripheral attentional resources, reallocating them to the stimulus at the point of fixation in an effort to improve processing. This would suggest that these inexperienced drivers suffer two forms of degradation in visual processing, with an increase in fixation durations that they try unsuccessfully to offset with a reallocation of extrafoveal attention. Alternatively one could suggest that it is the more experienced drivers who are likely to be more afflicted by a degradation of peripheral attention. The reallocation of peripheral attention to the point of fixation may actually be a strategy that is developed with driving experience, and could therefore be the cause of the reduced fixation durations seen in highly experienced drivers. Whichever of these theories is true has important ramifications for training interventions. If experienced drivers show peripheral degradation but novices do not, then this suggests that reallocation of peripheral resources to the point of fixation is a valuable strategy that drivers find useful. We could therefore train novice drivers in this strategy. Alternatively if novice drivers suffer greater degradation in the peripheral field then this would suggest that we are underestimating the differences between experienced and inexperienced drivers if we concern ourselves solely with eye movement measures without considering peripheral attention.

We compared these alternate hypotheses using a hazard perception test (Crundall, Underwood, and Chapman, 1999, 2002). In the first study we looked for differences between experienced, novice, and nondrivers in their allocation of extrafoveal attention while watching hazard perception clips of up to 60 s in length. In every clip there was at least one specific hazard (such as a pedestrian stepping into the road in front of the viewer's perceived vehicle). Participants were asked to watch each clip and make two judgments at the end of each clip regarding the level of perceived difficulty to drive through that particular scenario and the danger that they thought was posed.

In addition brief peripheral lights were presented for 200 ms around the edges of the screen, with one peripheral light appearing every 5 s on average. Participants were required to respond to these lights with a button press. The results showed that experienced drivers spotted more peripheral target lights than nondrivers, with the performance of a group of novice drivers falling midway between the other two groups. We also found fixation durations to decrease with experience as noted in Chapman and Underwood (1998). The results suggested that the decrease in fixation durations demonstrated by experienced drivers is not due to the use of a strategy of reallocating extrafoveal attention to the point of fixation, as they outperformed the other participant groups in the peripheral detection task as well. We concluded that not only do inexperienced drivers suffer from increased processing demands in terms of longer fixation durations, but that they also suffered from a degradation of peripheral attention.

The second study (Crundall et al., 2002) confirmed these findings, comparing peripheral detection rates of experienced and learner drivers. On this occasion we asked participants to press a foot pedal to respond to hazards, and a mouse button to respond to the peripheral lights. This allowed us to chart the time course of detection of peripheral lights around the identification of a hazard. Figures 6.6A and 6.6B display the results. As can be seen in Figure 6.6A, peripheral target detection rates fell

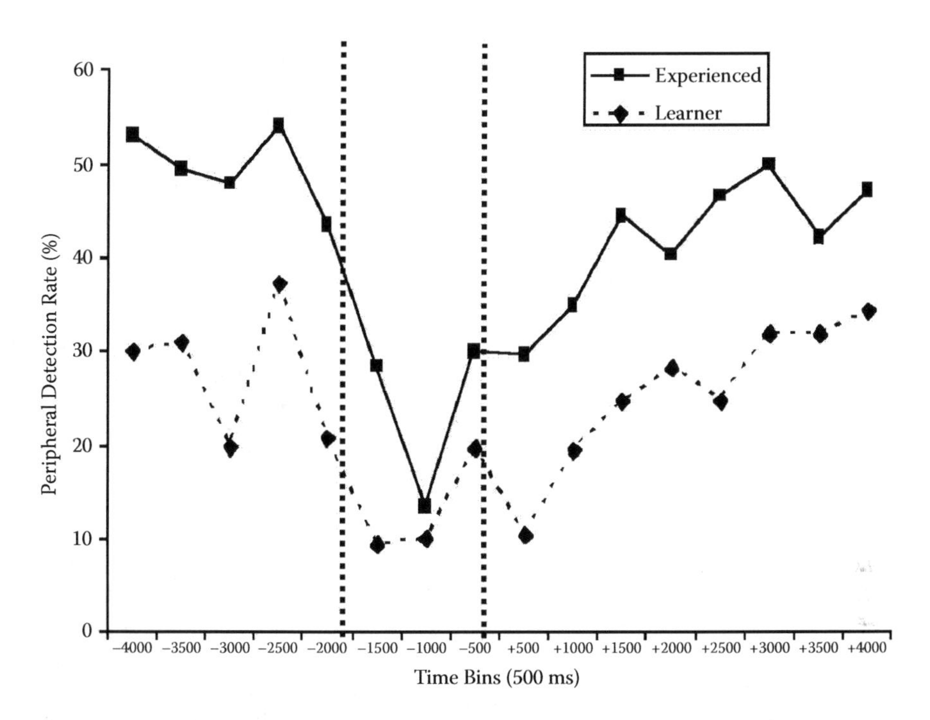

FIGURE 6.6A Peripheral attentional capture by hazards. Detection of peripheral targets decreases during a hazard. This graph displays 500 millisecond (ms) time bins. The first dashed line represents the average hazard onset, while the second dashed line represents the average hazard response.

dramatically if they occurred 1000 to 1500 ms before participants pressed the foot pedal to register a hazard. Response times to the appearance of hazards were 1453 ms on average (with no differences between the driver groups), which closely corresponds to the dip in peripheral detection rates. Figure 6.6B is a more fine-grained analysis with peripheral onsets grouped into periods of 200 ms. At about 900 to 1100 ms before the participants make a hazard response, both learners and experienced drivers have only a 10% chance of responding to a peripheral target. This means that the experienced drivers have suffered a greater degradation of peripheral attention than the learners, relative to their higher peripheral detection rates before the appearance of a hazard. Note also, however, that the time course of degradation is different for the two driver groups. The learner drivers suffer degradation as soon as they spot the hazard, and that degradation remains for a relatively long period of time, only picking up 700 ms after the hazard response. The experienced drivers, however, do not suffer any considerable degradation to peripheral target detection until approximately 400 ms after the hazard has appeared. At this point peripheral target detection virtually shuts down around 900 to 1100 ms before the hazard response, but then quickly returns to prehazard levels of detection, much faster than the learner drivers. This suggests that experienced drivers reallocate attention away from the periphery to the point of fixation later than learner drivers, but when they do

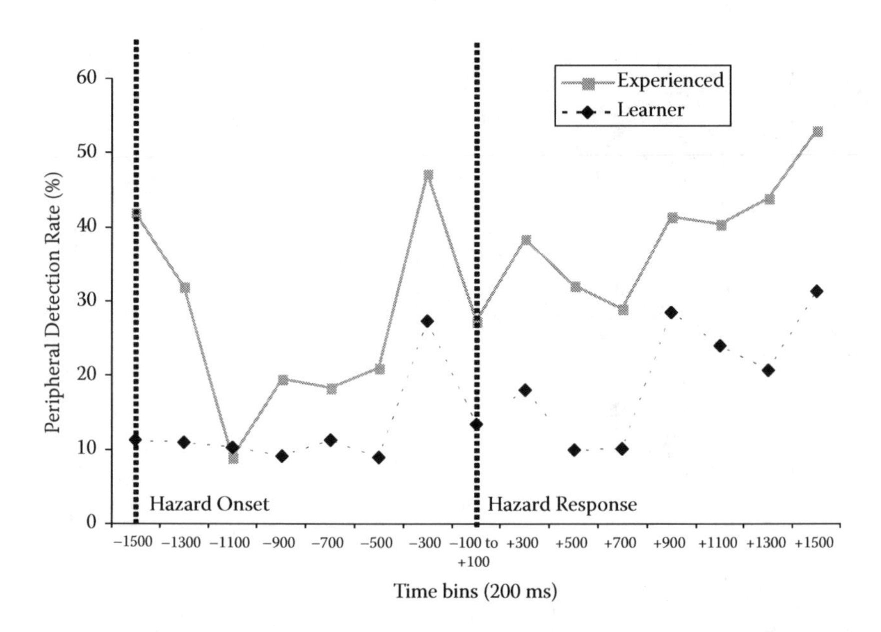

FIGURE 6.6B Peripheral attentional capture by hazards. Detection of peripheral targets decreases during a hazard. This graph displays 200 millisecond (ms) bins. The first dashed line represents the average hazard onset, while the second dashed line represents the average hazard response.

so, they invest a relatively higher proportion of attention. This allows them to process the hazard more quickly, after which their attentional resources are rapidly returned to the peripheral visual field.

This requires the conclusions of the first study (Crundall et al., 1999) to be modified somewhat. Whereas the first study concluded that as the experienced drivers spotted more peripheral targets, the reallocation of peripheral attention was therefore probably not a strategy used by experienced drivers to increase processing speed, the more fine-grained analysis of the second study found that experienced drivers could suffer greater costs of reallocation but over a very short time course. This suggests that reallocation of attention may be a skill that is learned by experienced drivers, which could account in part for their improved processing speeds and shorter fixation durations. This skill appears to be characterized by three key differences to the pattern of results produced by the learner drivers. First, they hold off reallocation for up to 400 ms, until the event has passed a criterion of hazardousness. Once reallocation has been triggered it is quick and almost total in magnitude, allowing the hazard to be processed quicker, and attention to be rapidly redeployed back to the peripheral field.

While peripheral attention is vital for the detection of hazards, it also has a role to play in basic lane maintenance. Land and Horwood (1995) used a simple driving simulator to investigate which portions of the lane markers are important to steering. They found that at high to moderate speeds, two sections of the road were vital to successful steering. One of these sections is a far preview of the road ahead, typically

1 s in front of the vehicle. The second section is closer to the vehicle and is necessary for feedback on position within the lane. Without this closer window steering is jerky and produces a highly varied lane position. However, Land and Horwood reported that their drivers rarely fixated this close segment of the road and instead were using peripheral vision to provide the lane position feedback allowing a smoother drive.

Land and Horwood's (1995) (experienced) drivers may not have fixated this near segment, but there are many examples of on-road driving that have found inexperienced drivers to fixate closer to the car than more experienced drivers, and to have a greater number of fixations upon lane markers (e.g., Mourant and Rockwell, 1972). This particular feature of inexperienced drivers' search strategies can again be explained by failures in peripheral attention. If novice and learner drivers do not have the spare resources to maintain a wide functional field of view, then it is probable that they would not be able to take in information from lane markers through peripheral vision. As this information is vital to a smooth drive, however, it is likely that the inexperienced driver feels the need to foveate these sources of information. Thus, limited peripheral attention forces inexperienced drivers to look at things that they would not need to look at if they were more experienced.

Although inexperienced drivers may suffer from reduced peripheral attention, there is a corresponding decline at the opposite end of a related continuum: age. Using a specific instrument (the useful field of view test) researchers have charted the decline in peripheral attention with advancing age (e.g., Sekuler, Bennett, and Mamelak, 2000). There is evidence, however, that training can improve the spread of visual attention by freeing resources from the point of fixation. It is even possible that the age-related decline that has been noted can be offset by appropriate levels of training, with effects lasting at least 3 months (Richards, Bennett, and Sekula, 2006).

Does this mean that inexperienced drivers could be trained to increase their spread of peripheral attention? Possibly, though any effect is highly context specific. Reducing resources from the point of fixation requires training directed at improving the processing of foveated items. Some stimuli seen on the road are regular enough to allow this type of training. For instance, road signs are standardized, and experienced drivers will process these with a cursory fixation (or may even be able to process them with peripheral attention). An inexperienced driver, however, may find that more resources are required to process a novel road sign, which will probably induce a reallocation of attention from the periphery to the point of fixation. Training on these regular aspects of the driving task may improve novice drivers' spread of peripheral attention.

Many stimuli encountered on the roads are, however, of a less regular nature or are less predicted by the context. Hazard perception training is aimed at providing exposure to a wide range of these unexpected events, allowing new drivers to draw on experience when faced with similar hazards on the road (albeit from hazard perception training videos rather than from real experience). It is possible that with the recent introduction (2002) of the hazard perception test in the United Kingdom that we will begin to see inexperienced drivers fare better in their spread of peripheral attention when faced with hazardous situations (providing they are similar to the ones they are trained on!).

6.5 TRAINING AND VISUAL SEARCH IN DRIVING

Relatively few attempts have been made to directly influence novice drivers' patterns of visual search. One reason to exercise caution in attempting to train drivers' eye movements is the fact that eye movements are likely to be a consequence of other aspects of visual processing—thus, a driver may fixate a region until information from that region is fully processed. At that point they may move onto a new region searching for additional information. Differences in fixation durations between novice and experienced drivers in unfamiliar situations (e.g., Chapman and Underwood, 1998) may thus reflect the additional time a novice driver requires to process novel risk-related information. A visual search training intervention that encourages novice drivers to mimic the search strategies of experienced drivers may simply cause them to leave a region of fixation before the relevant information is fully processed. Such training can be seen to be potentially dangerous. If an intervention is aimed at reducing fixation durations it makes better conceptual sense to teach novice drivers about hazardous situations in the hope that this will allow them to process relevant visual information faster and, consequently, reduce fixation durations and allow an increased spread of search. We might thus expect interventions that are designed to improve hazard perception to potentially improve visual search. One of the first carefully conducted studies of hazard perception training was reported by McKenna and Crick in 1994. In this study novice drivers were trained in hazard perception by watching video clips of potentially hazardous situations. The videos were halted part way through and drivers were asked to predict what would happen next. This training improved scores on a subsequent hazard perception test suggesting that trained novices had indeed learned to process visual information about hazards faster. Unfortunately, no visual search measures were recorded directly from this paradigm. Verbal commentaries have also been found to improve subsequent hazard perception, either with the driver giving a commentary or listening to a commentary provided by an expert (e.g., Horswill and McKenna, 2004; McKenna, Horswill, and Alexander, 2006). Another study has used both group training in the classroom and personal on-road training, emphasizing scanning of critical areas and continued movement of the eyes (Mills, Hall, McDonald, and Rolls, 1998). This combination of training was found to reduce response times to hazards in a subsequent hazard perception test. A review by Deery (1999) considered a series of hazard perception training interventions, including the use of photographs, video clips, and simulated driving. Deery suggested that visual scanning and hazard prediction are both critical for successful hazard perception performance. Deery also suggested that training drivers to vary their distribution of attention between different visual tasks can improve dual task performance and hazard perception ability.

Studies of hazard perception training techniques provide a useful basis for training broader visual search strategies, but relatively few studies have actually measured the influence of such training on drivers' eye movements. The first large-scale study attempting to train and measure drivers' visual search strategies was reported by Chapman, Underwood, and Roberts (2002). Their training intervention took the form of a 1-hour video-based task designed to train three specific components—knowledge, anticipation, and scanning. Similar to previous hazard perception studies, it was assumed that watching videos of potentially dangerous situations while providing commentaries and listening to expert commentaries would improve novice drivers' knowledge of hazardous road situations and potentially process them faster. Anticipation training was provided using a "What happens next?" prediction test similar that used by McKenna and Crick (1994). The final component of this intervention was designed to more directly influence drivers' visual scanning. Here, drivers were shown videos of dangerous situations with multiple areas of potential hazards circled. Such videos were played initially at half speed to give drivers time to fully process information in all areas and were subsequently played at full speed to encourage a scanning strategy that was both wide and rapid. Chapman et al. (2002) monitored their novice drivers' eye movements while performing hazard perception tests before training, immediately after training, and in a long-term follow-up condition between 3 and 6 months after the training intervention. Figure 6.7 shows the fixation durations and spread of horizontal search over the three phases for trained drivers and a matched group of novice drivers who did not take part in the training.

The results from Chapman et al. (2002) strongly suggest that eye movement training can help novice drivers develop visual search strategies in filmed hazardous situations that are more like those of experienced drivers (Chapman and Underwood, 1998). Pollatsek, Narayanaan, Pradhan, and Fisher (2006) have found similar effects of hazard perception training upon subsequent eye movements in a driving simulator.

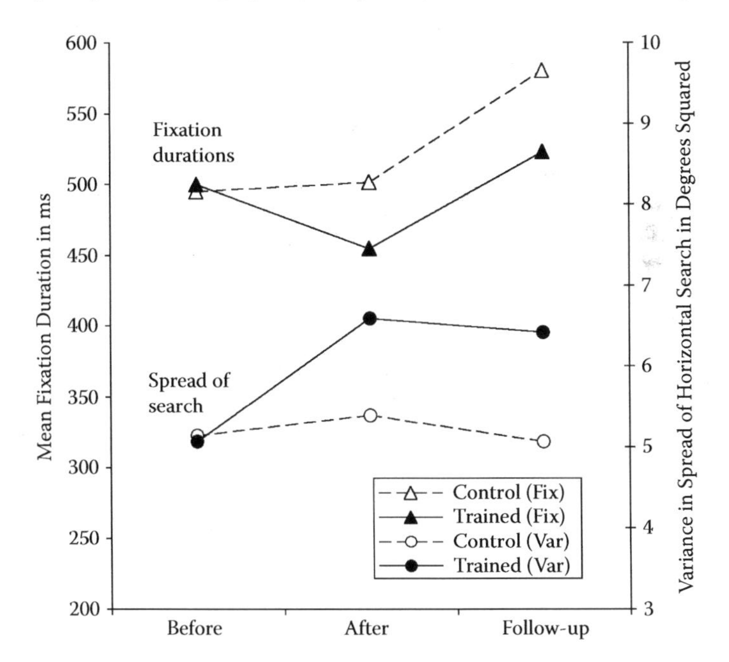

FIGURE 6.7 Training for scanning: laboratory study. Mean fixation durations in milliseconds (Fix) and mean horizontal spread of search in degrees squared (Var) for trained and untrained novice drivers while performing hazard perception tests at three points during their first year of driving after passing their practical driving test.

They gave their novice drivers aerial views of road scenarios and asked them to mark on the pictures the location of hazards that were obscured from the view of the driver. Trained drivers then drove in a simulator while their eye movements were recorded. They were found to subsequently show more successful visual search for potential hazards on those scenarios that were structurally the same as the overhead views that they had been trained with. Critically, there was also evidence that visual search in new scenarios had been affected by the training, suggesting transference of skill between driving situations. A follow-up study suggested that such training effects, like those shown by Chapman et al. (2002), were relatively long lasting (Pradhan, Fisher, and Pollatsek, 2006).

So far it has been seen that visual search training can transfer to eye movements during hazard perception tests and driving in a simulator. There is additional evidence that such training can also transfer into driving in the real world. Chapman et al. (2002) monitored the eye movements of their drivers while driving on real roads in an instrumented vehicle both before and after training. Figure 6.8 shows the same measures, mean fixation durations and spread of horizontal search during actual driving. Although there were no significant differences in fixation durations, this is hardly surprising. Data are averaged over approximately 15 minutes of driving, very little of which would be expected to include specific diving hazards. It is thus striking that significant differences in the spread of horizontal search were observed

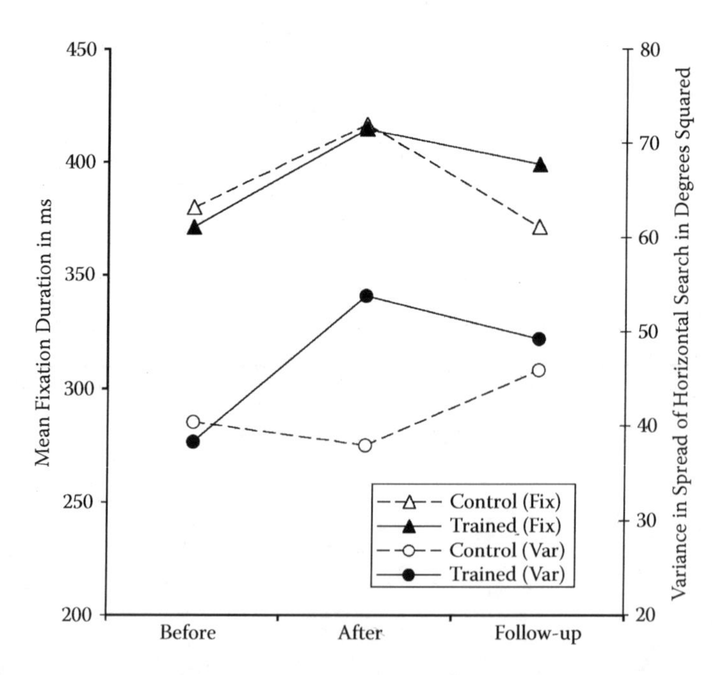

FIGURE 6.8 Training for scanning: on-road study. Mean fixation durations in ms (Fix) and mean horizontal spread of search in degrees^2 (Var) for trained and untrained novice drivers while driving on real roads at three points during their first year of driving after passing their practical driving test.

immediately after the intervention, although these differences were no longer significant in a follow-up 3 to 6 months after the intervention. This is partly because the untrained group of novices increased their spread of search during the 3-month follow-up period, suggesting that experience had taught them the importance of scanning widely. By training novice drivers early, to put this another way, we can see the effects of experience being accelerated at the very point when they have their greatest crash risk.

Pollatsek and his colleagues have recently followed up a variant of their aerial view training technique using eye tracking on real roads (Pradhan, Fisher, Pollatsek, Knodler, and Langone, 2006). Here they explored eye movements in a series of critical scenarios that either occurred naturally or were staged (for example, by deliberately parking vehicles in places where the driver's view would be obscured). They found significant training effects both for scenarios that were like those in the training phase and those that were less similar. The training effects were, however, smaller for less similar scenarios and they attribute this reduction to the fact that the modified training procedure used photographed situations that may have been very visually similar to those encountered during the actual drive (Pollatsek, Fisher, and Pradhan, 2006).

There is thus evidence that both types of training described can improve drivers' visual search strategies; that these benefits can be observed in hazard perception tests, simulated driving, and actual driving; and that some of these benefits are reasonably long lasting. Despite this promising conclusion, there are still a number of unanswered questions about the effectiveness of training. Although the risk awareness and perception training (RAPT) training described by Pollatsek, Narayanaan, et al. (2006) does appear to be successful in encouraging drivers to look to the location of potentially obscured objects, it is arguable that this is only one component of expert visual search in driving. In contrast, a problem with eye-movement-based training is that it may encourage a general scanning strategy that is not always appropriate; indeed there was evidence from Chapman et al. (2002) to suggest that changes in scanning strategy occurred in both dangerous and safer situations. This is of potential concern as there was no reason to expect changes in scanning for safer situations (c.f. Chapman and Underwood, 1998). Video-based hazard perception training can also be criticized on the grounds that a good driver could have avoided many hazards by an appropriate approach speed and direction. Video-based hazard training takes this opportunity out of the hands of the driver and simply allows them to focus on event unfolding in front of them. This may remove the need to change their visual search according to the demands of the current road situation. The ideal training regime would thus involve actual driving on real roads in which hazardous situations are encountered and dealt with; however, for obvious practical reasons, the opportunities for such training are extremely limited. A clear improvement over using training videos would be the use of simulated driving in which car control was required, but hazardous situations could be created for the driver to deal with in a safe, controlled environment.

Our suggestion, therefore, is that further efforts at training drivers' visual search patterns might be best concentrated on strategies that are appropriate in both general driving and in specific hazardous situations. The advent of low-cost, high-fidelity driving simulators suggests that these may be an ideal location for training in simulated hazardous situations. In addition to proposing the driving simulator as an ideal location for training, we have specific proposals for the components of training that are most likely to be effective in producing optimal visual search strategies in novice drivers. This proposal is described in the following section.

6.6 CONCLUSION: PROPOSED COMPONENTS OF NOVICE DRIVER TRAINING

Novice drivers are exceptionally liable to road traffic accidents, especially during the first few months of fully licensed driving. They are also vulnerable to distractions, and there is also a notable association between their accident liability and their failure to search the roadway adequately. They have a high proportion of crashes in which a scanning deficiency is identified as a factor. It is appropriate to consider the introduction of driver training that addresses this problem, and having considered existing interventions that have been described and evaluated, we can now identify the components of an effective package. We suggest that novices should receive an emphasis on anticipating the behavior of other road users and develop a mental model of the driving situation that will help with this process of anticipation. Drivers' responses to hazardous situations involves their management of attention—knowing what to attend to and knowing when to switch attention between objects. New drivers need to have control of their attention, not just to avoid distractions, but to scan the roadway for potential hazards without being totally captured by them. We conclude this discussion of the relationship between driving experience and visual attention with specific suggestions for the training of new drivers.

6.6.1 Predicting the Behavior of Other Road Users

Many real accidents are preceded by the unexpected behavior of other road users and could have been prevented if the driver had correctly predicted their behavior in advance. Realizing that an oncoming vehicle might be planning to turn in front of your car or that a pedestrian walking on the sidewalk might be about to step into the road requires very deep processing of the visual scene. In such cases a driver needs to be aware of potential sources of danger, fixate them at length, and return to them frequently. A driver can never be sure of the behavior of other road users, but knowledge about how potential hazards might develop will help the novice driver choose appropriate areas of the visual scene on which to concentrate their search. This may be one of the main benefits of expert commentaries, and important knowledge may be gained by simply watching a wide variety of hazardous scenarios. Additional training based on getting the driver to predict what happens next in a variety of scenarios may also be valuable.

6.6.2 Developing a Mental Model of the Situation

Some of the dangers to a driver cannot be seen until they become hazardous. For example, a child who steps out into the road from behind a parked ice-cream van becomes an immediate hazard. Experience allows a driver to identify and monitor these dangerous areas prior to the appearance of the hazard. This component is similar to the concept of anticipation as used by McKenna and colleagues (e.g., McKenna et al., 2006) and is clearly one of the main skills trained in RAPT training (Pollatsek, Narayanaan, et al., 2006). Training in these skills can be partly accomplished by standard hazard perception and what-happens-next tests, but there may be considerable additional benefit from the kind of multiple perspectives on road scenes shown in RAPT training.

6.6.3 Dividing and Focusing Attention

The ability to monitor multiple potential sources of hazards is essential when navigating congested urban roads. The driver must prioritize locations in the visual scene according to their importance and frequently monitor the most likely hazard spots, while inhibiting the impulse to fixate nonhazard-related information. Such an ability is similar to the concept of scanning training as proposed by Chapman et al. (2002); however, one particular focus is the need to not just scan the road continually, but particularly to disengage from hazards once they have been detected. Crundall, Underwood, and Chapman (1999, 2002) have observed a reduction in ability to detect peripheral targets in hazardous situations that may be particularly pronounced for inexperienced drivers. Training should thus focus on ensuring that once a hazard has been appropriately identified, attentional resources are redistributed and the remainder of the driving scene is also considered.

6.6.4 Hazard Management

One danger with emphasizing the role of eye-movement training in novice drivers is that it avoids the general issue that many hazards can be avoided by simply adopting a safe and defensive driving style. There is always a danger that any training intervention will encourage overconfidence in a driver. Although there is evidence that components of hazard perception training can be safely trained without encouraging an increase in risk-taking behaviors (McKenna et al., 2006), great care needs to be taken during any skill-based driver training. It is thus important that any training of visual search is integrated into a more general model of safe and responsible driving.

REFERENCES

Chapman, P., and Groeger, J. (2004). Risk and the recognition of driving situations. *Applied Cognitive Psychology*, *18*, 1231–1249.

Chapman, P., and Underwood, G. (1998). Visual search of driving situations: Danger and experience. *Perception*, *27*, 951–964.

Chapman, P., and Underwood, G. (1999). Looking for danger: Drivers' eye movements in hazardous situations. In. A.G. Gale et al. (Eds.), *Vision in vehicles VII* (225–232). Amsterdam: Elsevier.

Chapman, P., Underwood, G., and Roberts, K. (2002). Visual search patterns in trained and untrained novice drivers. *Transportation Research F: Psychology and Behaviour, 5*, 157–167.

Christianson, S., Loftus, E.F., Hoffman, H., and Loftus, G.R. (1991). Eye fixations and memory for emotional events. *Journal of Experimental Psychology: Learning, Memory, and Cognition, 17*, 693–701.

Cohen, A.S. (1981). Car drivers' patterns of eye fixations on the road and in the laboratory. *Perceptual and Motor Skills*, *52*, 515–522.

Crundall, D., Chapman, P., France, E., Underwood, G., and Phelps, N. (2005). What attracts attention during police pursuit driving? *Applied Cognitive Psychology*, *19*, 409–420.

Crundall, D., Chapman, P., Phelps, N., and Underwood, G. (2003). Eye movements and hazard perception in police pursuit and emergency response driving. *Journal of Experimental Psychology: Applied*, *9*, 163–174.

Crundall, D., Shenton, C., and Underwood, G. (2004). Eye movements during active car-following. *Perception, 33*, 575–586.

Crundall, D., and Underwood, G. (1998). Effects of experience and processing demands on visual information acquisition in drivers. *Ergonomics*, *41*, 448–458.

Crundall, D., Underwood, G., and Chapman, P. (1999). Driving experience and the functional field of view. *Perception*, *28*, 1075–1087.

Crundall, D., Underwood, G., and Chapman, P. (2002). Attending to the peripheral world while driving. *Applied Cognitive Psychology, 16*, 459–475.

Crundall, D., van Loon, E., and Underwood, G. (2006). Attraction and distraction of attention with outdoor media. *Accident Analysis and Prevention, 38*, 671–677.

Deery, H.A. (1999). Hazard and risk perception among young novice drivers. *Journal of Safety Research, 30*, 225–236.

Duncan, J., Williams, P., and Brown, I. (1991). Components of driving skill: Experience does not mean expertise. *Ergonomics, 34*, 919–937.

Easterbrook, J.A. (1959). The effect of emotion on cue utilization and the organization of behavior. *Psychological Review, 66*, 183–201.

Endsley, M. (1995). Measurement of situation awareness in dynamic systems. *Human Factors, 37*, 65–84.

Eriksen, B.A., and Eriksen, C.W. (1974). Effects of noise letters on the identification of a target letter in a nonsearch task. *Perception and Psychophysics, 16*, 143–149.

Eriksen, C.W., and Murphy, T.D. (1987). Movement of attentional focus across the visual field: A critical look at the evidence. *Perception and Psychophysics, 42*, 299–305.

Gugerty, L. (1997). Situation awareness during driving: explicit and implicit knowledge in dynamic spatial memory. *Journal of Experimental Psychology: Applied, 3*, 42–66.

Hella, F., Laya, O., and Neboit, M. (1996). *Perceptual demand and eye movements in driving*. Paper presented at International Conference on Traffic & Transport Psychology (ICTTP) '96, Valencia, Spain.

Horswill, M.S., and McKenna, F.P. (2004). Drivers' hazard perception ability: Situation awareness on the road. In S. Banbury and S. Tremblay (Eds.), *A cognitive approach to situation awareness* (pp. 155–175). Aldershot, UK: Ashgate.

Kass, S.J., Cole, K.S., and Stanny, C.J. (2007). Effects of distraction and experience on situation awareness and simulated driving. *Transportation Research F: Psychology and Behaviour, 10*, 321–329.

Klauer, S.G., Dingus, T.A., Neale, V.L., Sudweeks, J.D., and Ramsey, D.J. (2006). *The impact of driver inattention on near-crash/crash risk: An analysis using the 100-car naturalistic driving study data.* Washington, DC: National Highway Traffic Safety Administration.

Kramer, T.H., Buckhout, R., and Eugenio, P. (1990). Weapon focus, arousal and eyewitness memory: attention must be paid. *Law and Human Behavior, 14,* 167–184.

LaBerge, D. (1995). *Attentional processing: The brain's art of mindfulness.* Cambridge, MA: Harvard University Press.

Land, M.F., and Horwood, J. (1995). Which parts of the road guide steering? *Nature, 377,* 339–340.

Land, M.F., and Lee, D.N. (1994). Where we look when we steer. *Nature, 369,* 742–744.

Lestina, D.C., and Miller, T.R. (1994). Characteristics of crash-involved younger drivers. In *38th Annual Proceedings of the Association for the Advancement of Automotive Medicine* (pp. 425–437). Des Plaines, IL: Association for the Advancement of Automotive Medicine.

Loftus, E.F., Loftus, G.R., and Messo, J. (1987). Some facts about "Weapon Focus." *Law and Human Behavior, 11,* 55–62.

Maycock, G., Lockwood, C.R., and Lester, J. (1991). The accident liability of car drivers. (Report 315.) Crowthorne, UK: Transport and Road Research Laboratory.

McCartt, A.T., Shabanova, V.I., and Leaf, W.A. (2003). Driving experience, crashes and traffic citations of teenage beginning drivers. *Accident Analysis and Prevention, 35,* 311–320.

McKenna, F.P., and Crick, J.L. (1994). *Hazard perception in drivers: A methodology for testing and training* (Transport Research Laboratory Report 313). Crowthorne, UK: Transport Research Laboratory.

McKenna, F.P., Horswill, M.S., and Alexander, J. (2006). Does anticipation training affect drivers' risk taking? *Journal of Experimental Psychology: Applied, 12,* 1–10.

McKnight, J.A., and McKnight, A.S. (2003). Young novice drivers: Careless or clueless? *Accident Analysis and Prevention, 35,* 921–925.

Mills, K.L., Hall, R.D., McDonald, M., and Rolls, G.W.P. (1998). *The effects of hazard perception training on the development of novice driver skills* (Research report). London: Department for Transport, UK.

Mourant, R.R., and Rockwell, T.H. (1972). Strategies of visual search by novice and experienced drivers. *Human Factors, 14,* 325–335.

Muira, T. (1979). Visual behaviour in driving. *Bulletin of the Faculty of Human Sciences, Osaka University, 5,* 253–289.

Neyens, D.M., and Boyle, L.N. (2007). The effect of distractions on the crash types of teenage drivers. *Accident Analysis and Prevention, 39,* 206–212.

Pradhan, A.K., Fisher, D.L., and Pollatsek, A. (2006). Risk perception training for novice drivers: Evaluating duration of effects of training on a driving simulator. *Transportation Research Record,* 1969, TRB National Research Council, Washington, DC, 58–64.

Pradhan, A.K. Fisher, D.L., Pollatsek, A., Knodler, M., and Langone, M. (2006). Field evaluation of a risk awareness and perception training program for younger drivers. In *Proceedings of the Human Factors and Ergonomics Society 50th Annual Meeting* (2388–2391). Santa Monica, CA: Human Factors and Ergonomics Society.

Pollatsek, A., Fisher, D. L., and Pradhan, A. (2006). Identifying and remedying failures of selective attention in younger drivers. *Current Directions in Psychological Science, 15,* 255–259.

Pollatsek, A., Narayanaan, V., Pradhan, A., and Fisher, D.L. (2006). The use of eye movements to evaluate the effect of a PC-based Risk Awareness and Perception Training (RAPT) program on an advanced driving simulator. *Human Factors, 48,* 447–464.

Rahimi, M., Briggs, R.P., and Thom, D.R. (1990). A field evaluation of driver eye and head movement strategies toward environmental targets and distracters. *Applied Ergonomics, 21,* 267–274.

Recarte, M.A., and Nunes, L.M. (2003) Mental workload while driving: Effects on visual search, discrimination, and decision making. *Journal of Experimental Psychology: Applied, 9,* 119–137.

Richards, E., Bennett, P.J., and Sekula, A.B. (2006). Age related differences in learning with the useful field of view. *Vision Research, 46,* 4217–4231.
Sekuler, A.B., Bennett, P.J., and Mamelak, M. (2000). Effects of aging on the useful field of view. *Experimental Aging Research, 26,*103–120.
Sereno, A.B. (1992). Programming saccades: The role of attention. In K. Rayner (Ed.), *Eye movements and visual cognition: Scene perception and reading*, 89–107. New York: Springer Verlag.
Shinar, D., McDowell E.D., and Rockwell, T.H. (1977). Eye movements in curve negotiation. *Human Factors, 19*, 63–71.
Sperling, G., and Weichselgartner, E. (1995). Episodic theory of the dynamics of spatial attention. *Psychological Review, 102*, 503–532.
Underwood, G. (2007). Visual attention and the transition from novice to advanced driver. *Ergonomics*, *50*, 1235–1249.
Underwood, G., Chapman, P., Bowden, K., and Crundall, D. (2002). Visual search while driving: Skill and awareness during inspection of the scene. *Transportation Research F: Psychology and Behaviour, 5*, 87–97.
Underwood, G., Chapman, P., Brocklehurst, N., Underwood, J., and Crundall, D. (2003). Visual attention while driving: sequences of eye fixations made by experienced and novice drivers. *Ergonomics, 46,* 629–646.
Underwood, G., Crundall, D., and Chapman, P. (2002). Selective searching while driving: The role of experience in hazard detection and general surveillance. *Ergonomics*, *45*, 1–15.
Underwood, G., Crundall, D., and Chapman, P. (2007). Driving. In F. Durso (Ed.), *Handbook of Applied Cognition* (2nd ed., 391–414). Chichester, UK: John Wiley.
Underwood, G., and Everatt, J. (1992). The role of eye movements in reading: Some limitations of the eye-mind assumption. In E. Chekaluk and K.R. Llewellyn (Eds.), *The role of eye movements in perceptual processes* (111–170). Amsterdam: North-Holland.
Underwood, G., and Everatt, J. (1996). Automatic and controlled information processing: The role of attention in the processing of novelty. In O. Neumann and A.F. Sanders (Eds.), *Handbook of perception and action*, 3, 185–228. London: Academic Press.
Underwood, G., Phelps, N., Wright, C., van Loon, E., and Galpin, A. (2005). Eye fixation scan paths of younger and older drivers in a hazard perception task. *Ophthalmic and Physiological Optics, 25,* 346–356.
Williams, A.F. (2003). Teenage drivers: patterns of risk. *Journal of Safety Research, 34,* 5–15.

7 The Environment

Roadway Design, Environmental Factors, and Conflicts

Marieke H. Martens, Rino F.T. Brouwer, and A. Richard A. van der Horst

CONTENTS

Reflection ... 118
7.1 Introduction ... 118
7.2 Roadway Design ... 118
7.2.1 Road Pavement ... 119
7.2.2 Road Markings ... 121
7.2.3 Road Environment ... 122
7.2.4 Road Alignment ... 123
7.3 Adverse Weather Conditions ... 124
7.3.1 Adverse Weather Conditions and Accident Risk ... 124
7.3.2 Behavioral Adaptation ... 125
7.4 Public Lighting ... 128
7.4.1 Definition of Public Lighting ... 128
7.4.2 Freeway Lighting ... 129
7.4.3 Rural Roads ... 130
7.4.4 Public Lighting and Driver Workload ... 132
7.5 Driver Conflicts, Gap Acceptance, and Accident Analysis ... 132
7.5.1 Video Observations ... 133
7.5.2 DOCTOR Conflict Quantification ... 134
7.5.3 Practical Implications and Modeling ... 136
7.6 Conclusions ... 139
7.6.1 Road Design ... 139
7.6.2 Adverse Weather Conditions ... 140
7.6.3 Public Lighting ... 141
7.6.4 Conflicts and Accidents ... 141
References ... 142

REFLECTION

The main goal of this chapter is to provide an overview of the type of effects of road design, public lighting, and adverse weather conditions on driving behavior and traffic safety. This overview is far from complete, since decades of research and publications are available. Our main goal was to get the reader to understand the primary processes and provide a basis for further detailed literature searches or new experimental studies. The section about conflicts between road users and video observations illustrates that focusing on accident analyses is too limited. Observing behavior and conflicts is extremely valuable in understanding safety and traffic accidents, again offering the possibility to improve road design.

7.1 INTRODUCTION

Driving behavior is the interaction between the road environment, the vehicle, the driver, and surrounding traffic, and driving is not without danger. Although the chances to be involved in an accident are slim, accidents and conflicts do happen on a daily basis. A lot of people join the traffic circus daily and since both vehicle and driver are prone to error, incidents and accidents are bound to happen. There are several ways to improve road safety, such as driver education, road safety campaigns, enforcement, and driver support systems. However, a general idea is that the largest safety improvements can be realized when roads are properly designed. Roads should not only be designed from a "vehicle perspective" but also from a driver's perspective, that is, the (im)possibilities of people, or human factors knowledge, should be taken into account. This chapter will concentrate on the interaction between road design, environmental factors, and driving behavior, and conflicts between different road users.

7.2 ROADWAY DESIGN

Roadway design mainly influences driving behavior at the maneuvering (e.g., deciding to overtake while driving) and/or control level (steering, braking, etc.), and not so much the planning level (route choice). Certainly there are drivers who will adjust their route to avoid, for example, driving on narrow roads but still the design of the other roads they drive on will affect whether they will overtake or speed. A curved road will generally make drivers more cautious in overtaking and a narrow road will generally slow traffic down. Before turning to such effects of roadway design and environment on driving behavior, it is worthwhile to take a step back and see what we want to accomplish by (re)designing a road, and that is to positively affect driving behavior.

In most countries there are different road categories such as motorways, rural roads, urban roads, and so forth. What should be achieved by road layout and the road environment is that drivers know what type of road they are driving on. This way they can at least infer at each point in time the road category they are driving on, what they can expect (e.g., type of road users, oncoming traffic), and what is expected of them (e.g., maximum driving speed, overtaking or not). A road then becomes self-explaining. A traffic environment that provokes the right expectations will reduce potential errors. Furthermore it would increase the possibility for long-term effects and could avoid dangerous side effects like those with speed humps. Fildes and Jarvis (1994) emphasize the importance of such changes in road design, which are a mixture of sensorial and cognitive aspects.

The concept of self-explaining roads (SER) has been discussed by Theeuwes and Godthelp (1992; see also Aarts, Davidse, Louwerse, Mesken, and Brouwer, 2005). The traffic environment should provoke the right expectations concerning the presence and behavior of other road users as well as the demands with regard to their own behavior. In order to reach this goal, distinct road categories must be used, each requiring their own specific driving behavior.

Experiments in the Netherlands (Riemersma, 1988a; Theeuwes, 1994; Kaptein and Theeuwes, 1996) have shown that official road categories did not correspond with the subjective road categories that road users have. This may lead to driving behavior that is not appropriate for the traffic situation. Only motorways form a clearly distinct road category, where drivers have a good idea of what to expect and what is expected of their own driving behavior. However, studies have shown that it is difficult for road users to understand the road category by its design characteristics and that it is difficult to assess what road design characteristics define subjective road categorization (e.g., Gundy, 1994, 1995; Gundy, Verkaik, and de Groot, 1997; Brouwer, Janssen, and Muermans, 2000). Despite these difficulties, it is crucial to keep a simple rule in mind that similar road categories should elicit similar driving behavior and expectations, which could better be achieved when they are similarly designed.

A road consists of a number of dependent dimensions that have an influence on driving:

1. Pavement (type and width)
2. Markings
3. Environment
4. Alignment

7.2.1 Road Pavement

7.2.1.1 Roughness of Road Surface

Roughness of road surface is a measure for the amount and kind of deviations from a smooth road surface. There is longitudinal roughness (e.g., bumps in asphalted concrete in front of traffic lights), transverse roughness (e.g., aquaplaning), road-surface irregularities (e.g., holes in the road surface), and roughness caused by road material (e.g., a brick road). Irregular road surfaces will result in an increased amount of noise and vibration compared to smooth road surfaces, thereby decreasing driver comfort. An effect on driving speed is not the result of the roughness of the road surface per se, but rather an effect of a reduction in driver comfort. A road surface can be described in terms of material and structure, microroughness, and color (Wildervanck, 1987).

Van de Kerkhof and Berénos (1989) indicated that driving speed on asphalted concrete is higher than speed on brick roads, since the surface of a brick road is much rougher than that of asphalted concrete. Van de Kerkhof (1987) stated that roughness of a road surface is the most important factor in determining driving speed, and that it can explain 91% of the variation in driving speed. The second most important factor was the amount of buildings and the third factor was the repeating character of objects along the roadside, with a more irregular character leading to lower speed. Slangen (1983) also indicated a reduction in driving speed on roads with a rough road surface (14%–23% reduction). This relationship between speed and amount of roughness has also been found by others (Karan, Haas, and Kher, 1977; Anund, 1993). However, if the road surface is too rough, this may result in damage to vehicles and in increased accidents due to loss of vehicle control. A subtle measure to reduce speed would be to use a road surface that has a microrough structure that only causes an increased noise level inside the car.

Cooper, Jordan, and Young (1980) found increases in driving speed up to 2.6 km/h after resurfacing three test sites, where the profile of the road surface was improved. Te Velde (1985) found that if a smooth road surface was followed by a rough surface, this resulted in a mean reduction of driving speed of 5%. There was no immediate increase in speed if a rough surface was followed by a smooth road surface. Makking and De Wit (1984) found a reduction in driving speed with a transition in the road from a concrete road part to a brick one, but they indicated it is not known if this reduction will continue after driving a brick road for a while. Sometimes this effect of speed reduction for rough road surfaces is not found, for instance, in a study by Michels and Van der Heijden (1978). Probably, other road characteristics could have influenced speed behavior in the other direction.

7.2.1.2 Lane Width

It is easy to assume that lane width plays a role in driving speed. After all, on narrow lanes other traffic is nearer and there is less space to keep a certain distance to obstacles along the side of the road. Also, a driver needs to put more effort in lane keeping on narrow lanes. This may also lead to decreased speeds if the driver wants to reduce the effort.

This *a priori* notion has also been found in the literature. Yagar and Van Aerde (1983) found a reduction in speed of 5.7 km/h for every meter of reduction beyond 4 m. A positive relationship between vehicle speed and lane width was also found by Vey and Ferreri (1968). A point of attention is that accidents may result when road users do not adapt their behavior enough to match the increased difficulty of driving on a narrow lane (Jacobs, 1976; DeLuca, 1985; Lamm, Choueiri, and Mailander, 1989). So reducing lane width is an effective measure to decrease speed as long as the safety effect is not washed out by a decrease in lane keeping performance or an increase in workload. However, the risk of running off the road (e.g., due to fatigue) may also increase.

7.2.1.3 Lateral Clearance

Lateral clearance indicates the space between obstacles to the left and the right side of the road or the space that is visually available between obstacles on either side of the sidewalk. In this case, obstacles can be front gardens, overgrowth, lampposts, ditches aside the road, parked cars, and so forth (Van de Kerkhof, 1987). Although lateral clearance covers more than pavement width, under some conditions, for instance, a barrier just along the side of the road, pavement width and lateral clearance refer to the same space.

A reduction of lateral clearance from 30 m to 15 m decreases speed by only 3%. However, when lateral clearance is decreased to 7.5 m, a speed reduction of 16% was found (see Van der Heijden, 1978). This indicates that reducing lateral clearance only results in larger speed reductions beyond a certain point. Its effect also depends on the kind of shoulder (soft, hard) and the amount of danger associated with leaving the road (for instance, hitting a tree). Therefore, there is a relationship between the distance of the car to obstacles in the shoulder and driving speed. With obstacles directly along the side of the road (reduced lateral clearance), driving speed reduces about 13% compared to obstacles placed 1 m away from the edge of the road (Knoflacher and Gatterer, 1981).

7.2.2 Road Markings

Road markings are line treatments on the road surface that provide guidance and regulatory warning information to the driver. Side markings and center line markings can serve as a cue to show the proper path to follow and transverse road markings can serve as a possible warning. For lane keeping and anticipation on the course of the road, road markings are extremely important. Also, enhanced road markings on highways were found to lower workload, and to help drivers keep their lateral position and speed (Horberry, Anderson, and Regan, 2006). A literature review by the Organisation for Economic Co-operation Development (OECD, 1990) suggests that the presence of road markings on two-lane rural roads produces a safety benefit. However, this finding is not in accordance with Elvik, Borger, and Vaa (1996). In a summary of studies from different countries, they did not find reductions in accidents as a consequence of the presence of center line and edge line markings. However, for crashes associated with drunk driving, road markings have been found to reduce the risk (Noordzij, 1996). Even though the general idea is that road markings are important for driving safety, their presence can also cause an increase in speed, especially at night due to increased visual guidance. In a study by Van der Horst (1983), a reduction in lane width from 4.6 m to 3.6 m by placing a central area between the two driving lanes led to an increase in driving speed of 7.5 km/h.

Besides the most common type of road markings—center line and edge line markings placed along the road axis—other types of pavement markings can be found. Transverse marking patterns, as well as decreased spacing between the center line markings, can decrease speed, since this leads to the illusion that the driving speed is higher than it actually is or even that the car is accelerating. Transverse road markings are especially suitable to reduce speed near a dangerous situation, for instance, just before a dangerous crossing, a roundabout, or a bend. Since the markings increase in number, this will probably alert the driver and decrease speed in that manner. Fildes, Fletcher, and Corrigan (1987) found that the use of herringbone road markings along the side of the road (that increase in frequency when approaching a dangerous location) led to a reduction in mean driving speed. Similar findings have been found in other studies (Denton, 1971, 1973; Rockwell and Hungerford, 1979; Agent, 1980). Specific road markings are also found to increase safety on horizontal curves (Charlton, 2007). In a driving simulator study, different types of road markings were used to affect drivers' speed and lane position as they drove through curves. Of the road marking treatments, only rumble strips produced any appreciable reductions in speed. A herringbone road marking was found to produce significant improvements in drivers' lane positions, effectively flattening the drivers' paths through the curves. A treatment combining the herringbone with chevron and repeater arrow signs produced a reliable reduction in speed as well as improved lane positions.

Besides reductions in mean driving speed, reductions in speed variance are also reported (Denton, 1973). However, there is some uncertainty about the durability of these speed reductions. Havell (1983) suggests that effectiveness of such measures can be maintained for months, whereas others suggest the benefits fade in a matter of days or weeks (e.g., Maroney and Dewar, 1987). Also, rumble strips can be used to enlarge the effect of road markings. A rumble strip is a strip with a rather rough structure that can be placed on the road in either a lateral or a transverse direction. By driving on these strips, noise and vibrations will be produced, so driving comfort will be decreased.

7.2.3 Road Environment

There are some aspects in the road environment (e.g., amount of trees or overgrowth near a road, buildings, etc.) that play a role in driving behavior.

Trees close to the road decrease speed; on the other hand, they indicate the line of the road and may therefore increase speed. A line of trees that is not in parallel with the road may even confuse the driver (De Ridder and Brouwer, 2002).

On urban roads, Van de Kerkhof (1987) found an effect of the presence of buildings on driving speed. Buildings that were positioned next to the investigated road surface and are visible for the driver reduced driving speed. Smith and Appleyard (1981) also reported that the distance of housing to the road was positively correlated with speed. A driving simulator study (Perdok, 2003) showed that the closer buildings or trees were located to the road, the lower the speed, and that houses led to lower speeds compared to more industrial areas (high constructions) or more rural areas. Also, limitations in sight distance have been found to reduce speed, and strong transitions in lateral clearance or road width will also decrease speed and may even lead to traffic jams (see Janssen, Kaptein, Hogema, and Westerman, 1995).

Research has shown that restrictions in the amount of information available in the visual periphery may lead to an underestimation of the driving speed (Brandt, Wist, and Dichgans, 1975; Dichgans and Brandt, 1978). Salvatore (1968) had people drive in a car and provided either 25 degrees of frontal information or 25 degrees of peripheral information, all this with three different driving speeds. Subjects had to estimate the speed they were driving. It turned out that 25 degrees of peripheral visual stimulation led to a more accurate speed estimation than 25 degrees of frontal information. This can be explained by the fact that the angular velocity is much larger in the peripheral field. This finding suggests that enlarging the amount of information in the visual periphery may even lead to an overestimation in speed, possibly resulting in speed reductions.

Research by Yamanaka and Kobayashi (1970) shows that people consider speeds exceeding 2 rad/s in the visual periphery (at about 30 degrees left and right of the fovea) to be very disturbing. Road users usually choose their speed and position on the road in such a way that the angular speed of visual objects in the visual periphery does not exceed this value of 2 rad/s (Blaauw and Van der Horst, 1982; Van der Horst and Riemersma, 1984). These results seem to suggest that increasing the density of information in the visual periphery can help decrease driving speed. The layout of the environment should be designed in such a way that exceeding the speed limit leads to exceeding this value of 2 rad/s.

7.2.4 Road Alignment

The amount of curvature in roads actually affects speed in a number of ways. First of all, driving through curves requires some extra effort in lane keeping. Besides this, curves result in a reduction in the visibility distances along the road axes, limiting anticipation of the course of the road and upcoming traffic situations. Several studies found a significant relation between driving speed and visibility along the road axes, with reduced visibility resulting in reduced driving speed (Michels and Van der Heijden, 1978; Bald, 1987; Brenac, 1989). Reducing the visibility distance produces higher uncertainty about the course of the road. Several researchers point out that road curvature mainly predicts the amount of speed reduction, but does not predict actual driving speed (Kanellaidis, Golias, and Efastathiadis, 1990; Reinfurt, Zegeer, Shelton, and Newman, 1991). Yagar and Van Aerde (1983) indicate that there is only an effect of visibility along the road axis with visibility distances less than 500 m. Although Taragin (1954) found a close linear relationship between operating speed and the degree of curvature, curvature had a much greater effect on speed than sight distance. This was confirmed by McLean (1979), although Watts and Quimby (1980) claimed the opposite. The minimum sight distance was not necessarily related to the degree of curvature, but curves of larger radii did tend to have longer sight distances.

People often underestimate the sharpness of the curve and enter the curve with a speed that is too high, requiring abrupt braking behavior inside the curve. In a simulated driving task, Shinar, McDowell, and Rockwell (1974) found that subjective judgments of curve characteristics bear little relationship to the physical characteristics of curves. Riemersma (1988b) found that three subjective counterparts of distance, radius, and deflection angle of curves were not related to the objective characteristics on a one-to-one basis. This suggests that curve radii should not only be based on design speed. Marconi (1977) indicates that with deviations in the horizontal longitudinal profile, speed reductions will result since drivers have to put extra effort in lane keeping and uncertainty is increased due to reduced visibility distances. A phenomenon often found is that one enters a bend with a too high speed and only decelerates inside the bends. In that case large corrections are needed. Tenkink and Van der Horst (1991) showed that with tight curves, the amount of speed reduction drivers disposed was not sufficient to maintain the amount of line crossings at a low level. This indicates that people do not reduce the speed as much as would be necessary to guarantee safety. Advisory speeds, in comparison to general speed signs, work fairly well in changing driving speed. However, road users have to understand the reason for the warning or the restriction. According to Tenkink (1988), advisory speeds near curves work very well, but only if the reason of the advisory speed is explained. This is confirmed by Webb (1980), who found that when the reason for a speed restriction is not understood, advisory speed limits have only a marginal effect. Marconi (1977) found that advisory speeds work to some extent, in that they result in a more optimal traffic flow, but the reductions in speed are not always as large as aimed for. Rutley (1975) found that indicating the maximum speed at which drivers could comfortably negotiate a bend led to a mean speed closer to the advisory speed given by the sign. Drivers with low speeds increased their speed and drivers with high speeds decreased their driving speed toward the advised speed. Zwahlen (1987) suggests that warning signs should be placed before the beginning of the curve approach. Road markings and signs on the road surface are particularly helpful for improving safety in curve negotiation. Milosevic and Milic (1990) found that a warning sign and a speed limit sign helped drivers adjust their speed at the central point of a small radius curve more accurately.

A possibility to increase driving safety in case of tight curves by means of road markings is to introduce an "illusive curve phenomenon" (Shinar, 1977). This road marking will provide an image of a sharp curve, so drivers anticipate the curve more appropriately. Reinfurt et al. (1991) emphasize that although low cost measures like signing, marking, and delineation are attractive measures in an attempt to reduce driving speeds in curves, they cannot make up for intrinsic deficiencies of a poorly designed curve.

7.3 ADVERSE WEATHER CONDITIONS

7.3.1 Adverse Weather Conditions and Accident Risk

Adverse weather conditions (such as rain, snow, hail, wind, and fog) have a significant negative impact on road traffic safety. There are several studies that show that icy roads, snowy roads, higher wind speeds, and rainfall increase accident rates (Shankar, Mannering, and Barfield, 1995; Malmivu and Peltola, 1997; Khattak and Knapp, 2001; Andrey, Mills, Leahy, and Suggett, 2003). Andrey and colleagues (2003) claim that collision risk usually increases during precipitation, that snowfall has a greater negative impact than rainfall, and that the collision risk is highest for freezing rain and the first snowfalls of the season. It is interesting that snowfall has been shown to lead to more collisions, injuries, and vehicle damage, but to fewer fatalities (Andrey et al., 2003; Eisenberg and Warner, 2005). Precipitation in the form of rain and snow generally results in more accidents compared with dry conditions (Codling, 1974; Satterthwaite, 1976; Sherretz and Farhar, 1978; Brodsky and

Hakkert, 1988; Fridstrøm, Ifver, Ingebrigsten, Kulmala, and Thomsen, 1995; Levine, Kim, and Nitz., 1995; Changnon, 1996; Andreescu and Frost, 1998; Edwards, 1999; Eisenberg, 2004). There is evidence that wet or snowy weather, particularly if coupled with severe storms, can deter motorists from venturing onto the road (Knapp and Smithson, 2000). In a case-control study, risk factors (motor vehicle accidents on highways) were identified by comparing a driver who completed a trip with an accident and a driver who completed a trip without being involved in a traffic accident. Adverse weather conditions (rain, fog, wet pavement) showed a clear association with the risk of an accident (Hijar, Carrillo, Flores, Anaya, and Lopez, 2000).

All these studies show the negative impact of adverse weather conditions on traffic safety. Underlying causes of decreased safety are reduced friction, poor visibility, strong lateral deviations, a combination of factors, and even stress. In a survey study, driving on icy roads or in heavy rain or snow were the main factors causing driver stress (Hill and Boyle, 2007). Females, older drivers, and drivers who reported being involved in a higher number of crashes reported higher levels of stress under these conditions. Although one may blame these weather conditions and accept the direct link between weather conditions and safety, the question is whether the accidents may be the result of drivers not compensating for these conditions by properly adjusting their driving behavior to the changed conditions.

7.3.2 Behavioral Adaptation

Drivers need to adjust their behavior to changes in driving conditions. Drivers are normally inclined to react to changes in the traffic system, whether they are in the vehicle, in the road environment, or in their own skills or states, and the reaction occurs in accordance with their own motives (Summala, 1996). This principle of behavioral adaptation (see Näätänen and Summala, 1974, 1976; Wilde, 1974, 1975, 1976), also called risk compensation, is a central theme in the discussion of the effects of weather on safety.

Some studies have found that even in case of very slippery road conditions, drivers still drive faster than the speed limit and consider it safe driving (Heinijoki, 1994). Even safety measures such as studded tires may have limited effect because drivers compensate for this improved friction (risk compensation). Drivers with studded tires drive somewhat faster in curves, which to some extent decreases this safety improvement (Rumar, Berggrund, Jernberg, and Ytterbom, 1976; Summala and Merisalo, 1980). It has even been shown that on slippery road surfaces, over half regard the friction as quite normal (Heinijoki, 1994). Although the average speed on a slippery road is somewhat lower than in good winter conditions and the standard deviation of speed is also lower due to a drop in the highest speeds (Saastamoinen, 1993), the adaptation has been proved insufficient to fully compensate for the reduced friction (Malmivuo and Peltola, 1997). In general, headways are not substantially affected by winter conditions.

In the case of fog, visibility distances are found to affect free-driving speeds (Hogema and Van der Horst, 1994a). Also in fog, drivers still choose speeds that do not compensate for the reduced visibility. Especially in the visibility range between 40 and 120 m, speeds are too high to allow for a successful stop for a stationary obstacle. Although loop detection data (Hogema and Van der Horst, 1994a) and literature (White and Jeffery, 1980; Hawkins, 1988) showed shorter following distances in fog, this only seemed to be the result of reduced speeds (Hogema and Van der Horst, 1994b). There does not seem to be a direct relationship. This indicates a fixed headway strategy regardless of the visibility condition. Constant headway as a function of speed has also been found by others in clear visibility conditions (Colbourn, Brown, and Copeman, 1978; Van Winsum, 1993). Dense fog restricts perception of the situation ahead of the lead vehicle, and thus inhibits anticipation to the behavior of the lead vehicle.

The outcome of experimental research showed that drivers clearly perceive weather-related risks and do adjust their behavior, but not as much as the weather conditions require. In order to support drivers in adverse weather conditions different systems have been developed to warn the driver.

In Finland, there is a TWIS (Traffic Weather Information Service) that produces forecasts that classify the driving conditions in a specific region as normal, poor, or very poor. Kilpeläinen and Summala (2007) studied the effect of drivers' perception of the weather conditions on driving behavior and drivers' use of the TWIS system. The parameters studied were air temperature, wind speed, relative humidity, precipitation, and dew point temperature. About 16% of the drivers reported to have actively acquired traffic-related weather information for the trip, with 13% from the radio and 9% from television. This was substantially lower than was found in earlier telephone interviews, in which more than 40% of the drivers reported to have seen TWIS forecasts on television at least once a day and 32% reported to have heard them on the radio (Antilla, Nygård, Rämä, 2001). In general, Kilpeläinen and Summala (2007) stated that about 3% of the drivers claimed to have adjusted their travel plans according to the weather information, but this was 16% for the drivers who actively requested information. Reported behavioral changes included allowing more time for the trip, altering time of departure, and changing the route. The drivers' own weather ratings had a significant effect on speed choice, overtaking frequency, and headways. Interestingly, there was no effect of information acquisition on the actual driving behavior. This may have been because of the nonspecificity of the system.

A more specific warning system that only presents messages during driving in case of specific weather conditions may have better results. Cooper and Sawyer (1993) showed that a fog warning system on the London motorway in which the text FOG was presented on roadside matrix signals in case of actual fog only reduced speeds by 3 km/h. Rämä and Kulmala (2000) evaluated two types of variable message signs (VMS) to warn for slippery road conditions. One warned for slippery road conditions by means of a pictogram of a sliding car (actually informing what the problem is), and the other by showing a minimum headway sign (actually informing how to respond). The pictogram of the sliding car was always on in case of slippery or possibly slippery conditions due to ice or snow. The minimum headway sign was adapted to the road surface condition, taking into account vehicle type and speed of the vehicle. The slippery road condition sign decreased driving speeds by 1–2 km/h at a distance of 500–1100 m after the signs, with more substantial effects at night. The minimum headway sign decreased the proportion of short headways. The slippery road condition sign alone did not seem to affect headways.

Gupta, Bisantz, and Singh (2002) investigated the effect of an in-vehicle system to warn for an imminent skid or rollover in case of snowy conditions in a driving simulator. Two different alerting devices were used (one with low and one with high sensitivity) and two types of auditory alarms (one with a binary signal [message or no message] and one with a signal that increased in intensity with increasing risk). Velocity at the instance of a skid was higher for the low-sensitivity alarm compared to the high-sensitivity alarm. Five seconds after the alarm onset there was a larger decrease in velocity for participants in the low-sensitivity condition. This can be due to the fact that there was more time available for participants in the high-sensitivity alarm condition to reduce their speeds before the onset of a skid. The lower-sensitivity alarm tended to lead to increased trust and better performance as measured by steering wheel deviation. Overall, there was no indication that alarm sensitivity affected driver control as measured by yaw angle, slip angle, and lateral acceleration. Low-sensitivity alarm combined with a ramp alarm condition had fewer skids than the high-sensitivity ramp alarm condition. Overall, the presence of the alerting system tended to have a positive effect on performance compared to a control group. However, in the high-sensitivity condition, participants also produced greater steering wheel deviations—a trait that is not desirable in slippery conditions.

An automatic fog signaling and warning system in the Netherlands showed reduced speed limits based on the available visibility distances. In case of fog, a danger sign, the word MIST, and flashers on a VMS with reduced speed limits were shown. Traffic data were collected for a period of more than 2 years using inductive loop detector measurements at an individual vehicle level. This allowed studying speed, time headway, following distance, and time-to-collision. Speed was most strongly influenced by visibility, lane, and flow (Hogema, Van der Horst, and Bakker, 1994). The mean speed decreased when the visibility distance was reduced, speeds were higher in the fast lane (left), and when the flow increased, the mean speed decreased. With the system, speeds were lower than without the system. In the fast lane, the system caused a larger speed reduction than in the slow lane. As a consequence, the difference in speed between both lanes decreased due to the system. Also, the mean speed reduction for cars was larger than for trucks, which resulted in a decrease of the speed difference between both vehicle categories. A more restrictive sign resulted in a lower mean speed in the same visibility conditions. Even though the mean speeds were still higher than indicated by the matrix signs, the warning system yielded a speed reduction of 8 to 10 km/h. Based upon the relationships found in the literature between the mean driving speeds and the number of accidents, a speed reduction of 5 km/h would already yield a decrease of 15% of the number of accidents. In extremely dense fog (visibility range of less than 35 m), the system seemed to have an adverse effect on speed. The system still displayed 60 km/h, whereas without the system the mean speed was considerable lower than that. Therefore, it was recommended that the fog signaling system be based on the same set of speed limits as is used in the motorway control and signaling system (MCSS): 50, 70, and 90 km/h. This is currently implemented. There was hardly any effect of the system on following distance but the percentage of time-to-collision smaller than 5 and 10 seconds was lower with the system, but rather small. Both the number and the severity of fog accidents were reduced after the system was implemented.

7.4 PUBLIC LIGHTING

Since about 90% of the information presented to the driver is visual, the assumption was that introducing public lighting would also increase traffic safety because of increased visibility. In urban areas, it is common to find streetlights that generally have a beneficial effect on the visibility available to the drivers (Mortimer, 2001). Average daytime legibility and recognition distances are about 1.8 times longer than the average nighttime legibility and recognition distances (Zwahlen and Schnell, 1999). Urban streetlights are considered particularly useful in increasing the visibility of pedestrians and other objects on the road. In an experimental setting, individual drivers needed to detect a simulated hazard on the road surface of an urban freeway (Janoff and Staplin, 1987). The results demonstrated significant decrements in drivers' ability to detect the target as alternative reduced lighting tactics are implemented. However, the standard deviation for detection distance was greater with increased lighting. Subjective ratings of target visibility under test conditions were in general agreement with the detection response data.

Although the main reason for introducing public lighting is traffic safety, public lighting is also supposed to make the driving task less strenuous for the driver and is supposed to increase social safety as well. Over the years, however, the ideas about the application of lighting have changed due to, for example, environmental reasons. Public lighting is reduced unless there is a clear benefit with respect to traffic safety (or sometimes due to negative effects on social safety).

It is difficult to assess the effect of public lighting on traffic safety because of the complexity of the situation. In a before–after study, comparing accident rates before the introduction of public lighting with the accident rates after placing public lighting, other things may change as well. Examples are traffic volume, percentage of heavy goods vehicles, weather conditions, and so forth. These should all be taken into account. The remainder of this section describes the reported effects of public lighting on traffic safety.

7.4.1 DEFINITION OF PUBLIC LIGHTING

When discussing the effect of public lighting on traffic safety, it has to be kept in mind that there are different versions of public lighting. Public lighting can be used on motorways for orientation purposes only, mainly aimed at pointing out the direction of the road. The term public lighting is also used for alighting an entire stretch of road or for marking specific areas such as intersections with a single streetlight. Therefore, it is important to know what is meant exactly with *public lighting* when discussing experimental results or accident statistics.

For instance, accident data have shown an approximate reduction in night accidents of 50% after public lighting was introduced (Walker and Roberts, 1976). In this case, public lighting was only present at intersections. The average night accident rate per million entering vehicles was 1.89 before lighting the intersections and 0.91 after lighting. Bruneau and Morin (2005) evaluated the safety aspects of roadway lighting at rural and near-urban three-way and four-way junctions by comparing unlit intersections with those lit. Rural lighting of an intersection reduced the night accident rate by 30%–40%. In both examples, lighting specific areas already resulted in large safety improvements without public lighting the entire stretch of road.

Streetlights may also mark specific areas under dark conditions. However, streetlamps may also create glare and can interact with vehicle headlighting in a complex manner in certain conditions that can lead to a temporary reduction in the visibility of a pedestrian.

7.4.2 Freeway Lighting

Freeway lighting is often used at interchanges and along straight sections with high traffic volumes. Bruneau, Morin, and Pouliot (2001) indicated that "[p]ast analyses of motorway lighting reveal that full motorway lighting reduced the night-time accident rate. … The benefits of lighting along straight sections of motorways are similar in the case of accidents with injuries, yielding an average reduction of 38%. … Of the five studies found that examined interchange lighting, 3 revealed lighting to be significantly safer than darkness at interchanges. … A comparison of urban motorways further indicated that interchanges are safer when the motorway is continuously lit as opposed to lighting at interchanges only. … All together 22 results were found and almost half of them are significant. All the published studies tend to indicate that lighting reduces motorway accidents."

The objective of a study by Lamm, Kloeckner, and Choueiri (1985) was to assess the effectiveness of freeway lighting. In the case of freeway lighting, an entire stretch of road is completely illuminated. Lamm and his colleagues conducted a case study on traffic accident characteristics. For this, they used the data of a suburban freeway area in Germany between 1972 and 1981. The study revealed that the effects of lighting on suburban freeway accident rates were positive. There was a reduction in accidents, and these positive results of continuous freeway lighting were lost in the case of partial lighting, especially after switching off lights at night between 10:00 p.m. and 5:30 a.m. for the purpose of saving energy. In this case, partial lighting refers to no lighting in periods of low traffic volume and not to lighting of exits and entries.

A meta-analysis of 37 studies evaluating the safety effects of public lighting is reported by Elvik (1995). The 37 studies contain a total of 142 results, although a critical comment here is that it is unclear whether the type of public lighting was really comparable over studies. The studies included were reported from 1948 to 1989 in 11 different countries. The safety effects of public lighting were, however, sensitive to accident severity and type of accident. It was concluded that the best current estimates of the safety effects of public lighting are, in rounded values, a 65% reduction in nighttime fatal accidents, a 30% reduction in nighttime injury accidents, and a 15% reduction in nighttime property-damage-only accidents.

Roadway lighting design has evolved over the years from the illumination method, which is based on the amount of light falling on the road surface, to the luminance- and visibility-based methods that are in use today (Khan, Senadheera, Gransberg, and Stemprok, 1999). Visibility of an object on the roadway is directly related to the contrast between the object and its surroundings. In nighttime driving situations, the pavement acts as the background for most objects on the road. Therefore, reflectance characteristics of the pavement are important in visibility-based roadway lighting design processes.

Bruneau, Morin and Pouliot (2001) assessed the safety effect of continuous and interchange lighting on motorways by comparing the night/day accident rate ratio method. Various sources of data were used for calculating night/day accident rate ratios (e.g., accident databases). The categories of accidents they used were fatal and injury accidents, property damage only, and all accidents. They found that continuous lighting reduces the overall accident rate by 33% in comparison with interchange lighting alone and by 49% compared with dark motorways. By breaking down their data Bruneau, Morin, and Pouliot showed that these accident reductions appeared to be still valid regardless of traffic flow.

The traditional policy has been to install lighting as soon as the average traffic flow exceeded a certain threshold. However, over the years environmental and energy consumption issues have gained importance. In the Netherlands, Dynamic Public Lighting was regarded to be the optimal solution between safety, costs, and ecology. With Dynamic Public Lighting, the amount of lighting is adapted to the traffic and weather conditions in such a manner that the amount of lighting is sufficient to ensure safe and efficient traffic flow, while avoiding unnecessarily high levels of illumination (Folles, Ijsselstijn, Hogema, and Van der Horst, 1999). Besides the normal level of illumination (100% = 1 cd/m^2), a reduced (20%) and an increased level (200%) can be employed. On the basis of inductive loop data, drives with an instrumented vehicle, a road user survey, video recordings, and an accident analysis data were collected in a before period (no road lighting on the test section), a nil period (normal road lighting, 100%), and an after period (dynamic lighting: 20%, 100%, or 200%). Data were also gathered from a control section with normal lighting all the time. In addition to lighting condition, precipitation (dry versus rain) and traffic intensity were the main research factors. Results showed that when the visibility conditions were improved by means of road lighting, the driving speed was slightly increased. Since the number of critical situations (small time headways, time-to-collision) did not seriously increase, there are no indications for a reduction of traffic safety. In rain, speed choice under 100% lighting conditions is more similar to behavior in dry conditions than speed choice under 20% or 0%. However, differences between 100% and 200% were not found. At entries, a different level of road lighting did not affect the differences in driving speed between entering traffic and traffic on the main road. In case of rain, behavior is more comparable to behavior with dry weather in cases where the road is lit.

All the published studies tend to indicate that lighting reduces motorway accidents. Moreover, the significant results are similar to those not supported by a validation test.

7.4.3 Rural Roads

On rural roads, specifically vulnerable areas are normally illuminated, such as roundabouts and intersections. At minor side roads, there is often one single streetlight to indicate the presence of the side road. Looking at driving behavior and workload, a field study assessed how much lighting is actually required on rural roads if specific locations are already illuminated (Martens, 2007). How is mental workload and driving behavior affected by reducing the luminance level of public lighting or even turning off public lighting in the approach zones to roundabouts but not at the roundabouts themselves. Martens conducted a field study with an instrumented vehicle during nighttime driving. Participants drove a specific trajectory on an 80 km/h road, with four roundabouts with public lighting that could be dimmed or even switched off. Every participant drove the trajectory with full luminance level (100% luminance level on the approach zone and 100% at the roundabout), with a decreased luminance level (decreased to 20% luminance level in the approach zone and to 50% on the roundabout), and with lighting switched off (switched off in the approach zone and decreased to a luminance level of 50% on the roundabout). Driving behavior (speed, standard deviation of speed, lateral position, swerving within a lane, braking behavior, and steering behavior) was compared among the three luminance levels, as was subjective workload and the answers to the questionnaires. No effect was found of dimming or even switching off the public lighting on the speeds or standard deviation of speeds on the approach trajectory or on the roundabout. In the condition with the lighting being switched off, there was a slightly stronger but not a relevant maximum deceleration than in the dimmed condition or in the full lighting condition. It didn't produce an effect of lateral displacement within the lane nor for swerving within the lane. There were no effects in subjective workload, nor did a driving instructor find any difference in driving performance between the three conditions. In the questionnaires, some people indicated they preferred the public lighting being switched off, while others indicated that they needed the lighting to estimate the distance to the roundabout.

Another field study on switching off public lighting on a rural road while keeping specific areas such as roundabouts and minor side roads illuminated showed that on trajectories on which a lot of streetlamps were switched off, there was only one trajectory with a somewhat higher driving speed with all streetlamps on (Martens, 2005). There was no difference in the standard deviation of the speed between the two lighting conditions. This study also revealed the importance of having proper control conditions. In the limited lighting condition, participants drove somewhat more to the right side of the road. This could have led to the conclusion that reduced public lighting leads to a different lateral position. However, the before–after results corresponded to the before–after behavior found on the control location, so no extra risks are introduced due to lighting restrictions. The swerving was somewhat less with less road lighting (between 2 and 5 cm). There were no differences in steering effort or in subjective workload. The questionnaires revealed some critical notes, such as the opinion that parts of the road were not well lit or that the roundabouts were not clearly designed. There were no or hardly any differences in remarks between the two lighting conditions. The general conclusion from this study was that in terms of driving behavior, subjective experience, and workload, there is no reason to assume that decreasing the number of streetlamps according to the specific plans in that area increases safety risks compared to the current situation. It should be mentioned that this holds for the situation in which roundabouts and intersections are always lit and there are no cyclists on the road. For roads on which street lamps will also be removed on roundabouts and on intersections or for roads where cyclists are using the same carriageway, different conclusions may apply.

7.4.4 Public Lighting and Driver Workload

It may very well be that under some conditions, drivers are able to keep up good driving performance but at the expense of a higher workload. When the driving task is mentally loading, the driver can compensate by reducing the attention paid to other tasks. This mechanism is the basis of the secondary task paradigm. By introducing an extra task that is not related to the driving task, the performance on this secondary task is an indicator of the "spare capacity" the driver has available. A secondary task can also be applied to increase the overall task load. With such a loading secondary task, effects of independent variables on workload can occur that are not found in the absence of this secondary task. In a field experiment by Hogema and Veltman (2003), the secondary task applied was a new version of the continuous memory task (CMT) that was successfully applied in earlier research (Van Breda and Veltman, 1998). The results showed that mean speed marginally decreased without public lighting (111 km/h and 113 km/h, respectively). The secondary task itself also affected driving behavior, with higher speeds with and without the CMT, respectively (113.1 and 110.6 km/h, respectively). However, with relatively high traffic volumes, there was no effect of public lighting on speed. Without lighting, the steering reversal rate (SRR) was higher than with lighting, indicating more strenuous steering. The blink frequency was somewhat higher with road lighting than without lighting, which indicates a lower visual effort due to road lighting.

7.5 DRIVER CONFLICTS, GAP ACCEPTANCE, AND ACCIDENT ANALYSIS

As we have also seen in the previous studies, the assessment of the effect of specific measures is often based on accident analyses. One of the main drawbacks of reactive traffic safety assessments (accident analyses) is that this approach only shows a fraction of the total number of events. Traffic (un)safety is characterized by a much broader set of events than accidents alone, ranging from undisturbed passages, normal interactions, and conflicts to collisions. In experimental studies, driving behavior is observed, being mostly undisturbed passages and normal interactions This broad set of events is shown as a continuum of traffic events, which describe the traffic process (Figure 7.1; Hydén, 1987). Another drawback of accident analyses is that accidents are underrepresented in accident statistics, mostly showing the more serious accidents. And if they are represented in statistics, police reports do not always contain the information researchers are interested in from a traffic safety perspective and "objective" eyewitness testimonies are biased due to subjective interpretation.

In several projects, the limitations of accident analyses and statistics have been overcome by developing a method for investigating conflicts and traffic accidents in more detail. The method should be informative, provide detailed information, be

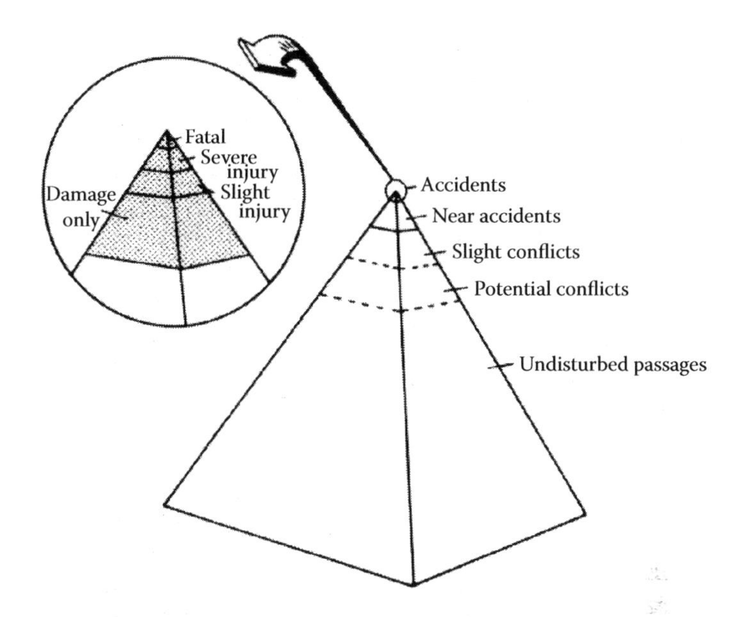

FIGURE 7.1 The pyramid continuum of traffic events from undisturbed passages to fatal accidents (Hydén, 1987).

easily executable, and have a high link with what is actually happening on the road. A very valuable tool in this is the method of video observations.

7.5.1 Video Observations

The method of using video recordings for actually assessing what type of behavior is shown at a specific location is extremely useful as a research tool. Video observations are considered to provide more insight into the circumstances and chains of events (Noordzij and Van der Horst, 1993; Svensson, Hakkert, and Van der Horst, 1996). The aim of this tool is to provide more knowledge on the mechanism that differentiates normal driving behavior from conflicts, and conflicts from accidents. Furthermore it provides more details about an accident, making it possible to validate and calibrate methodologies and techniques for research on driver behavior with respect to severe events (accidents) and less severe events. Also, video recordings allow traffic events to be analyzed by quantitative measures. Video observations allow a close view on what is happening at black spots (dangerous stretches of road where accidents frequently occur), but they can also help in understanding the parameters for interaction between road users at intersections (for instance, as input for behavioral models) or they can evaluate design modifications by means of before and after studies.

Video observations can be used in different time spans, depending on the purpose for the video observations. In order to get a better understanding of the interaction between road users, observing one to several days (with a bird's-eye view) may be sufficient. For these situations, much information can be gathered in a short period of time. When looking at before-and-after studies, longer observations are required to come to reliable and valid conclusions; that is, in order to rule out random factors like rain or traffic intensity.

When being interested in observing real accidents (or conflicts), one should realize that:

1. The low frequency of accidents makes it difficult to store accidents by video observation. Therefore, it is necessary to select intersections with a more than average amount of accidents (so-called black spots).
2. A bird's-eye point of view is required to determine output parameters such as position, relative position, and the relevant derivatives.

7.5.2 DOCTOR Conflict Quantification

Traffic conflict techniques (TCT) enable an objective and quantitative assessment of traffic events such as conflicts (as result from the video observations). In 1977 at the first International Traffic Conflicts Workshop, a group of researchers assembled a general definition of a conflict: A conflict is an observational situation in which two or more road users approach each other in space and time to such an extent that a collision is imminent if their movements remain unchanged.

The Dutch TCT is called DOCTOR (Dutch Objective Conflict Technique for Operation and Research). DOCTOR was developed by the Institute for Road Safety Research (SWOV) and TNO Human Factors. This TCT was primarily a result of an international calibration study that took place in Malmö under the auspices of the International Cooperation on Theories and Concepts in Traffic Safety (ICTCT) in order to compare existing techniques (Grayson, 1984). A comparison with videotaped conflicts and accidents indicated that severity scores, performed by individual observers, were mainly correlated to TTC and type of accident (Van der Horst, 1984).

According to DOCTOR, a conflict is defined as "a critical traffic situation in which two or more road users approach each other in such a way that a collision threatens, with realistic chance of injury or material damage if their course and speed remain unaltered." The severity scores in the DOCTOR technique are applied if the available space for a maneuver is less than needed for a normal reaction, which is a critical situation (Van der Horst and Kraay, 1986). The severity of the conflict is then scored on a scale from 1 to 5, taking into account the probability of a collision and the extent of the consequences if a collision had occurred.

The probability of a collision is determined by the time-to-collision (TTC) and/or the postencroachment time (PET; Van der Horst, 1990). The TTC is the remaining time until two road users on a collision course collide if course and speed remain unaltered. The TTC is a continuous function of time as long as the road users are on their collision course. The TTC_{min} is the lowest attained value of a collision course, which is a good indicator for the maximum probability of a collision. A low value of TTC corresponds to a high probability of an accident. The TTC value differentiates between encounters and conflicts, and between avoidable and unavoidable accidents. In urban areas, a TTC value of 1.5 seconds or lower constitutes a potentially dangerous situation. The deficiency of this concept is that TTC can only be applied in case of a collision course. The PET value is a measure that includes the "close misses." It

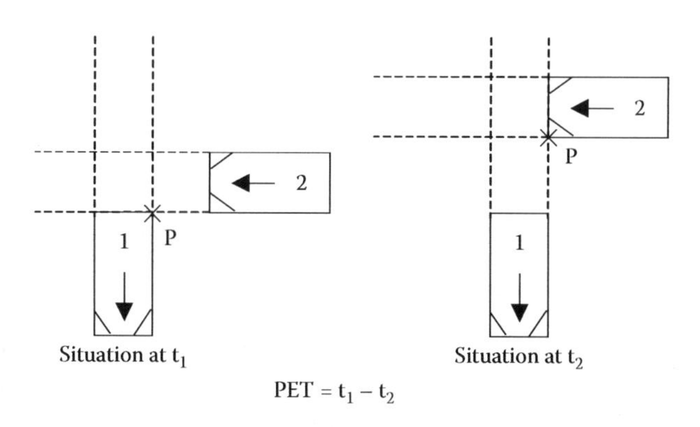

FIGURE 7.2 Definition of postencroachment time.

is defined as the time between the moment that the first road user leaves the path and the moment that the second reaches the same path (see Figure 7.2). The PET value indicates the extent to which they missed each other. In urban areas, PET values of 1 second and lower are indicated as possibly critical.

The severity of the consequences if a collision had occurred is mainly dependent on the potential collision energy and the vulnerability of the road users involved. Affecting factors are the relative speed, available and necessary space to maneuver, the angle of approach, the type and condition of road users, and so forth. The mass and maneuverability of the vehicles are critical; that is, consequences of colliding with a tram (streetcar) are very different from colliding with a pedestrian. For obtaining a relatively unambiguous estimate of the injury severity and additional information for analysis and diagnosis, several aspects are scored and registered on the DOCTOR observation sheet. For this methodology a manual (Kraay, Van der Horst, and Oppe, 1986) has been developed in which DOCTOR is described in detail.

The large Dutch project Integral Approach of the Analysis of Traffic Accidents (IAAV) was initiated to develop an integral multidisciplinary approach on the investigation of traffic accidents (Hoogvelt et al., 2007). As part of this, video observations of both collisions and traffic conflicts were made (Van der Horst, Rook, Van Amerongen, and Bakker, 2007). Long-term video observations were made to collect data on the precrash phase of real accidents (what exactly happened just before the collision) and traffic conflicts. This is unique in the sense that neither the analysis of accident statistics nor the study of road users' behavior includes direct observation of accidents. When conducting road safety research, there is a real need to have reliable information about the precrash phase of an accident. Direct observation of actual collisions would lead to a better understanding of the causes of the underlying processes and safety problems (Noordzij and Van der Horst, 1993). For a detailed description of the long-term video recordings in the IAAV study, the reader is referred to Van der Horst et al. (2007).

From the video recordings, collisions were manually selected. Apart from the real accidents, a limited number of conflicts and some observations about the functioning of the intersection were collected. In case of a conflict, these situations have been analyzed according the criteria of the DOCTOR methodology. Moreover, an arbitrary day had been taken for each location for which potential conflict situations have been selected.

In total, 16 collisions could be identified from video from which 10 were car–car collisions, 2 single–car accidents, 2 car–bicyclist/moped collisions, 1 single-bicycle accident, and 1 single-snowmobile accident. From the analysis of the video recordings, it was learned that there were basically two main problems at one specific intersection: the left-turn maneuver from the minor road and the bicycles crossing the main road from and to the bicycle track. Especially, the left-turn maneuvers from the minor road to the main road appeared to be problematic due to the limited sight of traffic on the main road, the frequent traffic on the main road, and its relatively high speed. It happened frequently that minor-road car drivers stopped with their car front already partly on the main-road carriageway, and in this way directly interacting with traffic from the left. Sometimes the minor-road left-turning cars tried to use the same gaps as bicycles from the cycle track crossing the main road. The gap acceptance problems for minor-road traffic also occurred for bicyclists to the minor road from the separate bicycle track. The relatively high speed of the main-road traffic was contributing to the task difficulty of crossing or merging traffic. The collisions as occurred at this intersection during a 22-month period of video observations properly reflect these findings. The results for the other three locations can be found in Van der Horst et al. (2007). In general, it can be concluded that traffic conflicts and deviant behavior gave good insight into potential safety problems at intersections from a road user's perspective, well in line with the results from the analysis of the collisions. Remarkably, in most cases, another road user was indirectly involved, either as a distracting or as a contributing element, for example, by occluding the view of one of the road users involved.

7.5.3 Practical Implications and Modeling

Video observations can also be used to analyze different road layouts. In the 1980s, a study was conducted to explore the determinants of gap acceptance by bicyclists intersecting an artery with two-way traffic (Van der Horst and ten Broeke, 1984). Video recordings were made on the intersection, focusing on two types of left-turn maneuvers, one from the cycle track along the main road and one from a minor street. Two different layouts of the main road were studied, one with 4.5-m wide lanes (before period) and one with 3.6-m wide lanes (after period). In the after period, there was a free area in the middle of the road with a refuge. About one-half of the crossings resulted in an accepted gap (mean traffic volumes were 600 veh/h for each direction). What was interesting to see is that crossing the first traffic stream resulted in acceptance of a gap that was larger than the gap accepted for crossing the second traffic stream (e.g., 10.6 s for intersection 1 for the first stream and 6.6 s for the second stream; 10 s for the first stream on intersection 2 and 7.8 s for the second stream). The countermeasures in the after period influenced the decision making by bicyclists resulting in a lower critical gap, mainly due to different crossing times (gap of 8.3 s and 5.7 s for intersection 1, and 7.1 s and 5.6 s for intersection 2). The mean PET is hardly discriminating between the conditions. Nevertheless from PET distributions, indications can be derived by means of defining the probability of minimal PET values. With respect to this, the situation in the after period is better for intersecting the second traffic stream.

Real-driving behavior data, either from experiments, video observations, or loop detection data, can be used to model driving behavior. Spek, Wieringa, and Janssen (2006) emphasized the necessity to have models that allow designers or researchers to estimate the likeliness of an intersection accident as a function of intersection approach speed. This is necessary for forensic institutes for accident speed reconstructions and for cause attribution in cases in which one driver violated the right-of-way while the other exceeded the speed limit. In the so-called critical gap models, decision making at intersections without signals is described as a process in which time gaps between successive major stream vehicles whose duration exceeds a certain value—the critical gap—are accepted, while shorter gaps are rejected (see also Troutbeck and Brilon, 1999). In these models, the critical gap value is considered dispersed among individual drivers while the distribution of values is considered unique for each intersection. These models assume that drivers are sensitive to the time it takes for approaching traffic to reach the intersection, whereas approach time perception and gap acceptance behavior are influenced by the speed of the approaching vehicle. In order to take this into account, Spek et al. (2006) constructed an alternative gap acceptance model, assuming that humans construct gap time indirectly by dividing perceived distance over perceived approach speed. Both perceived distance and perceived approach speed were modeled as exponential functions of their physical values, following Stevens' power law (Stevens and Stevens, 1975).

Data from various experiments have been used to test the model. Among others, the model has been applied to data from the experimental gap acceptance data from the Hancock and Caird (1993) driving simulator experiments (Hancock et al., 1991). In one of their experiments, young drivers were instructed to drive up to a rural, nonregulated intersection and turn left if they felt it was safe to do so. Each subject did this 49 times and crossed an oncoming stream of traffic of uniform speed with different time gaps and different speeds. Figure 7.3 shows for every condition the percentage of subjects that considered a condition to be safe for crossing. Hancock and Caird (1993) featured eight such plots, one for each combination of age group and vehicle type. Figure 7.3 presents their averaged percentages, as measured manually from these eight plots and as used in Spek et al. (2006).

The model described by Spek et al. (2006) fitted the data very well; more than 97% of the variation was accounted for. The incorporation of perceived speed or speed perception itself is not sufficient for pure gap time estimation. The probability of a conflict turning into a collision increases with speed.

This model has also used data of video observations at three- and four-legged intersections without signals made by Brilon and Weinert (2001). All of these intersections featured two major streams of vehicles, which had priority over all conflicting minor streams. From these observation data, a database of almost 30,000 decisions was derived. As only a specific selection of passages from major vehicles was taken into account, the results for the left-turn maneuver may have been subject to bias. Therefore, although results were presented for all maneuvers, conclusions were based on results for the merge and oncoming maneuvers only. The model of Spek et al. (2006) fits the observations reasonably well, although the fraction of

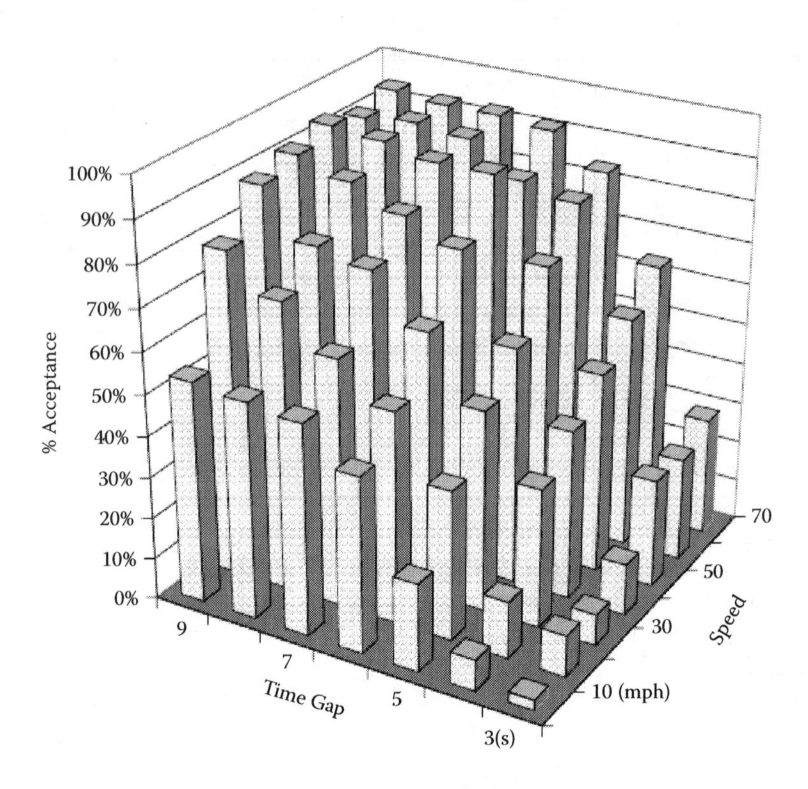

FIGURE 7.3 Percentage of subjects in the study of Hancock and Caird (1993) that found a combination of time gap and speed to be safe for crossing. (Reproduced with permission.)

variance explained was considerably less than for the simulator data (R2 = 0.76 for all observations). Presumably, behavior in the real world is influenced by factors that are not incorporated in the model. For instance, drivers may reject a shorter but otherwise acceptable gap if he sees a larger gap coming along down the traffic stream. This latter effect did not play a role in the simulator experiment, as subjects were confronted with gaps of equal size. When different maneuvers are distinguished, the relevance of approach speed becomes more explicit. Figure 7.4 shows the modeled effect of speed on the number of conflicts and collisions. Note that the data in Figure 7.4 apply to the driving simulator data. The result for the left-turn maneuver is likely to be influenced by the traffic stream in the far lane. In this respect, it is worth noting that the fit for left-turn maneuvers was worse than for the maneuvers with a continuous stream of oncoming traffic. The difference between the merge maneuver on the one hand and the left turn and oncoming maneuvers on the other hand may be understood as follows. For the latter, some fixed minimum time gap is needed to cross the major stream; there is no logical reason for a driver from the minor road to adapt this time gap criterion to the speed of the approaching vehicle. In contrast, the merge maneuver implies not only entering the stream but also adaptation to the speed of the vehicle that closes the accepted gap. Thus, to merge before a faster vehicle, a larger gap is needed.

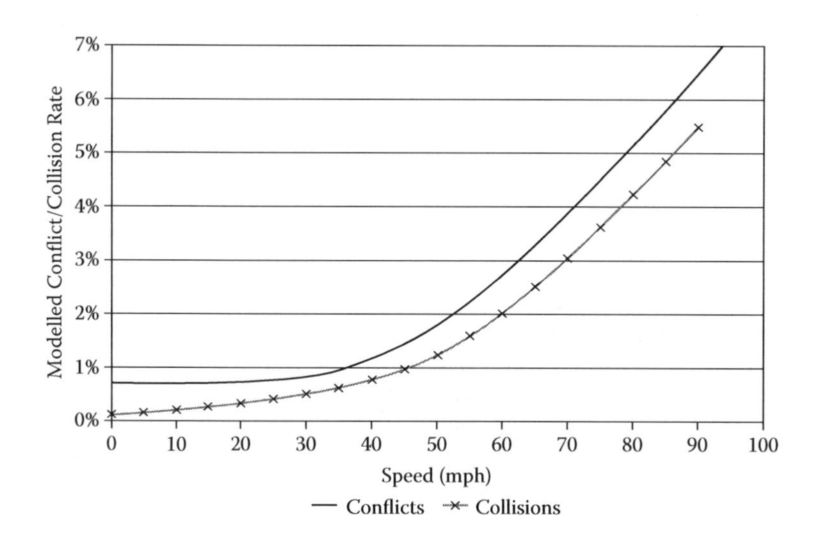

FIGURE 7.4 Conflicts, gap acceptance, and accident analysis.

In summary, by using data from detailed experiments, real-life data, and data from video observations some of the disadvantages of accident analysis can be overcome. By combining these data into models, better estimations and predictions can be done about the contributory effects of design and driver–driver interaction on traffic conflicts and accidents.

7.6 CONCLUSIONS

7.6.1 Road Design

Roadway design has an important influence on driving behavior. Roads should be self-explaining in the sense that their layout should explain how fast one can drive, if there is opposing traffic that has priority, and so forth. A traffic environment that provokes the right expectations will reduce potential errors.

Similar road categories should elicit similar driving behavior and correct expectations can only be achieved when they are similarly designed and if differences between road categories are maximized.

A road consists of a number of dependent dimensions that have an influence on driving:

1. Pavement (type and width)
2. Markings
3. Environment
4. Alignment

In road pavement, the type of pavement used and the width of the pavement are the most important factors in determining driver behavior. With type of pavement, the roughness of road surface is important. This is a measure for the amount and kind of deviations from a smooth road surface, with longitudinal roughness, transverse roughness, road surface irregularities, and roughness caused by road material. Irregular road surfaces will result in an increased amount of noise and vibration compared to smooth road surfaces, thereby decreasing driver comfort. An effect on driving speed is not the result of the roughness of the road surface per se, but rather an effect of a reduction in driver comfort, thereby reducing driving speeds. Buildings and obstacles close to the roadside and an irregular pattern of objects also reduce speed.

In narrow lanes, other traffic is near, there is less space to keep a certain distance to obstacles along the side of the road, and it requires more effort to stay in the lane. Therefore, lane width is also an important factor in driving behavior. However, one should keep in mind that there is also a risk associated with narrow lanes. When road users do not sufficiently adapt their behavior, traffic safety may suffer.

Road markings provide guidance and regulatory warning information to the driver. Side markings and center line markings can serve as cues to show the proper path to follow and transverse road markings can serve as a possible warning. For lane keeping and anticipation on the course of the road, road markings are extremely important. Road markings may also be used to visually narrow the lanes without any physical narrowing. Road markings are of great importance for driving safety, but their presence can also cause an increase in speed, especially at night due to increased visual guidance.

The amount of curvature in roads actually affects speed in a number of ways. First of all, driving through curves requires some extra effort in lane keeping. Besides this, curves result in a reduction in the visibility distances along the road axes, limiting anticipation of the course of the road and upcoming traffic situations.

7.6.2 Adverse Weather Conditions

Adverse weather conditions (such as rain, snow, hail, wind, and fog) have a significant negative impact on road traffic safety. Underlying causes of decreased safety are reduced friction, poor visibility, strong lateral deviations, a combination of factors, and even stress. When touching upon the topic of weather and driving behavior, many issues play a role, such as temperature, precipitation (and therefore reduced vision or slippery roads), relative humidity, wind speeds, solar radiation, and so forth. Since it is not possible to touch upon all these topics, the effects of the most common adverse weather conditions will be discussed. The effect of wind on driving behavior is a topic that is underrepresented in the literature.

Although one may blame these weather conditions and accept the direct link between weather conditions and safety, the question is if the accidents may be the result of drivers not compensating for these conditions by properly adjusting their driving behavior to the changed conditions. Drivers need to adjust their behavior to changes in driving conditions. Drivers are normally inclined to react to changes in the traffic system, whether they are in the vehicle, in the road environment, and in their own skills or states, and the reaction occurs in accordance with their own motives. This is the principle of behavioral adaptation, also called risk compensation. A good example is the effect of studded tires. In case of winter-road conditions, a large safety effect would be expected. But this safety measure has a limited effect because drivers compensate for this improved friction. Speed adaptation has been proved insufficient to compensate for the reduced friction and headways are not substantially affected by winter conditions. This also holds for fog. Even though visibility distances are found to affect free-driving speeds, especially in the lower visibility range, speeds are too high to allow for a successful stop for a stationary obstacle. Effects of visibility on headway or following distance are the direct result of speed.

In order to have drivers adjust their driving behavior according to the road conditions, countermeasures have been implemented. This could be a specific weather forecast on the radio or on the television, with the results varying from changing travel plans to taking more time for the trip, altering time of departure, and changing the route. The lack of actual strong effects on driving behavior would be the nonspecificity of the system. A more specific warning system that only presents messages during driving in case of specific weather conditions (either on the road or in-vehicle) will have better results, although the driver's own judgment of the weather conditions will affect driving behavior the most.

7.6.3 Public Lighting

Another countermeasure that is often used to increase traffic safety is public lighting. Although the main reason for introducing public lighting is traffic safety, public lighting is also supposed to make the driving task less strenuous for the driver (comfort, even without safety benefits), and is supposed to increase social safety and traffic throughput. Over the years, however, the ideas about the application of lighting have changed due to, for example, environmental reasons. Public lighting is reduced unless there is a clear benefit with respect to traffic safety (or sometimes due to negative effects on social safety). However, it is difficult to assess the exact effect of public lighting on traffic safety because of the complexity of the situation. Public lighting is used as a general term, but public lighting can be used on motorways for orientation purposes only, mainly aimed at pointing out the direction of the road. Public lighting is also used for alighting an entire stretch of road or for marking specific areas such as intersections with a single streetlight. Although, in general, accident data have shown a reduction in night accidents, public lighting can also increase speeds. It is generally believed that public lighting is especially safe at intersections or roundabouts.

7.6.4 Conflicts and Accidents

When studying traffic safety and driving behavior, conflicts are especially interesting. Although the analysis of accidents is very informative for studying traffic safety, conflicts and gap acceptance studies provide much more data and therefore much more data points to learn from. Also, accidents are underrepresented in accident statistics, which mostly show more serious accidents, and accident analyses are always afterward, not giving a clear picture about the effect of driver input on the outcome of the accident. By investigating conflicts, for instance, by using video recordings for actually assessing what type of behavior is shown at a specific location, an extremely useful research tool has been found since video observations are considered to provide more insight into the circumstances and chains of events. If specific conflicts have been recorded, these conflicts can be analyzed by means of different traffic conflict techniques that enable an objective and quantitative assessment of traffic events. Also, modeling these types of conflicts and accidents allows designers or researchers to estimate the likeliness of an intersection accident as a function of specific behavior such as intersection approach speed.

REFERENCES

Aarts, L.T., Davidse, R.J., Louwerse, W.J.R., Mesken, J., and Brouwer, R.F.T. (2005). *Herkenbare vormgeving en voorspelbaar gedrag* (Rapport R-2005-17). Leidschendam, Nederland: Stichting Wetenschappelijk Onderzoek Verkeersveiligheid.

Agent, K.R. (1980). Transverse pavement markings for speed control and accident reduction. *Transportation Research Record, 773,* 11–14.

Andreescu, M., and Frost, D.B. (1998). Weather and traffic accidents in Montreal, Canada. *Climate Research, 9,* 225–230.

Andrey, J., Mills, B., Leahy, M., and Suggett, J. (2003). Weather as a chronic hazard for road transportation in Canadian cities. *Natural Hazards, 28*(2–3), 319–343.

Antilla, V., Nygård, M., and Rämä, P. (2001). *Liikennesää-tiedotuksen toteutuminen ja arviointi talvikaudella 1999-2000.* Tiehallinnon selvityksiä (14/2001). Helsinki, Finland: Tiehallinto, Liikenteen palvelut.

Anund, A. (1993). *Effect of road surface on vehicle speed. Pavement technique and pavement surface—Consequences for traffic and environment.* Seminar held October 1992, Esbo, Finland.

Bald, S. (1987). *Untersuchungen zu Determinanten der Geschwindigkeitswahl. Bericht 1: Auswertung von Geschwindigkeitsprofilen auf Außerortsstraßen* (Bericht zum Forschungsprojekt 8525/3). Bergisch Gladbach, BRD: Bundesanstalt für Straßenwesen.

Blaauw, G.J., and Van der Horst, A.R.A. (1982). *Lateral positioning behavior of car drivers near tunnel walls* (Report IZF 1982 C-30). Soesterberg, The Netherlands: TNO Institute for Perception.

Brandt, T., Wist, E.R., and Dichgans, J. (1975). Foreground and background in dynamic spatial orientation. *Perception and Psychophysics, 17,* 497–503.

Brenac, T. (1989). Speed, safety and highway design. *Récherche Transports Securité, 5,* 69–74.

Brilon, W., and Weinert, A. (2001). *Ermittlung aktueller grenz- und folgezeitlucken fur auerortsknoten ohne lichtsignalanlagen.* Straßenbau und Straßenverkehrstechnik (Vol. 828).

Brodsky, H., and Hakkert, S. (1988). Risk of a road accident in rainy weather. *Accident Analysis and Prevention, 20,* 161–176.

Brouwer, R.F.T., Janssen, W.H., and Muermans, R.C. (2000). *Duurzaam veilige wegcategorieen en wegkenmerken: De invloed van de omgeving op de categorisatie van wegbeelden* (Report TNO TM-2000-C012) [Sustainable safe road categories and characteristics: The influence of the environment on the categorization of road images]. Soesterberg, The Netherlands: TNO Human Factors.

Bruneau, J.-F., and Morin, D. (2005). Standard and nonstandard roadway lighting compared with darkness at rural intersections. *Transportation Research Record, 1918,* 116–122.

Bruneau, J.-F., Morin, D., and Pouliot, M. (2001). Safety of motorway lighting. *Transportation Research Record, 1758,* 1–5.

Changnon, S. (1996). Effects of summer precipitation on urban transportation. *Clamatic Change, 32,* 481–494.

Charlton, S.G. (2007). The role of attention in horizontal curves: A comparison of advance warning, delineation, and road marking treatments. *Accident Analysis and Prevention, 39*(5), 873–885.

Codling, P. (1974). Weather and road accidents. In J. Taylor (Ed.), *Climatic resources and economic activity* (pp. 205–222). London: David and Charles Holdings, Newton Abbot.

Colbourn, C.J., Brown, I.D., and Copeman, A.K. (1978). Drivers' judgments of safe distances in vehicle following. *Human Factors, 20*(1), 1–11.

Cooper, B.R., and Sawyer, H. (1993). *Assessment of M25 automatic fog-warning system* (Final report, project report 16). Crowthorne, UK: Transport Research Laboratory.

Cooper, D.R.C., Jordan, P.G., and Young, J.C. (1980). *The effect on traffic speeds of resurfacing a road* (Supplementary Report 571). Crowthorne, Berks, UK: Transport and Road Research Laboratory, Department of Transport.

DeLuca, F.D. (1985, March). *Effects of lane width reduction on safety and flow.* Proceedings of a Conference on Effectiveness of Highway Safety Improvements, Tennessee Highway Division of American Society of Civil Engineers.

Denton, G.G. (1971). *The influence of visual patterns on perceived speed* (TRRL Report LR409). Crowthorne, Berks, UK: Transport and Road Research Laboratory, Department of Transport.

Denton, G.G. (1973). *The influence of visual pattern on perceived speed at Newbridge MB Midlothian* (TRRL Report LR531). Crowthorne, Berks, UK: Transport and Road Research Laboratory, Department of Transport.

De Ridder, S.N., and Brouwer, R.F.T. (2002). *Effecten van omgevingskenmerken op rijgedrag* [The influence of the road environment on road behavior] (Report TNO TM-02-C065). Soesterberg, Netherlands: TNO Human Factors.

Dichgans, J., and Brandt, T. (1978). Visual-vestibular interaction: Effects of self-motion perception and postural control. In R. Held, H.W. Leibowitz, and H.L. Teuber (Eds.), *Handbook of sensory physiology, VIII: Perception* (pp. xx–xx). Berlin: Springer-Verlag.

Edwards, J.B. (1999). Speed adjustment of motorway commuter traffic to inclement weather. *Transportation Research Part F, 2,* 1–14.

Eisenberg, D. (2004). The mixed effects of precipitation on traffic crashes. *Accident Analysis and Prevention, 36*(4), 637–647.

Eisenberg, D., and Warner, K.E. (2005). Effects of snowfalls on motor vehicle collisions, injuries, and fatalities. *American Journal of Public Health, 95*(1), 120–124.

Elvik, R. (1995). Meta-analysis of evaluarions of public lighting as accident countermeasure. *Transportation Research Record, 1485*, 112–123.

Elvik, R., Borger, A., and Vaa, T. (1996). *Utkast til reviderte tiltakskapitler i trafikksikkerhetshåndboka: Trafikkregulering.* Oslo, Norway: Transportøkonomisk Institutet.

Fildes, B.N., Fletcher, M.R., and Corrigan, J. McM. (1987). *Speed perception 1: Drivers' judgements of safety and speed on urban and rural straight roads* (Report CR 54).

Fildes, B.N., and Jarvis, J.R. (1994). *Perceptual countermeasures: Literature review* (Research Report CR4/94). Canberra, Australia: Federal Office of Road Safety, Department of Transport and Communication.

Folles, E., Ijsselstijn, J., Hogema, J.H., and Van der Horst, A.R.A. (1999). *Dynamische openbare verlichting (DYNO): Covernota* [Dynamic public lighting: cover report]. Rotterdam, The Netherlands: Ministry of Transport, Directorate General.

Fridstrøm, K., Ifver, J., Ingebrigsten, S., Kulmala, R., and Thomsen, L.K. (1995). Measuring the contribution of randomness, exposure, weather, and daylight to the variation in road accident counts. *Accident Analysis and Prevention, 27*(1), 1–20.

Grayson, G.B. (1984). *The Malmö study: A calibration of traffic conflict techniques* (Report R-84-12). Leidschendam: Institute for Road Safety Research, SWOV.

Gundy, C.M. (1994). *Cognitive organisation of roadway scenes: An empirical study* (Rapport R-94-86). Leidschendam, The Netherlands: Stichting Wetenschappelijk Onderzoek Verkeersveiligheid.

Gundy, C.M. (1995). *Cognitieve organisatie van wegbeelden, deel II: Een empirisch onderzoek naar wegen binnen de bebouwde kom* (Rapport R-95-75). Leidschendam, Nederland: Stichting Wetenschappelijk Onderzoek Verkeersveiligheid.

Gundy, C.M., Verkaik, R., and de Groot, I.M. (1997). *Cognitieve organisatie van wegbeelden, deel III* (Rapport R-97-27). Leidschendam, Nederland: Stichting Wetenschappelijk Onderzoek Verkeersveiligheid.

Gupta, N., Bisantz, A.M., and Singh, T. (2002). The effects of adverse condition warning system characteristics on driver performance: An investigation of alarm signal type and threshold level. *Behaviour and Information Technology, 21*(4), 235–248

Hancock, P.A., and Caird, J.K. (1993). *Factors affecting older drivers' left turn decisions.* Washington, DC: Transportation Research Board, National Research Council.

Hancock, P.A., Caird, J.K., Shekhar, S., and Vercruyssen, M. (1991). Factors influencing drivers' left turn decisions. In *Human Factors Society 35th Annual Meeting* (pp. 1139–1143). Santa Monica, CA: Human Factors Society.

Havell, D.F. (1983). *Control of speed by illusion at Fountains Circle, Pretoria* (Report RF/7/83). Republic of South Africa: National Institute for Roads and Transport Technology.

Hawkins, R.K. (1988). Motorway traffic behaviour in reduced visibility conditions. In A. Gale and M.H. Freeman (Eds.), *Vision in vehicles II.* Amsterdam, The Netherlands: Elsevier.

Heinijoki, H. (1994) *Kelin kokemisen rengaskunnon ja rengastyypin vaikutus nopeuskäyttäytymiseen* (Report 19) [Influence of the type and conditions of tires and drivers' perceptions of road conditions on driving speeds]. Helsinki, Finland: Finnish Road Administration.

Hijar, M., Carrillo, C., Flores, M., Anaya, R., and Lopez, V. (2000). Risk factors in highway traffic accidents: A case control study. *Accident Analysis and Prevention, 32*(5), 703–709.

Hill, J.D., and Boyle, L.N. (2007). Driver stress as influenced by driving maneuvers and roadway conditions. *Transportation Research Part F: Psychology and Behaviour, 10*(3), 177–186.

Hogema, J.H., and Van der Horst, A.R.A. (1994a). *Driving behaviour in fog: Analysis of inductive loop detector data* (Report TM 1994 C-6). Soesterberg, The Netherlands: TNO Human Factors.

Hogema, J.H., and Van der Horst, A.R.A. (1994b). *Driving behavior in fog: A simulator study* (Report TNO TM-94-C007). Soesterberg, The Netherlands: TNO Human Factors.

Hogema, J.H., Van der Horst, A.R.A., and Bakker, P.J. (1994). *Evaluation of the A16 fog-signalling system with respect to driving behaviour* (Report TM 1994 C048; in Dutch). Soesterberg, The Netherlands: TNO Human Factors.

Hogema, J.H., and Veltman, J.A. (2003). *Werkbelasting en rijgedrag tijdens duisternis: Tweede veldexperiment* [Workload and driving behavior: Second field experiment] (Report TNO TM-03-C018). Soesterberg, The Netherlands: TNO Human Factors.

Hoogvelt, R.B.J., Ruijs, P.A.J., Van der Horst, A.R.A., Van der Wijlhuizen, G.J., Verschragen, E.J.G., and Van der Wolf, F.E.C. (2007). *Integrale Aanpak Analyse Verkeersongevallen: overkoepelend rapport* (TNO rapport 07.OR.IS.024/RH). Delft, The Netherlands: TNO Industrie en Techniek, BU Automotive.

Horberry, T., Anderson, J., and Regan, M.A. (2006). The possible safety benefits of enhanced road markings: A driving simulator evaluation. *Transportation Research Part F: Traffic Psychology and Behaviour, 9*(1), 77–87.

Hydén, Ch. (1987). *The development of a method for traffic safety evaluation: The Swedish traffic conflicts technique* (Bulletin 70). Lund, Sweden: University of Lund, Lund Institute of Technology, Dept. of Traffic Planning and Engineering.

Jacobs, G.D. (1976). *A study of accident rates on rural roads in developing countries* (TRRL LR 732). Crowthorne, Berks, UK: Transport and Road Research Laboratory, Department of Transport.

Janoff, M.S., and Staplin, L.K. (1987). Effect of alternative reduced lighting techniques on hazard detection. *Transportation Research Record, 1149,* 1–7.

Janssen, W.H., Kaptein, N.A., Hogema, J.H., and Westerman, M. (1995). *"Quick scan" wegennet Rijkswaterstaat Zuid-Holland* ["Quick scan" main road network of South-Holland] (Report TNO TM-95-C042). Soesterberg, The Netherlands: TNO Human Factors.

Kanellaidis, G., Golias, J., and Efastathiadis, S. (1990, July). Drivers' speed behavior on rural road curves. *Traffic Engineering and Control,* 414–415.

Kaptein, N.A., and Theeuwes, J. (1996). *Effecten van vormgeving op categorie-indeling en verwachtingen ten aanzien van 80 km/h wegen buiten de bebouwde kom* [Effects of road design on categorization of 80 km/h roads] (Report TM-96-C010). Soesterberg, The Netherlands: TNO Human Factors Research Institute.

Karan, M.A., Haas, R., and Kher, R. (1977). Effects of pavement roughness on vehicle speeds. *Transportation Research Record, 602,* 122–127.

Khan, M.H., Senadheera, S., Gransberg, D.D., and Stemprok, R. (1999). Influence of pavement surface characteristics on nighttime visibility of objects. *Transportation Research Record, 1692*, 39–48.

Khattak, A.J., Kantor, P., and Council, F.M. (1998). Role of adverse weather in key crash types on limited-access roadways: Implications for advance weather systems. *Transportation Research Record, 1621,* 10–19.

Khattak, A.J., and Knapp, K.K. (2001). Snow event effects on interstate highway crashes. *Journal of Cold Regions Engineering*, *15*(4), 219–229.

Kilpeläinen, M., and Summala, H. (2007). Effects of weather and weather forecasts on driver behaviour. *Transportation Research* Part F, 288–299.

Knapp, K.K., and Smithson, L.D. (2000). Winter storm event volume impact analysis using multiple-source archived monitoring data. *Transportation Research Record, 1700,* 10–16.

Knoflacher, H., and Gatterer, G. (1981). *Der Einfluss seitlicher Hindernisse auf die* Verkehrssicherheit. Wien, Austria: Kuratorium für Verkehrssicherheit.

Kraay, J.H., Van der Horst, A.R.A., and Oppe, S. (1986). *Handleiding voor de conflictobservatietechniek DOCTOR* [Dutch Objective Conflict Technique for Operation and Research] (Report R-86-3). Leidschendam: Institute for Road Safety Research, SWOV.

Lamm, R., Choueiri, E.M., and Mailander, T. (1989). *Accident rates on curves as influenced by highway design elements—An international review and in-depth study. Proceeding of Road Safety in Europe, Gothenburg, Sweden, VTI-344 A*, pp. 38–54.

Lamm, R., Kloeckner, J.H., and Choueiri, E.M. (1985). Freeway lighting and traffic safety: A long-term investigation. *Transportation Research Record, 1027,* 57–62.

Levine, N., Kim, K., and Nitz, L. (1995). Daily fluctuations in Honolulu motor vehicle crashes. *Accident Analysis and Prevention, 27,* 785–796.

Makking, D.Th., and De Wit, T. (1984). Mythen, sagen en legenden in de verkeerstechniek [Myths and legends in traffic engineering]. *Verkeerskunde*, 35(9), 400–402.

Malmivuo, M., and Peltola, H. (1997). *Talviajan liikenneturvallisuus, tilastollinen tarkastelu 1991–1995 (6/1997).* Helsinki, Finland: Tielaitos, Tiehallinto.

Marconi, W. (1977). Speed control measures in residential areas. *Traffic Engineering, 37*(3), 28–30.

Maroney, S., and Dewar, R. (1987). Alternatives to enforcement in modifying the speeding behavior of drivers. *Transportation Research Record, 1111,* 121–126.

Martens, M.H. (2005). *Kunnen we met minder openbare verlichting toe? Een veldstudie in Drenthe* [Could we allow less public lighting? A field study in the province of Drenthe] (Report TNO DV3 C090). Soesterberg, The Netherlands: TNO Human Factors.

Martens, M.H. (2007). *Het dimmen van openbare verlichting langs provinciale wegen: Een veldstudie* [Reducing the luminance level of public lighting on rural roads: A field study] (Report TNO DV C227). Soesterberg, The Netherlands: TNO Human Factors.

McLean, J.R. (1979). An alternative to the design speed concept for low-speed alignment design. *Transportation Research Record, 702,* 55–63.

Michels, Th., and Van der Heijden, Th.G.C. (1978). De invloed van enkele wegkenmerken op de rijsnelheid op niet-autowegen [The influence of some road characteristics on driving speed on non-motorways]. *Verkeerskunde*, 6, 296–300.

Milosevic, S., and Milic, J. (1990). Speed perception in road curves. *Journal of Safety Research, 21*(1), 19–23.

Mortimer, R.G. (2001). The nighttime pedestrian collision: Human factors issues and a case study. In *Proceedings of the Human Factors and Ergonomics Society 45th Annual Meeting* (pp. 833–837). Santa Monica, CA: Human Factors Society

Näätänen, R., and Summala, H. (1974). A model for the role of motivational factors in drivers' decision making. *Accident Analysis and Prevention, 6,* 243–261.

Näätänen, R., and Summala, H. (1976). *Road-user behavior and traffic accidents.* Amsterdam, The Netherlands: North-Holland.

Noordzij, P.C. (1996). *Categorie, Vormgeving en Gebruik van Wegen.* Literatuurstudie (Deel 1: 80 km/uur-wegen) [Category, design and use of roads. Literature study (Part 1: 80 km/h roads)] (R-96-14). Leidschendam, The Netherlands: SWOV Institute for Road Safety Research.

Noordzij, P., and Van der Horst, A.R.A. (1993). Relationship between accidents and road user behavior: An integral research programme. In J.L. de Kroes and J.A. Stoop (Eds.), *Proceedings 1st World Congress on Safety of Transportation* (pp. 527–534). Delft, The Netherlands: Delft University Press.

Organisation for Economic Co-operation Development (OECD). (1990). *Behavioral adaptations to changes in the road transport system.* Paris: OECD Road Transport Research.

Perdok, J. (2003). *Ruimtelijke inrichting en verkeersgedrag. Technische rapportage aanvullende metingen: Simulatoronderzoek* [Environmental layout and driving behavior. Technical report additional measures: Driving simulator study]. Report MuConsult B.V., NO26, Amersfoort, The Netherlands.

Rämä, P., and Kulmala, R. (2000). Effects of variable message signs for slippery road conditions on driving speed and headways. *Transportation Research Part F, 3,* 85–94.

Rämä, P., Kulmala, R., and Heinonen, M. (1996). *Muuttuvien kelivaroitusmerkkien vaikutus ajonopeuksiin, aikavaleihin ja kuljettajien kasityksiin* [The effect of variable road condition warning signs] (Finnra Report 1). Helsinki: Finnish National Road Administration.

Reinfurt, D.W., Zegeer, C.V., Shelton, B.J., and Newman, T. (1991). *Analysis of vehicle operations on hHorizontal curves* (Paper 910802). Transportation Research Board, 70th Annual Meeting, Washington, DC.

Riemersma, J.B.J. (1988a). *Enkelbaans-dubbelbaans: Beleving van de weggebruiker* [An empirical study of subjective road categorization] (Report IZF 1988 C-4). Soesterberg, The Netherlands: TNO Institute for Perception.

Riemersma, J.B.J. (1988b). *De waarneming van Boogkenmerken* [The perception of curve characteristics] (Report IZF 1988 C-8). Soesterberg, The Netherlands: TNO Institute for Perception.

Rockwell, T.H., and Hungerford, J.C. (1979). *Use of delineation systems to modify driver performance on rural curves* (Report FHWA/OH/79/007). Washington, DC: Ohio Department of Transportation, Federal Highway Administration.

Rumar, K., Berggrund, U., Jernberg, P., and Ytterbom, U. (1976). Driver reaction to a technical safety measure: Studded tires. *Human Factors, 18,* 443–454.

Rutley, K.S. (1975). Control of drivers' speed by means other than enforcement. *Ergonomics, 18*(1), 89–100.

Saastamoinen, K. (1993). Kelin vaikutus ajokayttaytymiseen ja liikennevirran ominaisuuksiin [Effect of road conditions on driving behavior and properties of the traffic flow] (Finnra Report 80). Helsinki: Finnish National Road Administration.

Salvatore, S. (1968). The estimation of vehicular velocity as a function of visual stimulation. *Human Factors, 10*(1), 27–32.

Satterthwaite, S. (1976) An assessment of seasonal and weather effects on the frequency of road accidents in California. *Accident Analysis and Prevention, 8,* 96.

Shankar, V., Mannering, F., and Barfield, W. (1995). Effect of roadway geometrics and environmental-factors on rural freeway accident frequencies. *Accident Analysis and Prevention, 27*(3), 371–389.

Sherretz, L., and Farhar, B. (1978) An analysis of the relationship between rainfall and the occurrence of traffic accidents. *Journal of Applied Meteorology, 17,* 711–715.

Shinar, D. (1977). Curve perception and accidents on curves: An illusive curve phenomenon. *Zeitschrift für Verkehrssicherheit, 23,* 16–21.

Shinar, D., McDowell, E.D., and Rockwell, T.H. (1974). *Improving driver performance on curves in rural highways through perceptual changes* (Report Ohio-DOT-04-74). Columbus, OH: Ohio State University, Department of Industrial Engineering.

Slangen, B. (1983). *Verandering van de weg(-omgeving) kan leiden tot snelheidsverlaging* [Changes in road (environment) may lead to speed reduction]. *Wegen*, Oktober, 312–319.

Smith, D.T., and Appleyard, D. (1981). *Improving the residential street environment* (Report FHWA/RD-81/031). Washington, DC: Federal Highway Administration.

Spek, A.C.E., Wieringa, P.A., and Janssen, W.H. (2006). Intersection approach speed and accident probability. *Transportation Research Part F, 9,* 155–171.

Stevens, S., and Stevens, G. (1975). *Psychophysics: Introduction to its perceptional, neural and social prospects.* New York: Wiley.

Summala, H. (1996). Accident risk and driver behaviour. *Safety Science, 22*, 103–117.

Summala, H., and Merisalo, A. (1980). A psychophysical method for determining the effect of studded tires on safety. *Scandinavian Journal of Psychology, 21,* 193–199.

Svensson, Å., Hakkert, S., and Van der Horst, A.R.A. (1996). *Studying accidents in real-time: On the relationships between accidents and road-user behavior.* Proceedings of the ICTCT Workshop, Zagreb, Hungary.

Taragin, A. (1954). Driver performance on horizontal curves. *Highway Research Board, 33,* 446–466.

Tenkink, E. (1988). *Determinanten van rijsnelheid* [Determinants of drivers' speed] (Report IZF 1988 C-3). Soesterberg, The Netherlands: TNO Institute for Perception.

Tenkink, E., and Van der Horst, A.R.A. (1991). *Effecten van wegbreedte en boogkenmerken op de rijsnelheid* [Effects of road width and curve characteristics on driving speed] (Report IZF 1991 C-26). Soesterberg, The Netherlands: TNO Institute for Perception.

Te Velde, P.J. (1985). *De invloed van de onvlakheid van wegverhardingen op de rijsnelheid van personenauto's* [The influence of roughness of road pavement on driving speed of cars]. ICW Nota, 1599, Februari 1985.

Theeuwes, J. (1994). *Self-explaining roads: An exploratory study* (Report TNO-TM 1994 B-18). Soesterberg, The Netherlands: TNO Human Factors Research Institute.

Theeuwes, J., and Godthelp, J. (1992). *Self-explaining roads* (Report IZF 1992 C-8). Soesterberg, The Netherlands: TNO Human Factors Research Institute.

Troutbeck, R.J., and Brilon, W. (1999). Unsignalized intersection theory. In N. Gartner, C. Messer, and A.K. Raiti (Eds.), *Revised monograph on traffic flow theory* (chap. 8). McLean, VA: U.S. Department of Transportation, Federal Highway Administration.

Van Breda, L., and Veltman, J.A. (1998) Perspective and predictive information in the cockpit as a target acquisition aid. *Journal of Experimental Psychology, 4*(1), 55–68.

Van de Kerkhof, W. (1987). *De Invloed van Weg- en Omgevingskenmerken op de Rijsnelheid* [The influence of road and environment characteristics on driving speed]. Afstudeerverslag HTO voor Planologie, Verkeerskunde en Vervoerskunde, Richting Verkeerskunde, Mei 1987.

Van de Kerkhof, W., and Berénos, M. (1989). Stedebouwkundige factoren beïnvloeden rijsnelheid [Urban development factors affect driving speed]. *Verkeerskunde, 14*(1), 30–33.

Van der Heijden, Th.G.C. (1978). De Invloed van Weg- en Verkeerskenmerken op Rijsnelheden van Personenauto's [The influence of road and traffic characteristics in driving speeds of cars] (Nota 1106). Wageningen, Netherlands: Instituut voor Cultuurtechniek en Waterhuishouding.

Van der Horst, A.R.A. (1983). *Gedragsobservaties ten behoeve van (brom)fietsers: Demonstratieproject herindeling en herinrichting van stedelijke gebieden (in de gemeenten Eindhoven en Rijswijk)* [A behavioral study in relation to bicycle traffic: Demonstration project on redesigning urban areas (Eindhoven and Rijswijk)] (Report IZF 1983 C-11). Soesterberg, The Netherlands: TNO Institute for Perception.

Van der Horst, A.R.A. (1984). *The ICTCT Calibration Study at Malmö: A quantitative analysis of video-recordings* (Report IZF 1984-37). Soesterberg, The Netherlands: TNO Human Factors.

Van der Horst, A.R.A. (1990). *A time-based analysis of road user behavior in normal and critical encounters.* PhD thesis, Delft University of Technology, Delft, The Netherlands.

Van der Horst, A.R.A., and ten Broeke, W. (1984) *Hiaat-acceptatie door fietsers bij het oversteken van een verkeersader*. Een orientatie [Gap-acceptance by bicyclists intersecting an artery: A pilot study] (Report TNO IZF 1984 C-16). Soesterberg, The Netherlands: TNO Institute for Perception.

Van der Horst, A.R.A., and Kraay, J.H. (1986, September 8–10). The Dutch conflict observation technique DOCTOR. In *Proceedings ICTCT Workshop Traffic Conflicts and Other Intermediate Measures in Safety Evaluation*. Budapest: Institute for Transport Sciences KTI.

Van der Horst, A.R.A., and Riemersma, J.B.J. (1984). *Herindeling rijstroken Heinenoordtunnel:Effecten op rijgedrag?* [Redesign traffic lanes Heinenoord tunnel: Effects on driving behavior?] (Memo IZF 1984 M-28). Soesterberg, The Netherlands: TNO Institute for Perception.

Van der Horst, A.R.A., Rook, A.M., Van Amerongen, P.J.M., and Bakker, P.J. (2007). *Video-recorded accidents, conflicts and road user behaviour: Integral approach analysis of traffic accidents* (Report TNO TM-07-D154). Soesterberg, The Netherlands: TNO Human Factors.

Van Winsum, W. (1993). *Car-following behaviour in the driving simulator, determinants and consistency of time headway* (Report VK 93-05). Haren, The Netherlands: Traffic Research Centre, University of Groningen.

Vey, A.H., and Ferreri, M.G. (1968). The effect of lane width on traffic operation. *Traffic Engineering, 38*(8), 22–27.

Walker, F.W., and Roberts, S.E. (1976) Influence of lighting on accident frequency at highway intersections. *Transportation Research Record, 562,* 73–78.

Watts, G.R., and Quimby, A.R. (1980). *Aspects of road layout that affect driver's perception and risk taking* (TRRL Report 920). Crowthorne, Berks, UK: Transport and Road Research Laboratory, Department of Transport.

Webb, P.J. (1980). *The effect of an advisory speed signal on motorway traffic speeds* (TRRL Report 615). Crowthorne, Berks, UK: Transport and Road Research Laboratory, Department of Transport.

White, M.E., and Jeffery, D.J. (1980). *Some aspects of motorway traffic. Behaviour in fog* (TRRL Report 958). Crowthorne, Berks, UK: Transport and Road Research Laboratory, Department of Transport.

Wilde, G.J.S. (1974). Wirkung und Nutzen von Verkehrssicherheitskampagnen: Ergebnisse und Forderungen–ein Uberblick. *Zeitschrift für Verkehrssicherheit*, 20, 227–238.

Wilde, G.J.S. (1975). *Roads user behaviour and traffic safety: Toward a rational strategy of accident prevention.* Annual Convention of the Dutch Road Safety League, Amsterdam. Studies of Safety in Transport, Queen's University, Kingston, Ontario, Canada.

Wilde, G.J. (1976). Social interaction patterns in driver behavior: An introductory review. Human Factors, *18*, 477–492.

Wildervanck, C. (1987). Wegdek en rijgedrag, gedragseffecten van wegdekkarakteristieken vaak verrassend [Road surface and driving behavior, behavioral effects of road surface characteristics often surprising]. *Verkeerskunde*, *38*(2), 59–63.

Yagar, S., and Van Aerde, M. (1983). Geometric and environmental effects on speeds on 2-lane rural roads. *Transportation Research Record, 17A*(4), 315.

Yamanaka, A., and Kobayashi, M. (1970). *Dynamic visibility of motor vehicles* (International Automobile Safety Compendium, Paper 700393). Warrendale, PA: Society of Automobile Engineers.

Zwahlen, H.T. (1987). Advisory speed signs and curve signs and their effect on driver eye scanning and driving performance. *Transportation Research Record, 1111,* 110–120.

Zwahlen, H.T., and Schnell, T. (1999). Legibility of traffic sign text and symbols. *Transportation Research Record, 1692,* 142–151.

8 On Allocating the Eyes
Visual Attention and In-Vehicle Technologies

William J. Horrey

CONTENTS

Reflection ... 151
8.1 Overview ... 152
8.2 Benefits and Costs of In-Vehicle Technologies ... 152
8.3 Visual Demands of In-Vehicle Devices and Telematics ... 154
8.3.1 Factors That Impact In-Vehicle Glance Behavior ... 155
8.4 Modeling Drivers' Visual Attention Allocation While Interacting with In-Vehicle Devices ... 156
8.5 Focal and Ambient Visual Processing ... 160
8.6 Distributions of Glances ... 161
8.7 Conclusions ... 162
Acknowledgments ... 163
References ... 163

REFLECTION

Today, there is an abundance of new devices that we can look at or interact with while driving. Some of these (and this number is growing) are embedded in the vehicle while many others we bring into the vehicle ourselves. Although these devices are a great resource, offering information, connectivity, and entertainment, we must handle them appropriately lest we fail in our driving duties. Vision is a limited resource and, since we have difficulty looking at two places at the same time, we must allocate our attention in a manner that ensures a successful interaction. I learned firsthand how these in-vehicle devices might act as "attention sinks" when I bought my first satellite radio. While I enjoy countless hours of entertainment, I was amazed at how compelling it was to look down at the display for prolonged periods of time, waiting for the artist or album name to finish its crawl across the small screen.

As suggested by many researchers, complex in-vehicles tasks and devices tend to draw the eyes away from the road more often, whether through more frequent glances or longer individual glances—a fact that was abundantly clear with my satellite radio. However, in this chapter, I argue that simply understanding what happens to eye glance behavior with Display A or Task B is not sufficient in our understanding of driver–device interactions. Rather, exploring why glances are driven to different areas at different times will help our understanding and perhaps help predict how new devices will impact scanning. I try to illustrate how concepts borrowed from models of supervisory control can help in this endeavor. Finally, I discuss how the tail end of the distribution of in-vehicle glance durations can offer insight into high-risk situations, though we oftentimes tend to focus on the average behavior. For now, I try to resist the temptation of my satellite radio while driving, saving it instead for traffic lights or the like.

8.1 OVERVIEW

New technological innovations are rapidly bringing more and more embedded as well as portable devices into surface transport vehicles. Although there are many benefits to these technologies, oftentimes they compete with driving-related activities for drivers' limited visual resources. In this chapter, we discuss the general benefits and costs of these technologies, but focus on how these technologies impact drivers' visual scanning behavior. Understanding how visual scanning changes as a function of in-vehicle activities is an important precursor to understanding drivers' susceptibility to missed traffic events and other driving errors. While simulator-based and on-road studies offer much knowledge on the nature of driver scanning when interacting with in-vehicle devices, it is suggested that models of visual attention and supervisory control propose factors that will impact the distribution of visual attention and can aid in the determination of safety implications of emerging technologies. Finally, we discuss the importance of the underlying distribution of data in describing safety-critical phenomenon.

8.2 BENEFITS AND COSTS OF IN-VEHICLE TECHNOLOGIES

New technologies have exploded in the automotive domain in recent years. Vehicles are being equipped with new in-vehicle devices and telematics that are embedded in the vehicle systems. The potential for this technology is enormous and the demand is increasing. For the 2008 model year, it is projected that navigation systems will be offered as standard or optional equipment in more than 80% of vehicle models, touch-screen displays in more than 55%, and Bluetooth interfaces in nearly 70% (Telematics Research Group [TRG], 2007). These figures will only increase in coming years, as services expand and costs are reduced. Meanwhile, cell phones and other portable (nomadic) devices are becoming ubiquitous. For example, there are more than 239 million wireless subscribers in the United States alone (Cellular Telecommunications Industry Association [CTIA], 2007). Schnabel (2002) reports that the global market volume for in-vehicle telematics could reach US$400 billion by the year 2015. Thus, collectively, the amount of information that drivers can access while behind the wheel—whether driving-related or not—is expanding.

Drivers have increasing access to wireless applications that deliver navigation assistance, including real-time traffic information and alternate routing to ease traffic congestion; emergency and roadside assistance (made more effective using GPS information); and news, weather, financial information, and entertainment delivered through Web applications (e.g., Ashley, 2001). Concierge (location-based) services will provide drivers with all sorts of local information in support of shopping, service-seeking, sightseeing, or otherwise. For example, drivers in an unfamiliar neighborhood will be able to access information regarding nearby restaurants, service stations (including up-to-date gas prices), and points of interest. Onboard computers and sensors will also be able to provide drivers with real-time vehicle diagnostics. In addition to these information systems, vehicles will also include a variety of driver support systems such as collision and lane departure warning systems, vision enhancement systems, and other passive and active safety systems.

The increased connectivity that this new technology affords will create the potential for traffic-bound workers to be more productive. The Transportation Research Board (TRB, 2006) reports that in the United States, travel times for workplace commutes are increasing, along with trip length (distance). In addition, more people are driving alone than in previous years. Bluetooth, personal digital assistants (PDAs), cellular phones, and e-mail applications allow workers to remain in contact with their workplace or colleagues, thereby extending the workday into the "mobile office." Hahn, Tetlock, and Burnett (2000) estimated that—at the time—the economic benefit of in-vehicle cell phone use alone exceeded US$25 billion per annum. Of course, the application of these technologies is not limited to the private sector; commercial businesses stand to make significant gains from these technologies through more efficient delivery of products, real-time fleet tracking, and more effective communication with drivers.

Although these in-vehicle devices afford drivers with greater connectivity, information, and capacity to be productive, there are obvious safety concerns to the extent that these devices and associated in-vehicle activities detract from what should be a driver's primary goal: driving safely. Put simply, distraction or inattention occurs when drivers do not pay enough attention to the roadway or the task of driving. In the United States, studies estimate that 25%–50% of police-reported crashes involve some form of driver distraction or inattention (Wang, Knipling, and Goodman, 1996; National Highway Traffic Safety Administration [NHTSA], 1997; Ranney, Mazzae, Garrott, and Goodman, 2000). A recent on-road/naturalistic study suggests that driver inattention contributes to upward of 80% of crashes and 65% of near-misses (Dingus, Klauer, Neale, Petersen, Lee, and Sudweeks, 2006). Although there is no consensus on the precise crash statistics, there is general agreement that the problem of driver distraction is compounded by the arrival of new, sometimes unregulated, devices that drivers can deploy while on the road.

Much research has examined the impact of various in-vehicle activities on driver performance. For example, many studies have documented slowed response times to external events (such as roadside hazards), greater likelihood of missed events, as well as decreased lane-keeping ability and speed control (e.g., Young, Regan, and Hammer, 2003; Lee and Strayer, 2004). Some in-vehicle devices can create cognitive load and degrade performance for drivers—even without requiring that drivers look at the device (e.g., cell phone conversations; Caird, Scialfa, Ho, and Smiley, 2004; Horrey and Wickens, 2006); however, here we focus on those in-vehicle activities that do involve a visual component (e.g., interactions with dashboard displays or nomadic devices). To interact with these devices, the driver must scan back and forth from the road to the device in order to perform the task. Of course, the eyes can only focus on one location at a time and therefore eye movements work to bring different information into focus. This has two important implications: (a) models of visual scanning can help to characterize how drivers allocate their visual attention across multiple areas of interest (AOIs; e.g., road, instrument panel, telematic device) and (b) drivers are susceptible to missed information that is currently unattended (e.g., Simons, 2000).

In this section, we described the numerous benefits that new in-vehicle devices and technologies afford drivers, including increased connectivity with location-based services and increased productivity. Unfortunately, there are concerns as well, as visual in-vehicle devices compete with routine driving activities for limited visual resources (Wickens, 2002). As the eyes spend more time directed inward, drivers are vulnerable to missed traffic events and are thus at a greater risk of crash involvement (Wierwille and Tijerina, 1998). Although many in-vehicle activities require only a short amount of time to complete (e.g., checking the speedometer), new activities—particularly those that are complex or engaging—create greater concerns for safety.

8.3 VISUAL DEMANDS OF IN-VEHICLE DEVICES AND TELEMATICS

For many in-vehicle tasks (though not all), multiple interactions or steps are required to complete the task. For example, programming a new destination in a navigation system (the task) requires that drivers access different submenus as well as manually enter address information (each representing a step toward task completion). As such, one can look at in-vehicle task performance according to overall task engagement as well as each discrete step toward task completion.

With respect to eye glances, each step toward completing an in-vehicle task is generally reflected by the amount of time that the driver looks at the display or device during a single interaction. Thus, a glance represents the time between the transition of the eyes toward the in-vehicle display from somewhere else in the visual field (e.g., the road) until they transition away from the device (e.g., back to the road). A single glance toward a display may include multiple fixations and corresponding saccades (provided they do not leave the display to fixate in another area; e.g., instrument panel). Overall task engagement generally captures the extent of the interaction with the device, including all steps taken toward the end goal. A common measure, total glance time, is the cumulative glances to the display from the onset of the task until it is completed (in other words, the total eyes-off-the-road time). Other related measures such as glance frequency and percent dwell time also reflect this overall measure of task involvement.

While the total task time may reflect the overall difficulty or complexity of a task, measures of glance duration more adequately capture the driver's strategy for interacting with the device. That is, it captures how the driver chooses to parse the two (driving and in-vehicle) activities or how they choose to interrupt one task in favor of the other. Obviously, the nature of this strategy can have important implications for safety.

8.3.1 Factors That Impact In-Vehicle Glance Behavior

Traditional in-vehicle activities, such as monitoring instrument gauges (e.g., speed, fuel), generally require few in-vehicle glances and these glances tend to be short in duration (see Green, 1999a, for an excellent review). For example, glances to the speedometer can be as short as approximately 0.6 s and, on average, reading this information does not require much more than a single glance (M = 1.3 glances; Dingus, Antin, Hulse, and Wierwille, 1989). In contrast, more complicated tasks involving new display devices require longer and more frequent glances (e.g., a moving map display; Wierwille, Antin, Dingus, and Hulse, 1988). These effects are exacerbated when some degree of interaction is required (e.g., manually adjusting the zoom of the map).

The impact of increased complexity on in-vehicle glance behavior has also been examined. Lansdown (2002) varied task complexity through the number of required steps (or interactions) with an in-car entertainment system and found that as task complexity increased, so did the total amount of time spent looking at the display, the number of fixations on the display, and the mean glance duration (ranging from 0.5 to 1.6 s). Similarly, Victor, Harbluk, and Engström (2005) found that drivers made longer and more frequent in-vehicle glances for tasks of increased complexity (in this case, complexity was varied through the number of distractor elements in a visual search array). Horrey, Wickens, and Consalus (2006) varied in-vehicle task complexity through the amount of information to be processed by the driver. Drivers were asked to determine whether there were more even or odd numbers in a randomly generated string of 5 digits (simple) or 11 digits (difficult). They found that drivers spent more time looking away from the roadway when completing in-vehicle tasks of greater complexity and difficulty.

Thus, it appears that devices or activities that are more complicated will naturally lead to prolonged and more frequent in-vehicle glances. This complexity is often derived from increased required steps toward task completion or through more to-be-processed information displayed per unit time. Advanced in-vehicle devices promote increased complexity, given the limited real estate available on vehicle dashboards and consoles, coupled with the vast amount of textual or graphical information that can be displayed. The use of more sophisticated menu structures in this limited display space will become the typical solution to accommodate this information (auditory displays notwithstanding—a topic that is described in Spence and Ho, Chapter 10, this volume).

Although several studies have shown that more complex in-vehicle tasks (here a proxy for new, advanced in-vehicle systems) can lead to increased glance durations, oftentimes these glances still tend to fall within an acceptable range (Wierwille, 1993). That is, drivers elect to interrupt the in-vehicle task in order to scan back to the roadway. Others have found that regardless of complexity, in-vehicle glance durations tend to be fairly consistent—the differences instead being reflected in the number of glances (e.g., Gellatly and Kleiss, 2000; Hoffman, Lee, and McGehee, 2006). That is, drivers appropriately partition the in-vehicle activity into more manageable chunks, rather than completing it in fewer, albeit longer interactions.

Several additional factors may influence the extent of scanning to in-vehicle devices. For example, Lansdown (2002) showed that novice drivers tended to look at in-vehicle technologies (IVT) longer and more often than experienced drivers, suggesting that novice drivers may not be particularly well calibrated to the demands of driving and, hence, are more likely to scan away from the road for longer (possibly inappropriate) durations. Another factor, age, has shown some mixed results. For example, Tsimhoni, Smith, and Green (2004) showed that older drivers made shorter glances to an advanced in-vehicle display than younger drivers, but made more of them (resulting in a longer overall interaction). In contrast, Dingus, Hulse, Mollenhauer, Fleischman, McGehee, and Manakkal (1997) found that older adults had longer glances to a navigation display than younger drivers, though these differences were small (on the order of 0.3 s). It is possible that these mixed results are due to the influence of another important factor: exposure.

In a typical experimental paradigm, the in-vehicle device or task is unfamiliar to drivers (and this is true also for new and/or unfamiliar vehicles) and therefore the extent of scanning may be due to a novelty effect. In other words, drivers will scan more frequently to the display or device, simply because it may be new and interesting to them. Along these lines, there may also be age-related cohort effects given that younger adults may be more familiar with similar technologies and related-tasks. For example, cell phone and iPod ownership among younger drivers far outpaces that of older drivers (Center for Media Research [CMR], 2005). In one study, Dingus et al. (1997) found that, after six weeks of practice and exposure to a new in-vehicle navigation system, drivers had better strategies for interacting with them, including shorter glance durations and fewer long duration glances (over 2.5 s).

In summary, interactions with in-vehicle devices affect drivers' scanning behavior by diverting the eyes away from the roadway. Although drivers generally do not like to look away from the roadway for extended periods of time, complex devices and in-vehicle devices—those that are becoming more prevalent—can lead to increases in in-vehicle glance durations and glance frequency. As more time is spent looking inside the vehicle drivers are at increased risk of missing critical traffic information or events. To date, the long-term impact of new technologies on scanning behavior is not well known.

While simulator and on-road studies offer much knowledge on the nature of driver scanning when interacting with in-vehicle devices, controlled experimentation is often difficult, time-consuming, and expensive. Ideally, models of visual attention could be used to examine the impact of various in-vehicle tasks on scanning behavior and, consequently, safety. In the following section, we outline some models of visual attention and describe how they may be applied to the driving context.

8.4 MODELING DRIVERS' VISUAL ATTENTION ALLOCATION WHILE INTERACTING WITH IN-VEHICLE DEVICES

Wierwille (1993) describes a deterministic model of in-vehicle sampling behavior in which visual scanning is regulated by the amount of time required to extract information from an in-vehicle display. For example, if the information required for a particular task cannot be completely processed in under 1.6 s, the driver will return his or her eyes to the roadway momentarily before refocusing on the in-vehicle display. This cycle continues until the in-vehicle task is completed or abandoned by the driver. The threshold used in this model (1.6 s) is based on scan data for a number of different in-vehicle systems and controls (e.g., speedometer, climate control gauges; Dingus et al., 1989). According to Green (1999a; 2002), drivers report that they are comfortable looking away from the roadway for relatively short periods of time, ranging from 0.8 to 1.1 s, depending on the traffic situation.

Although drivers may break up the in-vehicle activities into more manageable chunks of time (Wierwille, 1993), there is evidence that the frequency of glances—in addition to glance duration—contributes to increased crash risk. Wierwille and Tijerina (1998), using data on frequency of system use, glance duration, glance frequency, and crash statistics, established a positive correlation between the visual demands of an in-vehicle system and crash risk. Thus, longer single glances or an increased number of short duration glances downward to deal with an in-vehicle activity will increase the likelihood that the driver will miss—or delay responding to—important traffic events. The finding that many new in-vehicle devices draw more frequent glances is a robust one (Dingus et al., 1989; Lansdown, 2002; Victor et al., 2005; Hoffman et al., 2006).

The deterministic model outlined by Wierwille (1993) is elegant in its simplicity—capturing how the eyes are directed while interacting with in-vehicle devices—however, it does not describe why visual attention is distributed in a particular way. Other models of visual attention, including supervisory control, may help to address the *why* underlying scanning in complex environments, such as driving.

In general, models of supervisory control characterize the eye as a single-server queue and visual scanning as the means of serving the queue (e.g., Senders, 1964; Carbonell, 1966; Moray, 1986). Single-server queue models tend to differ from visual search models (e.g., Wolfe, 1994) in that (1) the observer is monitoring a series of dynamic processes (e.g., gauges, displays), rather than searching for a static target; (2) the focus is on detecting critical events at relatively consistent locations, rather than finding critical targets at uncertain locations; (3) eye scanning patterns and proportion of fixations in different AOIs (e.g., roadway, in-vehicle device) are the variables of interest, more so than target detection times; and (4) there is an emphasis on knowing when to look where (given the different AOIs) as opposed to knowing simply where to look.

Knowing when to look where is a function of numerous factors. In early work on supervisory control, Senders (1964) described a model of visual scanning based on the concept of event or information bandwidth. In general, when relevant information occurs at a given location more frequently, observers will tend to sample this location more frequently, compared to other locations where information does not occur as frequently. As such, observers tend to sample where they *expect* to find information (e.g., Theeuwes and Hagenzieker, 1993). For example, the frequency of a driver's glances to the roadway will be based on uncertainty regarding the vehicle's current lane position and heading as well as potential obstacles. This uncertainty increases much faster in a driving environment with heavy traffic, winding roads, or increased wind turbulence, for example. With increased driving experience, drivers' expectations become better calibrated to actual information bandwidth (e.g., Underwood, Chapman, Bowden, and Crundall, 2002). With respect to in-vehicle devices, if information is presented on the display frequently or more information is presented, then the driver is likely to scan there more often.

Scanning in a complex environment is not a function of information bandwidth alone. The relative importance of the information plays a role as well. Carbonell (1966) suggested that optimal scanning strategies should attempt to maximize the benefits of perceiving certain information and minimize the costs of missing it. That is, the bandwidth of information as well as the expected value of perceiving this information at different AOIs dictates the optimal scan strategy. For example, in driving, the value of detecting a pedestrian that has stepped in front of a vehicle's path far outweighs the value associated with correctly perceiving information displayed on a roadside billboard. Naturally, different in-vehicle tasks lend themselves to varying degrees of value, and each task itself can vary considerably across drivers and contexts. Information presented on a navigation display will be highly valued when the driver is in an unfamiliar traffic environment; however, this information will be less valuable when the driver is in a very familiar environment (even though the bandwidth of information across the two contexts may be similar). Likewise, the value afforded to activities that allow drivers to be productive (i.e., the mobile office) will vary across driver and circumstances. This shifting prioritization of in-vehicle tasks remains a challenge in modeling task value. Ideally, driving should always be the top priority; however, as crash statistics, experimental data, and casual observation suggest, this may not always be the case.

Recently, Wickens and colleagues (Wickens, Goh, Helleburg, Horrey, and Talleur, 2003; Horrey et al., 2006) have elaborated on these earlier models of visual scanning to include additional factors. According to the SEEV model, the allocation of visual attention in a dynamic environment is driven by four factors: Salience, Effort, Expectancy, and Value. The final two factors, Expectancy and Value, are consistent with the factors described by Senders (1964) and Carbonell (1966). We describe the others below.

The salience or conspicuity of information occurring at a given location has been a fundamental aspect of models of visual search (e.g., Itti and Koch, 2001). Highly salient objects tend to capture attention while observers are searching for a target object (e.g., Wolfe, 1994). For example, observers' visual attention is much more likely to be drawn to a highly salient location (e.g., a brightly clad pedestrian) as opposed to a less salient location (e.g., a dark corner). Unfortunately, in some cases, salience does not correspond to task relevance or importance. As such, salient objects can easily become distractors for more important information. For example, a particularly bright or flashy billboard may draw the eyes away from a highly relevant, but darkly clad pedestrian. Similarly, in-vehicle devices or displays that are highly salient may inadvertently draw attention to themselves (consider a bright or flashing console display). Although systems aimed at advising the driver of critical safety information (e.g., collision warning systems) can benefit from using highly salient features, there is a risk that less important information could be equally salient (e.g., incoming text messages).

Effort is a factor that reduces the likelihood that visual scanning will occur (Sheridan, 1970). For example, effort may be manifested through movement and physical costs (e.g., scanning distances; Wickens, 1992), costs of manipulation or interactivity (e.g., Gray, 2000), or costs of cognitive load (e.g., Recarte and Nunes, 2000). As the required effort to access information increases, the likelihood of scanning or seeking out the pertinent information decreases. Alternatively, observers may seek to reduce the number of times they need to scan between the two locations—a phenomenon known as the "in the neighborhood effect." In driving, however, the influence of effort on scanning appears to be minimal given drivers' willingness to scan frequently to in-vehicle displays (Wierwille, 1993; Horrey et al., 2006).

The computational version of the SEEV model is described by Wickens and colleagues (Wickens et al., 2003; Horrey et al., 2006). The model generates predictions regarding the expected probability of attending a given AOI according to the value of all tasks supported by that AOI, the relevance of the AOI for those tasks, and the bandwidth of task-relevant information in that AOI. In their application, the authors used a "lowest ordinal algorithm" to determine the assignment of coefficients for the various components of the model. It is important to note that the model allows one to specify model parameters *a priori* based on display constraints, task demands, and assumed task priorities. Of note, the authors in these studies did not include the salience factor. Because of the transient nature of salient features in a dynamic environment, salience is more difficult to characterize in the computational model. For example, in contrast to other factors, salience may describe a particular object or event occurring at a given location (e.g., pedestrian on the roadside) as opposed to the location itself (i.e., the roadside). Please see Wickens et al. (2003) or Horrey et al. (2006) for additional details regarding the computational model.

The predictive capabilities of the computational version of the SEEV model have been examined in both the aviation and driving domains with positive results. For example, Horrey et al. (2006) examined drivers' visual attention allocation while in a driving simulator. In Experiment 1, task-relevant information bandwidth and task priority was varied systematically in the road environment and on an in-vehicle device (which supported an in-vehicle task). In-vehicle task complexity was varied in Experiment 2. Using the computational model, they found that task value was the strongest predictor of visual scanning. The effect of expectancy varied by task type: IVT task bandwidth (manipulated by frequency or complexity) had a greater impact on scanning than did a corresponding increase in road bandwidth—possibly because the latter task could be supported by ambient vision (a concept we will discuss in the next section). Overall, a high proportion of the variance was accounted for by the model parameters at both the individual and aggregated levels. Although this modeling approach appears promising in the evaluation of in-vehicle activities, further validation is warranted.

The practical upside of using models of visual attention is that they provide users with a tool to predict scanning with different types of in-vehicle devices and activities based on underlying factors and thereby predict vulnerability to missed roadway hazards. The SEEV model characterizes the allocation of visual attention, or more specifically, it describes the momentary allocation of *focal* visual attention (i.e., mapping onto the fovea). There are, however, certain tasks that can be performed without requiring that drivers directly fixate on a given location or AOI (e.g., lane keeping). In such cases, the predictive model may be less useful. In the next section, we describe two visual systems (focal and ambient), their functions, and the implications for models of scanning and display or device location.

8.5 FOCAL AND AMBIENT VISUAL PROCESSING

Several researchers have distinguished between focal and ambient visual systems (e.g., Leibowitz and Post, 1982; Previc, 1998), which differ in terms of function and underlying physiology. Previc (1998) describes the primary function of the focal (or focal extrapersonal) visual system as visual search and object recognition, including tasks that require high visual acuity (e.g., reading text). This system is predominantly ventrolateral, involving the parvo visual cortical pathways spanning the occipital–temporal cortex. Because focal vision is strongest in the fovea, it is tightly linked to eye movements. Put another way, visual scanning is a means of bringing relevant information into focal vision. As such, focal vision lends itself to models of visual attention, as described previously. Given that many in-vehicle activities involve accessing or resolving information present on in-vehicle displays, there is a natural competition with driving over limited focal resources (Wickens, 2002).

In contrast, the ambient (or ambient extrapersonal; Previc, 1998) visual system is involved in orienting in earth-fixed space, spatial orientation, and postural control in locomotion. Ambient vision is a predominantly dorsomedial neural activity. This system involves peripheral vision and relies on a few major visual cues, including horizontality, linear perspective, and motion flow; however, it is not fastidious about scene details inasmuch as orientation and stability can be maintained even when vision is degraded considerably. Because of its strong reliance on peripheral vision, ambient vision does not have a strong involvement in focal scan models; however, it does have important implications for some aspects of driving.

For example, ambient vision has been shown to be effective in supporting some driving activities, such as lane keeping (Summala, Nieminen, and Punto, 1996). Therefore, a driver may be able to use focal vision to read information from a road sign or IVT display while using ambient vision to support the control of vehicle heading. That being said, ambient vision is not very effective in supporting other critical tasks that require some degree of focal processing, such as hazard detection (e.g., Summala, Lamble, and Laakso, 1998; Horrey et al., 2006). Moreover, ambient vision is less susceptible to degraded resolution of visual information, such as when driving at nighttime. Unfortunately, this can cause drivers to overestimate their relative safety and, consequently, drive too fast for the conditions (Leibowitz and Owens, 1977). When interacting with in-vehicle devices, drivers may similarly become overconfident that ambient vision can "take care" of the driving task. Of course, as discussed above, ambient vision is quite limited in its ability to support several critical driving tasks.

Thus, while models of visual attention can be very useful in describing the nature of visual scanning in complex environments, the role of ambient vision should not be overlooked. Scanning models tend to characterize focal vision, although some activities can be performed without necessarily fixating them directly (i.e., with ambient

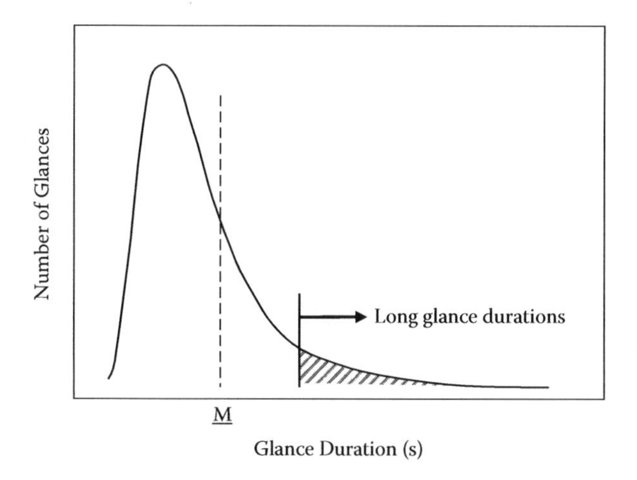

FIGURE 8.1 Log-normal distribution of glance durations.

vision). As such, models of scanning may overpredict the amount of scanning required for tasks that are well supported by ambient vision. Future efforts to unite both visual systems into a single model of visual attention would be advantageous.

8.6 DISTRIBUTIONS OF GLANCES

Finally, there is another important consideration when examining the safety implications of new in-vehicle devices, especially as they relate to driver glance behavior. As shown in Figure 8.1, glance durations are typically log-normally distributed, meaning that distribution is positively skewed (in which a lot of short glances, coupled with fewer long glances that extend the tail of distribution; e.g., Green, 2002). The nature of the underlying distribution of glances can have significant implications for safety. As Wickens (2001) suggests, the conditions or response characteristics that contribute to unsafe behavior or crash risk are not typical. That is, they do not reside at the mean of a given distribution. Rather, it is the more extreme conditions or responses—those that lie in the tails of the distribution—that contribute disproportionately to accidents. For example, it is the unusually slow response to a hazard that results in a collision, rather than the average response.

As such, it may be especially important to consider the nature of the distribution for in-vehicle glances. Much of the discussion of in-vehicle glance behavior, however, tends to focus on the average glance duration and other measures of central tendency (as these lend themselves to traditional statistical analysis). Several studies, however, have reported data on more extreme in-vehicle glances, employing measures of maximum glance duration, percentage of glances over a certain threshold, or the like (e.g., Dingus et al., 1997; Sodhi, Reimer, and Llamazares, 2002; Victor et al., 2005; Klauer, Dingus, Neale, Sudweeks, and Ramsey, 2006).

In an earlier paper, Horrey and Wickens (2007) demonstrated how a consideration of the more extreme values in the distribution of glances (versus the average glance duration) can lead to different conclusions related to the relative safety of in-vehicle devices. In this study, drivers' eye scanning behavior was recorded while performing a visual in-vehicle task that varied in complexity while driving in a simulator. An analysis of the mean glance duration for a simple versus complex in-vehicle task was not statistically significant (not to mention that the average values were well below the 1.6-s threshold). However, when the authors examined the proportion of glances that exceeded 1.6 s (i.e., those that fell in the tail end of the distribution), large and statistically significant display differences were observed. (Note: The authors used Wierwille's, 1993, 1.6-s threshold; other more extreme thresholds would have yielded similar results.) To emphasize, only 6% of glances in the simple task condition exceeded the 1.6-s threshold. In contrast, a whopping 20% of glances exceeded this limit in the complex task condition. Since long duration glances expose drivers to increased crash risk, it is clear that the complex task was far more problematic in this regard. Reaction times to hazard events and collision involvement, measured in the same experiment, corroborated this point. Thus, while a traditional analysis of mean values is certainly not inappropriate, alternate measures may yield important and safety-critical findings.

As Wierwille (1993) suggests, drivers tend to limit themselves to glances that are no longer than 1.6 s. Importantly, however, there is evidence that these complex displays and tasks have more extreme glance durations (i.e., those that lie in the tail end of the distribution) and that it is these glances that pose the greatest threat for safety (versus the average in-vehicle glance).

8.7 CONCLUSIONS

In-vehicle devices will become more and more prevalent in the coming years, not only for embedded in-vehicle systems and displays, but also for nomadic devices that drivers bring into the vehicles. These devices afford drivers the capacity for enhanced efficiency and productivity but also create potential problems arising from distraction and inattention (obviously a safety concern). Although we focus our discussion on the impact of in-vehicle devices that involve visual information, auditory interfaces may help offset these visual demands (e.g., Spence and Ho, Chapter 10, this volume).

The manner in which drivers interact with these devices can either increase or decrease crash risk. In general, drivers prefer to use a series of short glances to perform in-vehicle activities, returning their eyes to the road intermittently. However, as devices become more and more complex, these individual glances may increase in duration and drivers will glance downward more frequently. More importantly, complex in-vehicle tasks may inadvertently result in a larger number of excessively long glances, even though the average glance duration remains within safe or acceptable limits. As we discussed, it is important to consider the tail of the distribution of glances when assessing the relative safety of these device.

Models of visual attention and supervisory control have some promise in predicting the amount of visual scanning that new in-vehicle activities and devices can incur, based on several task properties and characteristics. However, given that these models tend to rely on focal visual attention, it is important to consider the influence of ambient vision in the context of driving and in-vehicle activities. Although the models are intended to predict drivers' vulnerability to missed traffic events, even an optimal scanning strategy is unlikely to capture all of these relevant features. For example, drivers are susceptible to "looked but did not see" crashes and inattentional blindness, resulting in missed traffic information even though the drivers were fixated on it (Herslund and Jørgensen, 2003; Wickens and Horrey, in press).

In recognizing the risks associated with some in-vehicle devices, design standards have been proposed to mitigate the amount of time drivers spend interacting with IVTs, such as the 15-s rule for total task time while the vehicle is stationary, which applies to navigation systems (from the Society for Automotive Engineers; Green, 1999b), and the 2–10–30-s guideline (which characterizes the maximum single glance time, total glance time, and total interaction time, respectively) recommended by the Alliance of Automobile Manufacturers (McGehee, 2001). However, these guidelines presuppose that drivers themselves will interact with these devices appropriately and that these shortened interactions will allow drivers to detect hazard events, which often occur suddenly and without warning. Moreover, nomadic and portable devices are more difficult to regulate.

Another important consideration is the location of the in-vehicle display or device relative to the outside world. Displays that are centrally located (e.g., one presented in a head-up display, or HUD) may yield a more parsimonious, more effective scan pattern by keeping focal vision closer to the road environment. Many studies have demonstrated the benefits of HUDs compared to traditional head-down displays, typically located in the dashboard or center mid-console (e.g., Hada, 1994; Kiefer, 1998; Horrey and Wickens, 2004).

ACKNOWLEDGMENTS

I am grateful to Mary Lesch, David Melton, Patrick Dempsey, and Cándida Castro for their helpful comments and suggestions regarding an earlier draft of this chapter.

REFERENCES

Ashley, S. (2001). Driving the info highway. *Scientific American, 285*, 52–58.

Caird, J.K., Scialfa, C.T., Ho, G., and Smiley, A. (2004). *Effects of Cellular Telephones on Driving Behaviour and Crash Risk: Results of a Meta-Analysis*. Edmonton, Alberta, Canada: CAA Foundation for Traffic Safety.

Carbonell, J.R. (1966). A queuing model of many-instrument visual sampling. *IEEE Transactions on Human Factors in Electronics, HFE-7*, 157–164.

Cellular Telecommunications Industry Association (CTIA). (2007). *CTIA's World of Wireless Communications*. Retrieved June 26, 2007, from http://www.ctia.org.

Center for Media Research (CMR). (2005, July 14). *Upscale Men Dominate Adult iPod Ownership*. Retrieved Sept 11, 2007, from http://www.centerformediaresearch.com/.

Dingus, T.A., Antin, J.F., Hulse, M.C., and Wierwille, W.W. (1989). Attentional demand requirements of an automobile moving-map navigation system. *Transportation Research, 23A*(4), 301–15.

Dingus, T.A., Hulse, M.C., Mollenhauer, M.A., Fleischman, R.N., McGehee, D.V., and Manakkal, N. (1997). Effects of age, system experience, and navigation technique on driving with an advanced traveler information system. *Human Factors, 39*, 177–199.

Dingus, T.A., Klauer, S.G., Neale, V.L., Petersen, A., Lee, S.E., and Sudweeks, J. (2006). *The 100-Car Naturalistic Driving Study, Phase II–Results of the 100-Car Field Experiment* (Report No. DOT HS 810 593). Washington, DC: National Highway Traffic Safety Administration.

Gellatly, A.W., and Kleiss, J.A. (2000). Visual attention demand evaluation of conventional and multifunction in-vehicle information systems. *Proceedings of the IEA 2000/HFES 2000 Congress*, 3:282–3:285.

Gray, W.D. (2000). The nature and processing of errors in interactive behavior. *Cognitive Science, 24*(2), 205–248.

Green, P. (1999a). *Visual and Task Demands of Driver Information Systems* (Report No. UMTRI-98-16). Ann Arbor, MI: University of Michigan Transportation Research Institute.

Green, P. (1999b). *The 15-Second Rule for Driver Information Systems*. Proceedings of the Intelligent Transportation Society of America Annual Meeting, Washington DC.

Green, P. (2002). Where do drivers look while driving (and for how long)? In R.E. Dewar and P.L. Olson (Eds.), *Human Factors in Traffic Safety* (pp. 77–110). Tucson, AZ: Lawyers and Judges.

Hada, H. (1994). *Drivers' Visual Attention to In-Vehicle Displays: Effects of Display Location and Road Type* (Report No. UMTRI-94-9). Ann Arbor, MI: University of Michigan Transportation Research Institute.

Hahn, R.W., Tetlock, P.C., and Burnett, J.K. (2000). Should you be allowed to use your cellular phone while driving? *Regulation, 23*(3), 46–55.

Herslund, M., and Jørgensen, N.O. (2003). Looked-but-failed-to-see-errors in traffic. *Accident Analysis and Prevention, 35*(6), 885–891.

Hoffman, J.D., Lee, J.D., and McGehee, D.V. (2006). Dynamic display of in-vehicle text messages: The impact of varying line length and scrolling rate. *Proceedings of the Human Factors and Ergonomics Society 50th Annual Meeting*, 574–578.

Horrey, W.J., and Wickens, C.D. (2004). Driving and side task performance: The effects of display clutter, separation, and modality. *Human Factors, 46*, 611–624.

Horrey, W.J., and Wickens, C.D. (2006). Examining the impact of cell phone conversations on driving using meta-analytic techniques. *Human Factors, 48*(1), 196–205.

Horrey, W.J. and Wickens, C.D. (2007). In-vehicle glance duration: Distributions, tails and model of crash risk. *Transportation Research Record, 2018*, 22–28.

Horrey, W.J., Wickens, C.D., and Consalus, K.P. (2006). Modeling drivers' visual attention allocation while interacting with in-vehicle technologies. *Journal of Experimental Psychology: Applied, 12*(2), 67–78.

Itti, L., and Koch, C. (2000). A saliency-based search mechanism for overt and covert shifts of visual attention. *Vision Research, 40*, 1489–1506.

Kiefer, R.J. (1998). Quantifying head-up display (HUD) pedestrian detection benefits for older drivers. *Proceedings of the 16th International Conference on the Enhanced Safety of Vehicles* (pp. 428–437). Washington, DC: National Highway Traffic Safety Administration.

Klauer, S.G., Dingus, T.A., Neale, V.L., Sudweeks, J.D., and Ramsey, D.J. (2006). *The Impact of Driver Inattention on Near-Crash/Crash Risk: An Analysis Using the 100-Car Naturalistic Driving Study Data* (Report No. DOT HS 810 594). Washington, DC: National Highway Traffic Safety Administration.

Lansdown, T.C. (2002). Individual differences during driver secondary task performance: Verbal protocol and visual allocation findings. *Accident Analysis and Prevention, 34*, 655–662.

Lee, J.D., and Strayer, D.L. (2004). Preface to the special section on driver distraction. *Human Factors, 46*(4), 583–586.

Leibowitz, H.W., and Owens, D.A. (1977). Nighttime driving accidents and selective visual degradation, *Science, 197*(4302), 422–423.

Leibowitz, H.W., and Post, R.B. (1982). The two modes of processing concept and some implications. In J. Beck (Ed.), *Organization and Representation in Perception* (pp. 343–363). Hillsdale, NJ: Lawrence Erlbaum.

McGehee, D.V. (2001). New design guidelines aim to reduce driver distraction. *Human Factors and Ergonomics Society Bulletin, 44*(10), 1–3.

Moray, N. (1986). Monitoring behavior and supervisory control. In K.R. Boff, L. Kaufman, and J.P. Thomas (Eds.), *Handbook of Perception and Human Performance* (Vol. 2, pp. 40.1–40.51). New York: John Wiley and Sons.

National Highway Traffic Safety Administration (NHTSA). (1997). *An Investigation of the Safety Implications of Wireless Communications in Vehicles* (Report DOT HS 808-635). Washington, DC: NHTSA.

Previc, F.H. (1998). The neuropsychology of 3-D space. *Psychological Bulletin, 124*, 123–164.

Ranney, T.A., Mazzae, E., Garrott, R., and Goodman, M.J. (2000). *NHTSA Driver Distraction Research: Past, Present, and Future.* Available at: www-nrd.nhtsa.dot.gov/departments/nrd-13/driver-distraction/welcome.htm.

Recarte, M.A., and Nunes, L.M. (2000). Effects of verbal and spatial-imagery tasks on eye fixations while driving. *Journal of Experimental Psychology: Applied, 6*, 31–43.

Schnabel, B. (2002). Telematics on the move. *Traffic Technology International,* Annual Review, 102–108.

Senders, J. (1964). The human operator as a monitor and controller of multidegree of freedom systems. *IEEE Transactions on Human Factors in Electronics, HFE-5*, 2–6.

Sheridan, T. (1970). On how often the supervisor should sample. *IEEE Transactions on Systems Science and Cybernetics, SSC-6*, 140–145.

Simons, D.J. (2000). Current approaches to change blindness. *Visual Cognition, 7*, 1–15.

Sodhi, M., Reimer, B., and Llamazares, I. (2002). Glance analysis of driver eye movements to evaluate distraction. *Behavior Research Methods, Instruments, & Computers, 34*(4), 529–538.

Spence, C., and Ho, C. (in press). Crossmodal information processing in driving. In C. Castro (Ed.), *Human Factors of Visual and Cognitive Demands in Driving*. Boca Raton, FL: CRC Press.

Summala, H., Lamble, D., and Laakso, M. (1998). Driving experience and perception of the lead car's braking when looking at in-car targets. *Accident Analysis and Prevention, 30*, 401–407.

Summala, H., Nieminen, T., and Punto, M. (1996). Maintaining lane position with peripheral vision during in-vehicle tasks. *Human Factors, 38*, 442–451.

Telematics Research Group (TRG). (2007). *Sneak Peek: Connectivity a "Must Have" for Model Year 2008.* Retrieved August 31, 2007, from http://www.telematicsresearch.com/.

Theeuwes, J., and Hagenzieker. M.P. (1993). Visual search of traffic scenes: On the effect of location expectations. In A.G. Gale et al. (Eds.), *Vision in Vehicles—IV* (pp. 149–158). Amsterdam: Elsevier.

Transportation Research Board. (TRB). (2006). *Commuting in America III: The Third National Report on Commuting Patterns and Trends* (NCHRP Report 550/TCRP Report 110). Washington, DC: National Academies.

Tsimhoni, O., Smith, D., and Green, P. (2004). Address entry while driving: Speech recognition versus a touch-screen keyboard. *Human Factors, 46*(4), 600–610.

Underwood, G., Chapman, P., Bowden, K., and Crundall, D. (2002). Visual search while driving: Skill awareness during inspection of the scene. *Transportation Research Part F, 5*, 87–97.

Victor, T.W., Harbluk, J.L., and Engström, J.A. (2005). Sensitivity of eye-movement measures to in-vehicle task difficulty. *Transportation Research Part F, 8*, 167–190.

Wang, J.-S., Knipling, R.R., and Goodman, M.J. (1996). The role of driver inattention in crashes: New statistics from the 1995 Crashworthiness Data System. *40th Annual Proceedings of the Association for the Advancement of Automotive Medicine*, 377–392.

Wickens, C.D. (1992). *Computational models of human performance* (Technical Report No. ARL-92-4/NASA-A3I-92-1). Savoy, IL: University of Illinois, Aviation Research Laboratory.

Wickens, C.D. (2001). Attention to safety and the psychology of surprise. *Proceedings of the 2001 Symposium on Aviation Psychology*. Columbus, OH: Ohio State University.

Wickens, C.D. (2002). Multiple resources and performance prediction. *Theoretical Issues in Ergonomics Science, 3*, 159–177.

Wickens, C.D., Goh, J., Helleburg, J., Horrey, W.J., and Talleur, D.A. (2003). Attentional models of multi-task pilot performance using advanced display technology. *Human Factors, 45*, 360–380.

Wickens, C.D., and Horrey, W.J. (2008). Models of attention, distraction and highway hazard avoidance. In M.A. Regan, J.D. Lee, and K. Young (Eds.), *Driver Distraction: Theory, Effects, and Mitigation*. Boca Raton, FL: Taylor & Francis.

Wierwille, W.W. (1993). Visual and manual demands of in-car controls and displays. In B. Peacock and W. Karwowski (Eds.), *Automotive Ergonomics* (pp. 299–320). Washington, DC: Taylor & Francis.

Wierwille, W.W., Antin, J.F., Dingus, T.A., and Hulse, M.C. (1988). Visual attentional demand of an in-car navigation display system. In A.G. Gale et al., (Eds.), *Vision in Vehicles— II* (pp. 307–316). Amsterdam: Elsevier.

Wierwille, W.W. and Tijerina, L. (1998). Modelling the relationship between driver in-vehicle visual demands and accident occurrence. In A.G. Gale et al. (Eds.), *Vision in Vehicles—VI* (pp. 233–243). Amsterdam: Elsevier.

Wolfe, J.M. (1994). Guided search 2.0: A revised model of visual search. *Psychonomic Bulletin and Review, 1*, 202–238.

Young, K., Regan, M., and Hammer, M. (2003). *Driver Distraction: A Review of the Literature* (Report No. 206). Victoria, Australia: Monash University Accident Research Centre.

9 Enhancing Safety by Augmenting Information Acquisition in the Driving Environment

Monica N. Lees and John D. Lee

CONTENTS

Reflection 167
9.1 Introduction 168
9.2 Older Driver Crashes and Their Causes 169
9.3 Younger Driver Crashes and Their Causes 170
9.4 Countermeasures That Enhance Information Acquisition 171
9.5 Acceptance and Reliance of Information Acquisition Systems 176
9.6 Conclusions 181
References 182

REFLECTION

According to the National Center for Statistics (NCSA), in 2006 over 6 million motor vehicle crashes occurred in the United States resulting in 42,642 deaths and approximately 2.6 million injuries. Many of these crashes result from a mismatch between the attentional and perceptual capabilities of drivers and the demands of the driving environment. A recent naturalistic study monitored 100 drivers in their own vehicles for 1 year and found approximately 85% of the crashes and near crashes resulted from some attentional failure, including fatigue and distraction (Klauer et al. 2006). These data demonstrate a fundamental problem that plagues driving safety: People have evolved to 2–10 mph locomotion, but not the demands of 20–100 mph locomotion. This mismatch leads to circumstances in which the demands of driving exceed the capacity of the driver to respond. Recent advances in sensor and computing technology may reduce these mismatches by augmenting the driver's ability to acquire relevant information. For example, radar-based sensors can scan the road ahead and detect cars that might pose a hazard to the driver, and algorithms can process these data to deliver a warning to help the driver avoid an impending collision. However, achieving the promise of such systems is far from certain.

9.1 INTRODUCTION

Experience with automation in a variety of domains demonstrates that how people respond to technology can have a dramatic influence on its effectiveness (Lee 2006b). If drivers find the system annoying, they might grow to distrust the system and reject its guidance (Parasuraman and Riley 1997). If drivers adapt their behavior and over-rely on the system, safety might actually decline (Parasuraman and Riley 1997). As an example, if the protection afforded by a collision warning system leads drivers to read the newspaper, a driver with the system might be more vulnerable compared to a driver without the system. How people respond to systems that enhance information acquisition from the driving environment depends on the characteristics of the driver and the technology (Lee 2006a).

Driving involves a very heterogeneous population. Compared to other domains where technology has been inserted to extend human capability, such as military aviation, driving involves people of greatly disparate ability, different objectives, expertise, and cognitive capacity. Age is the most obvious differentiator of drivers. Younger drivers lack experience, but have acute perceptual and motor processes. Older drivers have experience, but tend to see more poorly and respond more slowly than other age groups. More subtle differences between older and younger drivers concern how they deploy their attention to the driving environment. Drivers under the age of 25 and over the age of 65 tend to neglect hazards (Underwood et al. 2005) and are overrepresented in crashes, as shown in Figure 9.1 (Evans 2004). Emerging vehicle technology may be particularly beneficial to younger and older drivers if designed to address the factors that underlie the increased crash risk of these populations.

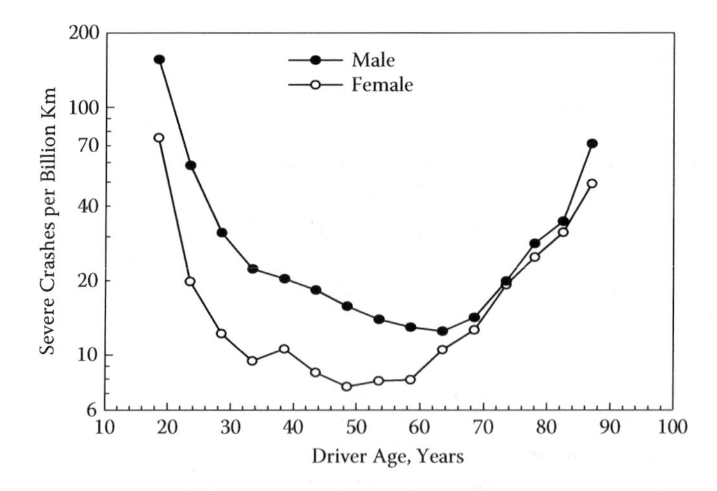

FIGURE 9.1 Both younger and older drivers are overrepresented in severe crashes. Reproduced with permission from Evans, L., *Traffic Safety* (2004). Bloomfield, MI: Science Serving Society. © Leonard Evans.

If emerging sensor and computing technology is to benefit drivers it is important to understand why younger and older drivers crash and how this might affect driver reliance on the technology. Comparing these populations in terms of top-down and bottom-up processing represents a useful approach for suggesting different technological countermeasures and for understanding how drivers will adapt to and accept information acquisition systems. Bottom-up perceptual and cognitive processes concern perception and responses driven by information in the environment. Top-down perceptual and cognitive processes are driven by driver knowledge and experience—information in the head rather than information in the environment (Theeuwes 1993; Wickens and Hollands 2000).

In general, older drivers have less effective bottom-up processes to accommodate the demands of the driving environment, whereas younger drivers have less effective top-down processes. Younger drivers have acute perceptual systems that effectively process information from the environment, but lack the experience needed to anticipate hazards and reconcile ambiguous information in the world. Older drivers are often able to use experience to anticipate conflicts, which can compensate for their less effective information acquisition and slower responses. Bottom-up and top-down processing deficiencies capture some of what contributes to increased crash risk for these populations; however, it is important to note that both top-down and bottom-up deficiencies exist among both older and younger drivers. For example, younger drivers that engage in a distracting activity are likely to experience diminished bottom-up processing. Likewise, cognitive impairments, such as Alzheimer's disease, can undermine the ability of older drivers to draw on their experience to support top-down hazard identification (Lundberg et al. 1997; Uc et al. 2004).

This chapter identifies characteristics associated with older and younger drivers' increased crash risk. These characteristics suggest how technology might be tailored to these groups to enhance information acquisition. We also explore factors that may influence how these populations accept and use emerging in-vehicle technology.

9.2 OLDER DRIVER CRASHES AND THEIR CAUSES

By the year 2030, fatal crashes involving drivers over the age of 65 are expected to increase by 155% and account for approximately 25% of all driver fatalities (Lyman et al. 2002). The substantial research on older drivers describes the factors that underlie their increased crash risk and can help identify technology that can benefit these drivers.

Older drivers are overrepresented in intersection and multiple-vehicle crashes. They have difficulty with complex traffic situations, intersections, merging into traffic, turning, and changing lanes (Preusser et al. 1998; Zhang et al. 1998; McGwin and Brown 1999a). Older drivers are more likely to be considered at fault compared to middle-aged drivers. For this driver group, at-fault crashes often involve inappropriate actions, improper turning, failure to obey traffic controls, and failure to yield right-of-way (Zhang et al. 1998; McGwin and Brown 1999a).

Increased risk for older drivers stems from age- and disease-related declines in visual and cognitive ability. These declines appear to influence driving through bottom-up processes, or the inability to detect, attend to, or extract relevant information from the driving environment. Several age-related declines have been reported in drivers over the age of 65 including (Preusser et al. 1998): (1) reduced visual field of view, (2) difficulty seeing in dim light, (3) difficulty in visually defining and separating objects, (4) diminished divided attention capacity in cluttered environments, and (5) impaired visual search that undermines the ability to attend to relevant information. Inefficient attention allocation seems particularly important in understanding older-driver crash involvement. For instance, an attention-related construct called the useful field of view (UFOV) is a better predictor of crash involvement, on-road driving test performance, and driving simulator performance for older drivers than other visual or cognitive function tests (Owsley et al. 1991; Ball and Owsley 1993; Goode et al. 1998; Myers et al. 2000). UFOV represents the area in which visual information can be extracted during a single glance without making eye or head movements (Sanders 1970). This construct reflects visual sensory function, visual processing speed, divided attention, and selective attention (Ball and Owsley 1993; Roenker et al. 2003). These perceptual and cognitive processes reflect diminished bottom-up processing capacity.

In-vehicle technology, such as collision warning systems that direct the driver's attention to an impending crash, could support bottom-up processes. Such technology might also help mitigate other age-related declines that increase accident involvement, such as delayed decision making, delayed action implementation, and longer processing times (Preusser et al. 1998). Night vision technology might counteract the sensitivity of older drivers to glare and diminish the processing demands of driving-related information (Bossi et al. 1997; Caird et al. 2001). Beyond helping drivers respond to specific situations, technology could enhance safety by reducing the overall demands of driving. For example, turn-by-turn navigation systems can make it easier for drivers to navigate and see critical features of the roadway, which could free attentional resources and enable older drivers to respond to hazards more quickly (Baldwin 2002). Navigation systems could further reduce driving demands by helping drivers to plan routes that are less demanding, such as routes that avoid traffic, freeways, and left turns.

9.3 YOUNGER DRIVER CRASHES AND THEIR CAUSES

Compared to other age groups, younger drivers are more likely to be involved in a fatal crash on a per mile basis. Sixteen-year-old drivers have a crash rate 10 times that of adults but experience a two-thirds reduction in their crash rate within the first 500 miles (McKnight and McKnight 2003). While crash rates for younger drivers rapidly decline during the first 6 months of driving, fatality rates remain relatively high during the first few years of driving (Mayhew et al. 2003).

Younger drivers are overrepresented in single-vehicle crashes and crashes occurring in rural locations. They are more likely to be considered at fault compared to middle-aged drivers and crashes are often attributed to alcohol impairment, fatigue, inattention, speeding, lack of control, misjudging stopping distance, inappropriate actions, improper turning, failure to obey traffic controls, failure to yield right-of-way, improper lane changes, and risk taking (Zhang et al. 1998; McGwin and Brown 1999a). The increased crash risk for younger drivers is often attributed to inexperience, lack of maturity, and risk taking (Deery 1999; McGwin and Brown 1999a; Mayhew et al. 2003). While traditional training programs have failed to show significant reductions in crash involvement, recent graduated driver licensing (GDL) programs show promise (Williams 2006). One reason for the reduced crashes associated with GDL is that it allows novice drivers to gain experience before they are exposed to demanding driving situations (Ferguson 2003). For example, delaying exposure to nighttime driving and situations with several passengers allows novice drivers to learn how to respond to the simpler situations of daytime driving without distractions.

In general, inefficient or immature top-down processes and engagement in inappropriate behavior underlie the increased risk for younger drivers. Novice drivers lack an overarching representation to combine information or to guide their expectations of what constitutes a hazard. Novice drivers are less accurate and are slower in perceiving hazards in filmed traffic situations (McKenna and Crick 1994; Horswill and McKenna 2004). Hazard detection ability represents the ability of the driver to anticipate potentially dangerous roadway situations and correlates with accident involvement and driver experience (McKenna and Crick 1994; Horswill et al., in press). Hazard detection depends upon high-level cognitive skills, efficient scanning, and awareness of one's surroundings (Horswill and McKenna 1999; Grayson and Sexton 2002). A critical skill that takes time for novice drivers to learn is the ability to distribute their attention so they are able to detect hazards in a timely fashion (Fisher et al. 2006; Pollatsek et al. 2006). The challenging task of hazard detection is more difficult at night and when drivers must also manage distractions (Lee 2007). Research suggests that repeated exposure of novice drivers to hazards or providing knowledge regarding how to identify hazards could improve hazard detection (McKenna and Crick 1994; Grayson and Sexton 2002; Fisher et al. 2006; McKenna et al. 2006). For instance, McKenna et al. (2006) demonstrated in a series of experiments that training novice drivers with commentated videos improved hazard detection and the ability to modulate speed according to the likelihood of hazards. The effect of such training shows that experience supports top-down processing that is needed to detect hazards and modulate behavior.

In-vehicle technology may help these drivers identify and combine relevant information and ultimately train top-down processing. For example, collision-warning systems might direct novice drivers' attention to hazards that they might not have otherwise noticed. Systems that alert drivers to sharp curves and intersection hazards could be particularly useful in helping novice drivers learn where to focus their attention (Tijerina et al. 1995; Lee 2007).

9.4 COUNTERMEASURES THAT ENHANCE INFORMATION ACQUISITION

Recent technological developments enable a range of countermeasures to enhance the acquisition of visual information. These developments include sensors, algorithms, and display systems that can substantially augment the way drivers perceive the road:

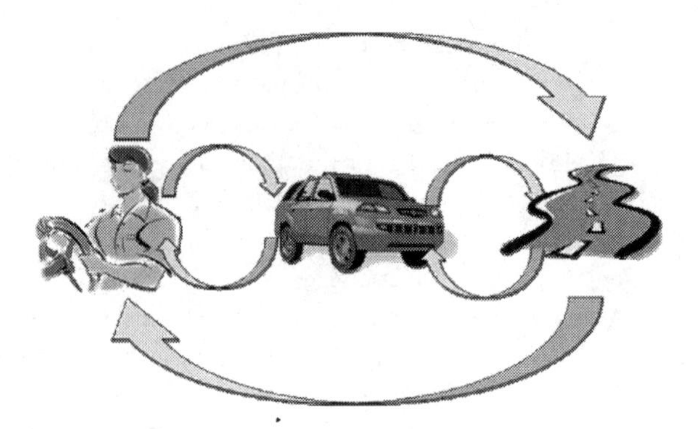

FIGURE 9.2 Sensors, databases, algorithms, and displays mediate driver interaction with the roadway and the vehicle.

- Infrared, laser, and radar sensors can see what drivers cannot.
- Global positioning systems (GPS) and map database systems can catalog information about unfamiliar roads and potential hazards.
- Sensors that assess driver state and estimate the focus of drivers' attention can identify what information the driver is acquiring.
- Algorithms that use powerful vehicle-based computers to combine data from multiple sensors can identify potentially hazardous situations or help the driver extract useful information.
- Haptic seats, spatially localized sound, head-up displays, and other interface systems can provide substantial flexibility in bringing information to the driver.

Figure 9.2 shows how sensors and algorithms can help support drivers in negotiating the roadway. The large arrows at the top and bottom of the figure indicate the potential for the driver to perceive and respond to roadway characteristics, which can be mediated by technology, particularly GPS and map database systems. The smaller inner arrows indicate the potential for the vehicle to sense and respond to the state of the driver, on the left, and to sense and respond to the state of the world, on the right.

These sensor, algorithm, and interface developments enable a wide range of interventions that can help people drive more safely. Initial application of this technology has been in helping drivers avoid crashing when a collision is imminent. With these systems, radar or laser rangefinders track the distance to other vehicles and warn the driver to either direct their attention to the threat or to guide the driver's response (e.g., brake to avoid colliding with a slowing lead vehicle). Directing attention to threats and guiding responses are two general ways technology can enhance information acquisition of drivers. Table 9.1 shows a range of general countermeasures that could enhance information acquisition in older and younger drivers.

These strategies range from mitigating imminent crashes to helping drivers learn and adapt their driving behavior. Systems that guide responses or direct attention enhance bottom-up perceptual processes, whereas those that enhance feedback focus on top-down processes. Strategies at the top of the list (i.e., those that guide response

TABLE 9.1
Countermeasures That Could Enhance Information Acquisition and Responses in Older and Younger Drivers

Behavioral Level	Intervention Type	Description	Example
Operational	Initiate response	Sensors detect a threat and the system initiates a braking or steering response. Such a system would give the driver additional time to intervene and may lessen the impact of an unavoidable crash.	Adaptive cruise control initiates braking to maintain a prespecified headway.
Operational	Supplement response	Sensors detect a threat and infer a response from the driver. The system augments the driver's braking or steering response to accommodate the threat.	Sensors detect an inevitable crash situation that exceeds the driver's capacity to respond and therefore amplifies the driver's braking response.
Operational/ tactical	Guide response	Sensors detect a threat and the system guides the driver's response by indicating how the driver should respond.	A forward-collision warning system uses a brake pulse to indicate that the driver should initiate braking. A curve-speed warning system issues a verbal warning indicating that the driver should slow down for an upcoming curve (e.g., curve ahead, slow down).
Operational	Direct attention	Sensors detect a crash situation or roadway hazard and the system directs the driver's attention to the hazard, but does not specify a response.	A collision warning system issues an alarm (e.g., an auditory or visual alarm) that indicates the direction or presence of a hazard.
Tactical/strategic	Enhance awareness	Sensors monitor the driver's state to estimate deficiencies in driver performance or behavioral aspects that may increase the likelihood of a crash.	The system monitors eye movements and, depending on scanning behavior, issues peripheral visual or auditory alarms to engage the driver in more appropriate scanning behavior.

TABLE 9.1 (CONTINUED)
Countermeasures That Could Enhance Information Acquisition and Responses in Older and Younger Drivers

Behavioral Level	Intervention Type	Description	Example
Tactical	Deliver information	GIS (geographic information system), GPS, and map data are combined to deliver information that drivers might otherwise have to search for in the environment, such as guidance and regulatory sign information. Such systems diminish information acquisition load and free attentional resources so that drivers extract other roadway information more effectively.	A GPS provides guidance or turn-by-turn route information so that drivers can focus primarily on driving rather than searching for signs and directions. A system that warns the driver of an upcoming stop sign at a controlled intersection may reduce the propensity of younger and older drivers to fail to obey traffic controls and to yield the right-of-way.
Tactical/strategic	Enhance feedback	Sensors monitor the driver's state in relation to the current roadway demands and provide feedback to the driver.	Using information about the driving environment and the drivers' state (e.g., eye movements). This indicates to the driver how a distraction slowed a response to a potential threat.
Tactical	Tune expectations	Repeated exposure of drivers to systems that direct their attention to hazards and enhance their awareness could train drivers to be more sensitive to roadway hazards and tune their expectations concerning what situations demand their attention.	Using information about the driver's eye movements the system could guide the driver to look at the edge of the crosswalk when the driver had not. Over time, such a system would teach a driver the location of roadway hazards.
Strategic/tactical	Calibrate capacity and demand	Repeated exposure of drivers to systems that indicate the demand of the roadway and the capacity of the driver could help drivers adapt their behavior to reduce risk.	A system alerts the driver to hazardous situations that the driver may be unaware of because the driver is glancing away from the road.

TABLE 9.1 (CONTINUED)
Countermeasures That Could Enhance Information Acquisition and Responses in Older and Younger Drivers

Behavioral Level	Intervention Type	Description	Example
Strategic/tactical	Postdrive feedback	The system provides the driver with information obtained during previous drives in which the driver made an error or the system provides suggestions on how to improve behavior.	At the end of the week the driver receives a report of risky situations and the associated consequences for his/her insurance premium.

and direct attention) might benefit older drivers most, whereas those that the bottom of the list might benefit younger drivers most. These countermeasures can be placed into three groups according to the behavioral level they address: operational, tactical, and strategic (Sheridan 1970; Michon 1989). The operational level concerns the lateral and longitudinal control of the vehicle and occurs at a timescale of milliseconds to seconds. Tactical control concerns lane and speed choice and occurs at a timescale of seconds to minutes. Strategic control concerns decisions regarding routes and travel patterns and occurs at a timescale of minutes to weeks. These three levels of control apply to activities that are critical to safe driving and can describe the behaviors addressed by a particular type of countermeasure. These levels can help guide understanding of how such countermeasures will improve driving safety (Lee 2006a).

Figure 9.3 presents a specific example of an information acquisition system that would use a head-up display to guide the drivers' attention to potential hazards in the roadway environment. Such a system could benefit older and younger drivers in different ways. The system might benefit older drivers by directing attention and highlighting information within the environment that the driver might otherwise overlook or fail to attend to. For younger drivers, the benefit of the system might be enhanced awareness of hazards. By highlighting such targets, younger drivers may be better able to identify particular sources of danger and to tune their ability to recognize future hazards. Similar systems could be developed to highlight information such as stop signs or a pedestrian crosswalk. Such systems could adapt to drivers by tracking their eye movements and only highlighting information when they fail to look at it. This could be a powerful way to help younger drivers develop the top-down processing skills needed to detect hazards in a timely fashion.

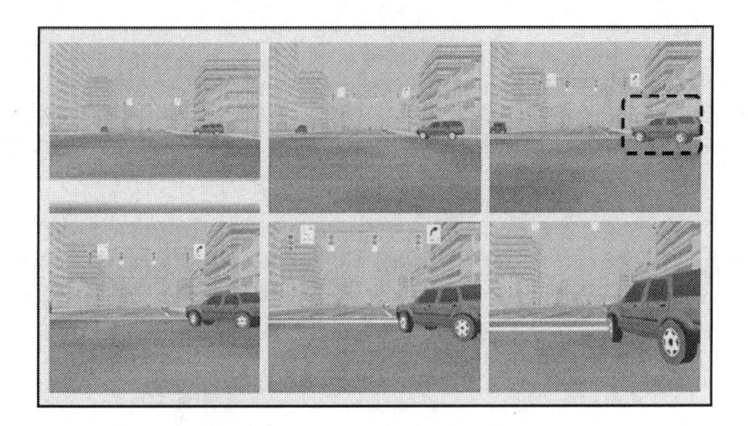

FIGURE 9.3 An example of a head-up display that highlights information within the driving environment. Such a system may benefit older and younger drivers through different mechanisms. For such a system, the dashed box would be more salient (e.g., a red flashing box).

9.5 ACCEPTANCE AND RELIANCE OF INFORMATION ACQUISITION SYSTEMS

Ideally, the combination of driver and technology will perform better than either alone. Yet, both drivers and vehicle technology are imperfect and technology that promises to enhance driver information acquisition may not necessarily provide the expected benefits (Parasuraman and Riley 1997). For example, if drivers fail to rely on the technology appropriately, the combination could perform worse than either the driver or the technology alone. Poorly designed technology can increase driver workload, distract drivers, and undermine performance (Lee 2007). It can also annoy drivers and cause them to reject it so that they derive no benefit from it. At the other extreme, drivers can become complacent and rely on it even when it is not appropriate.

When designing information acquisition systems it is imperative to consider the characteristics of the driver and the specific system (e.g., display modality, function, and algorithms). The same characteristics that increase crash risk for older and younger drivers may also influence the benefits derived by using the system, system acceptance, and reliance. For example, older drivers may fail to look both ways when determining whether to execute a left-hand turn (Preusser et al. 1998; McGwin and Brown 1999). An information acquisition system developed to modify this behavior by giving alerts that guide eye movements to critical areas of the road may help these drivers. Developing such a system requires consideration of these drivers' limitations. For example, the same factors that cause these drivers not to scan effectively or to attend to relevant stimuli may cause older drivers to miss peripheral visual alarms. For younger drivers, an information acquisition system could provide an important benefit by training the driver to be more sensitive to hazards (Lee 2007; McGehee et al. 2007). However, the diminished hazard recognition ability of these drivers may

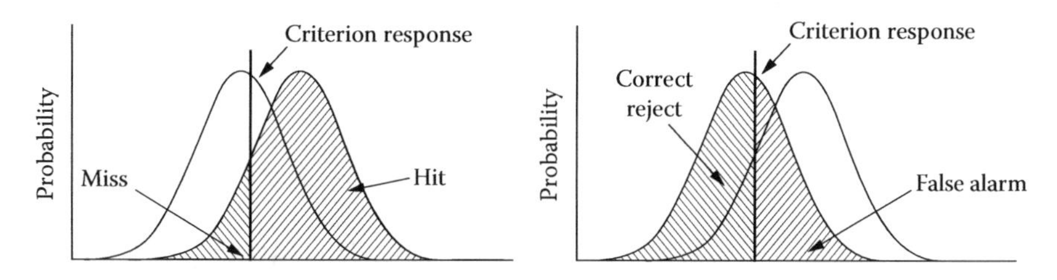

FIGURE 9.4 A graphical illustration of signal detection theory.

prevent them from recognizing the value of such information and may interfere with their ability to use the system to identify roadway hazards.

While information acquisition systems may benefit older and younger drivers, such systems are imperfect and experience failures that may jeopardize their benefits. Failures promote distrust in the system, which can impede appropriate compliance with and reliance on the system as suggested by diminished response frequency, fewer appropriate responses, slower reaction times, and degraded task performance (Bliss 2003). A system that adapts based on driver behavior may reduce alarm failures but at the same time may be more difficult for drivers to understand. Such a system may lack consistency or may be unable to inform the driver of what he or she did correctly in one circumstance and not in another. A framework for describing these imperfections is needed to understand how drivers might respond to imperfect information acquisition systems.

Signal detection theory (SDT) represents a standard method for classifying failures of both humans and technology (Green and Swets 1966). Figure 9.4 provides a graphical illustration of signal detection theory. The criterion response represents the threshold used by the human or system to discriminate between a true signal and noise. Four outcomes are possible using this classification:

1. Hit: the stimulus exists and is correctly detected
2. Correct rejection: the stimulus is absent and not detected
3. False alarm: the stimulus is absent but incorrectly detected
4. Miss: the stimulus is present but not detected

While insightful, this approach neglects the nature of the failure and the user's understanding of the failure, which can substantially influence both system acceptance and use. Such an approach may fail to characterize performance in a manner that explains why some drivers would trust and rely on a system and others would not. Lees and Lee (2007) extended the signal detection framework to understand how drivers might respond to systems that enhance the driver's information acquisition process, through intervention, alarms, and other information. The framework uses three dimensions assumed to underlie trust in automation: performance, process, and purpose (Lee and Moray 1992; Lee and See 2004). *Performance* or *utility* defines what the system does, and refers to the ability of the system to aid the driver with a particular task, such as information extraction. *Process* or *predictability* defines how the system works, and refers to how the algorithms and sensors monitor the roadway or driver state. *Purpose* or *intent* defines the reason the system exists, and refers to the designer's intent regarding what information the system should provide, and when or how the system should intervene or warn the driver. These three dimensions suggest that the driver's perception of an alarm or action depends on both its objective validity and on the state of the driver.

Table 9.2 presents a range of alarm types resulting from the dimensions of trust, which merit consideration in anticipating how older and younger drivers, who have very different experience and capacity, might accept and rely on different types of information acquisition systems. Next we examine how a subset of these alarms

TABLE 9.2
Alarms Based on the Performance, Process, and Purpose Dimensions of Trust That Could Affect the User's Perception of an Information Acquisition System

Alarm Type	Purpose, Process, and Performance	Example	Consequences
Accurate	Intended Predictable Useful	An alarm associated with a hazardous driving situation that the driver might not otherwise avoid.	May enhance driving performance and trust.
Unnecessary	Intended Predictable Nonuseful	An alarm associated with a situation judged hazardous by the designer, but not by the driver. The driver can understand what triggered the alarm.	May help drivers understand how the system works, but may annoy drivers if frequent.
Incomprehensible	Intended Useful Unpredictable	An alarm associated with a hazardous driving situation that the driver might not otherwise avoid, but is not recognized as such by the driver.	May diminish trust and compliance.
Unappreciated	Intended Unpredictable Nonuseful	An alarm that is associated with a situation judged hazardous by the designer, but is not understood or appreciated by the driver.	May diminish trust and compliance.
Fortuitous	Predictable Useful Unintended	An alarm that is inconsistent with the stated purpose of the system, but which helps the driver avoid hazards.	May enhance driving performance and trust, but could lead drivers to use the system differently than the designer intended.

TABLE 9.2 (CONTINUED)
Alarms Based on the Performance, Process, and Purpose Dimensions of Trust That Could Affect the User's Perception of an Information Acquisition System

Alarm Type	Purpose, Process, and Performance	Example	Consequences
Inadvertent/ nuisance	Predictable Unintended Nonuseful	An alarm triggered by events that do not pose a threat to the driver and were not intended by the designer, such as vehicles in the adjacent lane, roadside objects, and clutter on curves (Zador et al. 2000).	May help drivers understand how the system works, but may undermine system credibility.
Unforeseen	Useful Unintended Unpredictable	An alarm triggered in a manner that is inconsistent with the designer's intent and is not understandable to the driver, but is useful in avoiding a threat.	May enhance driving performance and trust, but could lead drivers to use the system differently than the designer intended.
False	Unintended Unpredictable Nonuseful	An alarm triggered by sensor noise or system malfunction that neither helps the driver understand the system or respond to threats.	May diminish trust and compliance.

Source: Lees and Lee 2007.

may manifest different behaviors, benefits, and consequences. Those considered are believed to have the greatest relevance for older and younger drivers.

Accurate—Are consistent with the driver's interpretation of the system's intent, the driver can understand why the alarm occurs, and the alarm helps the driver detect a threat. The source of the alarm and the consequences of not responding to the situation or alarm are immediate and transparent to any driver. These alarms may supplement a novice driver's lack of experience and exposure, mentoring novice drivers in how to acquire and use relevant information when assessing and managing roadway hazards. Over time, these alarms may accelerate the acquisition of top-down processing skills. For older drivers, these alarms may guide their attention to aspects of the environment they would have otherwise missed, and/or allow more time for information extraction, decision making, and response initiation.

Unnecessary—Reflect the designer's intent, but because of the peculiarities of the situation, drivers do not need the alarm but can understand what caused it to occur. These alarms may benefit both older and novice drivers through mechanisms similar to accurate alarms. With novice drivers, the big advantage is that these alarms may occur more frequently than accurate alarms, which are relatively rare. For novice drivers, these alarms may provide an outside source of expertise that can fine-tune hazard perception skills and top-down processing. This may be especially true for transparent systems that convey to the driver how the system is working. For example, a time headway display in which the driver can link the specific visual information to the current car following context may be more effective at increasing following distances than a system that provides an auditory alarm that occurs only in imminent crash situations.

Incomprehensible—Are intended by the designer and can help the driver avoid a collision, but the driver may not be able to understand why the alarm is delivered. These alarms promise the greatest benefit to drivers because the system can detect events the driver cannot; however, drivers may not see the value of these alarms and may fail to respond. Young drivers may be particularly vulnerable to this situation. Because they tend to be less sensitive to hazards than more experienced drivers, a warning system that alerts them to hazards has the potential to greatly enhance performance. At the same time, drivers may not detect the hazard even with the benefit of the alarm and so may disregard system information. Analysis of driver response to collision warning systems suggests that most drivers respond to alarms only after they have identified the reason for the alarm (Lee et al. 2002). Helping young drivers recognize the source of alarms may determine whether these alarms enhance driving safety.

Unappreciated—Are intended by the designer, but the driver may fail to understand why the alarm is delivered or the benefit of such an alarm. An information acquisition system developed to modify certain behaviors may be unappreciated by both younger and older drivers. This may be especially true if the driver cannot monitor what or how the system is working. For example, a system designed to randomly generate visual cues or auditory alarms to improve scanning behavior or gap acceptance may be unappreciated by older and younger drivers. Systems that alert based on the driver's behavior may be particularly prone to being unappreciated.

False—Are unintended by the designer, do not aid the driver in a given task, and appear random from the driver's perspective. These types of alarms can misdirect drivers by indicating an inappropriate response or by directing attention to unimportant information. Alarms that direct attention to the wrong location may be particularly detrimental for older drivers with diminished attentional flexibility or speed to recover.

Congruence between the driver's perception of the situation and the information provided by the system may also affect trust and acceptance of information acquisition systems. Drivers use multiple sources of information to determine risks and the actions required to avoid those risks, while automation detects and responds to only a subset of available cues and may fail to take into account the broader context of the situation (van der Hulst et al. 1999). Discrepancies between the system and the user can have detrimental effects on trust and use. When a person has little information about how a system works they tend to believe that it will outperform them, and reliance increases (Dzindolet et al. 2003). Reliance also increases if the user's knowledge of the system relates to general performance rather than the individual decisions made by the system. Exposure to errors the automation makes with simple tasks undermines the operator's trust and diminishes the operator's tendency to rely on automation with difficult tasks when the automation could enhance operator performance (Madhavan et al. 2006). These results suggest that the influence of in-vehicle technology depends on the complex interplay of the driving context and the experience and capacity of the driver.

To further complicate design, how the driver responds to technology that augments information acquisition depends on how the information is displayed. Information acquisition systems are likely to be susceptible to the same top-down and bottom-up deficiencies that contribute to crash involvement. Specifically, older adults may not be able to attend to system information and the roadway, and younger drivers may not be able to appreciate the risks that the system identifies. Effective design of such systems needs to consider the effects of modality in relation to the driving context, the type of information, and the user. Compared to other modalities, auditory alarms seize the user's attention and enable faster reaction times to time-critical situations (Hirst and Graham 1997). The driver can attend to auditory alarms without having to sacrifice visual attention or motor control (Graham 1999). At the same time, such alarms can be more annoying than other modalities and are public to both the driver and passengers, which could make the driver self-conscious. Visual displays may be particularly useful for presenting redundant information or continuous information, such as headway distance, or system status. However, such displays may fail to capture attention in time-critical situations or may be difficult for the driver to understand. Visual displays also have the potential to distract drivers and draw their attention away from the roadway. A visual display may also place greater demands on the driver, such as requiring drivers to interpret and extrapolate system information to the driving environment, which may be difficult for novice drivers. Haptic displays appear promising for systems that guide a specific response. Haptic displays, such as vibration in the seat, also promise to be less annoying for a given level of invalid alarms compared to auditory alarms (Lee et al. 2004). However, it may be difficult to provide spatial information (e.g., seat shakers offer little resolution beyond indicating right and left), and in certain cases, may lead the driver to initiate an inappropriate response (Suzuki and Jansson 2003). Design guidelines are beginning to catalog the range of display issues that need to be considered in the design of information acquisition systems (Campbell et al. 2007); however, the science of multisensory integration remains a relatively poorly understood aspect of human performance (Spence et al. 2001; Spence 2002; Lloyd et al. 2003; Ho et al. 2006; Campbell et al. 2007).

9.6 CONCLUSIONS

Failures of information acquisition, particularly those associated with inattention, account for a large proportion of crashes. Older and younger drivers crash at a relatively high rate and seem particularly vulnerable to information acquisition failures. The vulnerabilities of older and younger drivers differ, with older drivers having diminished bottom-up capacity associated with response speed, breadth of attention, and flexibility of attention. Younger drivers lack the experience needed to support effective top-down information acquisition. A variety of powerful sensor, algorithm, and interface technologies promise to enhance information acquisition. Although most research has focused on directing the drivers' response or guiding their attention to imminent collision events, several other strategies appear promising. The benefit that older and younger drivers might derive from these systems depends on the degree to which the systems compensate for the weaknesses of the bottom-up processing of older drivers and the deficiencies of top-down processing of younger drivers.

To a large extent, whether emerging technology provides the expected benefits depends on the factors governing driver reliance on the technology. Drivers' trust in the technology is one factor that guides reliance, and the dimensions of system purpose, process, and performance provide a useful extension to signal detection theory for anticipating how drivers might view alarms and system information. Designs for younger drivers need to consider that these drivers might not accept valid alarms because they do not have the experience needed to identify the hazard being warned. Designs for older drivers need to consider that invalid alarms could misdirect their attention and impair their response to actual threats. More generally, the types of alarms drivers consider invalid versus those that they consider acceptable depend on the capabilities of the system relative to the capability of the driver.

For both populations, an important challenge is to go beyond imminent crash warnings and consider how in-vehicle technology can enhance information acquisition at longer timescales, such as in tuning expectations and calibrating capabilities. In-vehicle technology is not simply a way to compensate driver limits. With older drivers, this concerns whether they will be as successful in adapting to their limits *and* to those of the technology as they are now in adapting to just their own limits. With younger drivers, this concerns whether they will be able to judge the combined limits of their own *and* those of the technology better than they are currently able to judge their own limits. In-vehicle technology changes the nature of the driving task and introduces new vulnerabilities and capacities that reflect the joint cognitive system composed of the driver and the technology (Woods and Dekker 2000). As a consequence, information acquisition in driving depends on a complex interaction of the driver, the in-vehicle technology, and the driving situation.

REFERENCES

Baldwin, C. (2002). Designing in-vehicle technologies for older drivers: Applications of sensory-cognitive interaction. *Theoretical Issues in Ergonomics Science, 3,* 307–329.

Ball, K., and Owsley, C. (1993). The useful field of view test: A new technique for evaluating age-related declines in visual function. *Journal of the American Optometric Association, 64,* 71–79.

Bliss, J. P. (2003). Investigation of alarm-related accidents and incidents in aviation. *International Journal of Aviation Psychology, 13,* 249–268.

Bossi, L. L., Ward, N. J., and Parkes, A. M. (1997). The effect of vision enhancement systems on driver peripheral visual performance. In Y. I. Noy (Ed.), *Ergonomics and Safety of Intelligent Driver Interfaces* (pp. 239–260). Mahwah, NJ: Lawrence Erlbaum Associates.

Caird, J. K., Horrey, W. J., and Edwards, C. J. (2001). Effects of conformal and nonconformal vision enhancement systems on older-driver performance. *Transportation Research Record, 1759,* 38–45.

Campbell, J. L., Richard, C. M., Brown, J. L., and McCallum, M. (2007). *Crash Warning System Interfaces: Human Factors Insights and Lessons Learned.* Seattle: Battelle.

Deery, H. A. (1999). Hazard and risk perception among young novice drivers. *Journal of Safety Research, 30,* 225–236.

Dzindolet, M. T., Peterson, S. A., Pomranky, R. A., Pierce, L. G., and Beck, H. P. (2003). The role of trust in automation reliance. *International Journal of Human-Computer Studies, 58,* 697–718.

Evans, L. (2004). *Traffic Safety.* Bloomfield Hills, MI: Science Serving Society.

Ferguson, S. A. (2003). Other high-risk factors for young drivers—how graduated licensing does, doesn't, or could address them. *Journal of Safety Research, 34,* 71–77.

Fisher, D. L., Pollatsek, A. P., and Pradhan, A. (2006). Can novice drivers be trained to scan for information that will reduce their likelihood of a crash? *Injury Prevention, 12,* 25–29.

Goode, K. T., Ball, K. K., Sloane, M., Roenker, D. L., Roth, D. L., Myers, R. S., and Owsley, C. (1998). Useful field of view and other neurocognitive indicators of crash risk in older adults. *Journal of Clinical Psychology in Medical Settings, 5,* 425–438.

Graham, R. (1999). Use of auditory icons as emergency warnings: Evaluation within a vehicle collision avoidance application. *Ergonomics, 42*(9), 1233–1248.

Grayson, G. B., and Sexton, B. F. (2002). *The development of hazard perception training* (TRL Report TRL558). Limited, Crowthorne, UK.

Green, D. M., and Swets, J. A. (1966). *Signal Detection Theory and Psychophysics.* New York: Wiley.

Hirst, S., and Graham, R. (1997). The format and presentation of collision warnings. In I. Y. Noy (Ed.), *Ergonomics and Safety of Intelligent Driver Interfaces* (pp. 203–219). Mahwah, NJ: Lawrence Erlbaum Associates.

Ho, C., Tan, H. Z., and Spence, C. (2006). The differential effect of vibrotactile and auditory cues on visual spatial attention. *Ergonomics, 49,* 724–738.

Horswill, M. S., Marrington, S. A., McCullough, C. M., Wood, J., Pachana, N. A., McWilliam, J., and Raikos, M. K. (in press). Older drivers' hazard perception ability. *Journal of Gerontology: Psychological Sciences.*

Horswill, M. S., and McKenna, F. P. (1999). The development, validation, and application of a video-based technique for measuring an everyday risk-taking behavior: Drivers' speed choice. *Journal of Applied Psychology, 84,* 977–985.

Horswill, M. S., and McKenna, F. P. (2004). Drivers' hazard perception ability: Situation awareness on the road. In S. Banbury and S. Tremblay (Eds.), *A Cognitive Approach to Situation Awareness* (pp. 155–175). Aldershot, UK: Ashgate.

Klauer, S. G., Dingus, T. A., Neale, V. L., Sudweeks, J. D., and Ramsey, D. J. (2006). *The Impact of Driver Inattention on Near-Crash/Crash Risk: An Analysis Using the 100-Car Naturalistic Driving Study Data.* Washington, DC: National Highway Traffic Safety Administration.

Lee, J. D. (2006a). Driving safety. In R. S. Nickerson (Ed.), *Review of Human Factors.* Santa Monica, CA: Human Factors and Ergonomics Society.

Lee, J. D. (2006b). Human factors and ergonomics in automation design. In G. Salvendy (Ed.), *Handbook of Human Factors and Ergonomics* (pp. 1570–1596). Hoboken, NJ: Wiley.

Lee, J. D. (2007). Technology and the teen driver. *Journal of Safety Research, 38,* 203–213.

Lee, J. D., Hoffman, J. D., and Hayes, E. (2004). Collision warning design to mitigate driver distraction. In *Proceedings of CHI 2004* ACM, New York, pp. 65–72.

Lee, J. D., McGehee, D. V., Brown, T. L., and Reyes, M. L. (2002). Collision warning timing, driver distraction, and driver response to imminent rear-end collisions in a high-fidelity driving simulator. *Human Factors, 44,* 314–334.

Lee, J., and Moray, N. (1992). Trust, control strategies and allocation of function in human-machine systems. *Ergonomics, 35,* 1243–1270.

Lee, J. D., and See, K. A. (2004). Trust in automation: Designing for appropriate reliance. *Human Factors, 46*(1), 50–80.

Lees, M. N., and Lee, J. D. (2007). The influence of distraction and driving context on driver response to imperfect collision warning systems. *Ergonomics, 50,* 1264–1286.

Lloyd, D. M., Merat, N., McGlone, F., and Spence, C. (2003). Crossmodal links between audition and touch in covert endogenous spatial attention. *Perception & Psychophysics, 65,* 901–924.

Lundberg, C., Johansson, K., Ball, K., Bjerre, B., Blomqvist, C., Braekhus, A., Brouwer, W. H., Bylsma, F. W., Carr, D. B., Englund, L., Friedland, R. P., Hakamies Blomqvist, L., Klemetz, G., Oneill, D., Odenheimer, G. L., Rizzo, M., Schelin, M., Seideman, M., Tallman, K., Viitanen, M., Waller, P. F., and Winblad, B. (1997). Dementia and driving: An attempt at consensus. *Alzheimer Disease & Associated Disorders, 11,* 28–37.

Lyman, S., Ferguson, S. A., Braver, E. R., and Williams, A. F. (2002). Older driver involvements in police reported crashes and fatal crashes: Trends and projections. *Injury Prevention, 8,* 116–120.

Madhavan, P., Dzindolet, M. T., and Lacson, F. C. (2006). Automation failures on tasks easily performed by operators undermine trust in automated aids. *Human Factors, 48,* 241–242.

Mayhew, D. R., Simpson, H. M., and Pak, A. (2003). Changes in collision rates among novice drivers during the first months of driving. *Accident Analysis & Prevention, 35,* 683–691.

McGehee, D. V., Raby, M., Carney, C. H., Reyes, M. L., and Lee, J. D. (2007). Extending parental mentoring using an event-triggered video intervention in rural teen drivers. *Journal of Safety Research, 38,* 15–227.

McGwin, G., and Brown, D. B. (1999). Characteristics of traffic crashes among young, middle-aged, and older drivers. *Accident Analysis and Prevention, 31,* 181–198.

McKenna, F. P., and Crick, J. (1994). *Hazard Perception in Drivers: A Methodology for Testing and Training*. Crowthorne, Berkshire, UK: Transportation Research Laboratory, Department of Transport.

McKenna, F. P., Horswill, M. S., and Alexander, J. L. (2006). Does anticipation training affect drivers' risk taking? *Journal of Experimental Psychology: Applied, 12,* 1–10.

McKnight, A. J., and McKnight, A. S. (2003). Young novice drivers: Careless or clueless? *Accident Analysis and Prevention, 35,* 921–925.

Michon, J. A. (1989). Explanatory pitfalls and rule-based driver models. *Accident Analysis and Prevention, 21,* 341–353.

Myers, R. S., Ball, K. K., Kalina, T. D., Roth, D. L., and Goode, K. T. (2000). Relation of useful field of view and other screening tests to on-road driving performance. *Perceptual and Motor Skills, 91,* 279–290.

Owsley, C., Ball, K., Sloane, M. E., Roenker, D. L., and Bruni, J. R. (1991). Visual/cognitive correlates of vehicle accidents in older drivers. *Psychology and Aging, 6,* 403–415.

Parasuraman, R., and Riley, V. (1997). Humans and automation: Use, misuse, disuse, abuse. *Human Factors, 39,* 230–253.

Pollatsek, A., Fisher, D. L., and Pradhan, A. (2006). Identifying and remedying failures of selective attention in younger drivers. *Current Directions in Psychological Science, 15,* 255–259.

Preusser, D. F., Williams, A. F., Ferguson, S. A., Ulmer, R. G., and Weinstein, H. B. (1998). Fatal crash risk for older drivers at intersections. *Accident Analysis and Prevention, 30,* 151–159.

Roenker, D. L., Cissell, G. M., Ball, K. K., Wadley, V. G., and Edwards, J. D. (2003). Speed-of-processing and driving simulator training result in improved driving performance. *Human Factors, 45,* 218–233.

Sanders, A. F. (1970). Some aspects of the selective process in the functional field of view. *Ergonomics, 13,* 101–117.

Sheridan, T. B. (1970). Big brother as driver: New demands and problems for the man at the wheel. *Human Factors, 12,* 95–101.

Spence, C. (2002). Multisensory attention and tactile information-processing. *Behavioural Brain Research, 135,* 57–64.

Spence, C., Kingstone, A., Shore, D. I., and Gazzaniga, M. S. (2001). Representation of visuo-tactile space in the split brain. *Psychological Science, 12,* 90–93.

Suzuki, K., and Jansson, H. (2003). An analysis of driver's steering behaviour during auditory or haptic warnings for the designing of lane departure warning system. *JSAE Review, 24,* 65–70.

Theeuwes, J. (1993). Visual selective attention: A theoretical analysis. *Acta Psychologica, 83,* 93–154.

Tijerina, L., Browning, N., Mangold, S. J., Madigan, E. F., and Pierowicz, J. A. (1995). *Examination of Reduced Visibility Crashes and Potential IVHS Countermeasures.* Washington, DC: U.S. Department of Transportation.

Uc, E. Y., Rizzo, M., Anderson, S. W., Shi, Q., and Dawson, J. D. (2004). Driver route-following and safety errors in early Alzheimer disease. *Neurology, 63,* 832–837.

Underwood, G., Phelps, N., Wright, C., van Loon, E., and Galpin, A. (2005). Eye fixation scanpaths of younger and older drivers in a hazard perception task. *Ophthalmic and Physiological Optics, 25,* 346–356.

van der Hulst, M., Meijman, T., and Rothengatter, T. (1999). Anticipation and the adaptive control of safety margins in driving. *Ergonomics, 42,* 336–345.

Wickens, C. D., and Hollands, J. G. (2000). *Engineering Psychology and Human Performance.* Upper Saddle River, NJ: Prentice Hall.

Williams, A. F. (2006). Young driver risk factors: Successful and unsuccessful approaches for dealing with them and an agenda for the future. *Injury Prevention, 12,* 4–8.

Woods, D. D., and Dekker, S. W. A. (2000). Anticipating the effects of technological change: A new era of dynamics for human factors. *Theoretical Issues in Ergonomics Science, 1,* 272–282.

Zador, P. L., Krawchuk, S. A., and Voas, R. B. (2000). *Automotive Collision Avoidance (ACAS) Program* (Final report). Washington, DC: National Highway Traffic Safety Administration.

Zhang, J., Fraser, J., Lindsay, J., Clarke, K., and Mao, Y. (1998). Age-specific patterns of factors related to fatal motor vehicle traffic crashes: focus on young and elderly drivers. *Public Health, 112,* 289–295.

10 Crossmodal Information Processing in Driving

Charles Spence and Cristy Ho

CONTENTS

Reflection 187
10.1 Driving and the Senses: An Introduction 188
10.2 Dividing Attention between Eye and Ear 188
10.2.1 Talking on the Mobile Phone 189
10.2.2 Talking to a Passenger 191
10.2.3 Listening to the Radio 192
10.3 Interim Summary 192
10.4 Multisensory Information Displays 192
10.5 Warning Signals: (Re)Capturing Driver Attention 193
10.5.1 Multisensory Warning Signals 194
10.6 Vigilance: Alerting the Sleepy Driver 195
10.7 Conclusions 196
References 196

REFLECTION

Recent discoveries in cognitive neuroscience are beginning to have an increasingly important impact on the design of everything from the food we eat to the cars we drive. The research in our laboratory in Oxford is based on trying to understand the rules used by the human brain to combine the various sensory cues that are available to it, and then seeing how those rules can be applied to help design "things" more effectively. In our chapter in this volume, we have tried to show some of the ways in which such cognitive neuroscience findings are increasingly coming to influence the design of multimodal (or multisensory) driver interfaces and warning signals. We are particularly excited by this research area just now because it seems to offer the very real opportunity to temper the rapidly emerging new technologies/interfaces that are coming online for car drivers with a better understanding of the constraints that researchers have identified with people's ability to monitor and process multiple sources of sensory information. We hope that in our chapter we have been able to convey some of our enthusiasm for, and belief in, the idea of neuroscience-inspired interface design. This approach to interface design holds the promise of being able to promote safer driving by enabling interface designers to develop multisensory displays that will more effectively/efficiently stimulate the senses of the driver.

10.1 DRIVING AND THE SENSES: AN INTRODUCTION

It has been widely reported in the literature that at least 90% of the information used by drivers is visual (e.g., Booher, 1978; Bryan, 1957). Even though robust empirical support for this particular claim is lacking, and indeed it is unclear how one could actually substantiate it (Sivak, 1996), most people would nevertheless intuitively agree that the majority of information used by drivers is visual. In fact, vision is typically the only sense whose acuity is explicitly tested in driving license tests (Booher, 1978; Sivak, 1996), and the very notion of a blind person driving is certainly more startling (although, surprisingly, not unheard of; see Nugent, 2006) than that of someone driving who is missing one of their other senses (for example, smell).

It is, however, important to note that several of the other human senses—principally audition, touch, proprioception/kinesthesis, and the vestibular system—also provide potentially important information to car drivers (see Figure 10.1). For example, drivers typically use the sound of their car horns to alert other road users of their presence (and often to their state of mind; see Graham, 1999). People can also use the sound made by their car engines when deciding whether to change gear, or when they may be pushing their car too hard (see Matthews and Cousins, 1980; McLane and Wierwille, 1975; though see also Booher, 1978). Meanwhile, the rumble strips on the approach to many roundabouts highlight the need for a driver to slow down by means of the combined audiotactile stimulation that they provide. Similar cues can also alert drivers to the fact that they have veered out of their lane (cf. Suzuki and Jansson, 2003). Finally, proprioceptive, kinesthetic, and vestibular cues (that generate information concerning the position and movement of the body in space) provide drivers with important cues concerning the acceleration/deceleration of their vehicle, and are also important for steering (see Gibson and Crooks, 1938; Kemeny and Panerai, 2003). In fact, it is the very absence of these cues (and the presence of realistic visual cues suggesting motion) that triggers the simulator sickness experienced by a proportion of the drivers in driving simulators.

10.2 DIVIDING ATTENTION BETWEEN EYE AND EAR

Given that the majority of the information used by drivers is visual, it has often been assumed that combining driving with an auditory task, such as listening to the car radio or having a conversation on the mobile (cellular) phone (or for that matter, with a passenger), should cause little, if any, decrement in driving performance. This view has gained support from the influential account of human information processing put forward by Christopher Wickens more than 25 years ago (e.g., see Wickens, 1980, 1992, 2002). According to Wickens' multiple resources theory (MRT), people have relatively independent pools of attentional resources for the processing of visual and auditory stimuli. Thus, according to Wickens, while giving a driver a visual display to inspect might well have a detrimental effect on their driving performance (due to competition for the same limited pool of attentional resources), any secondary task

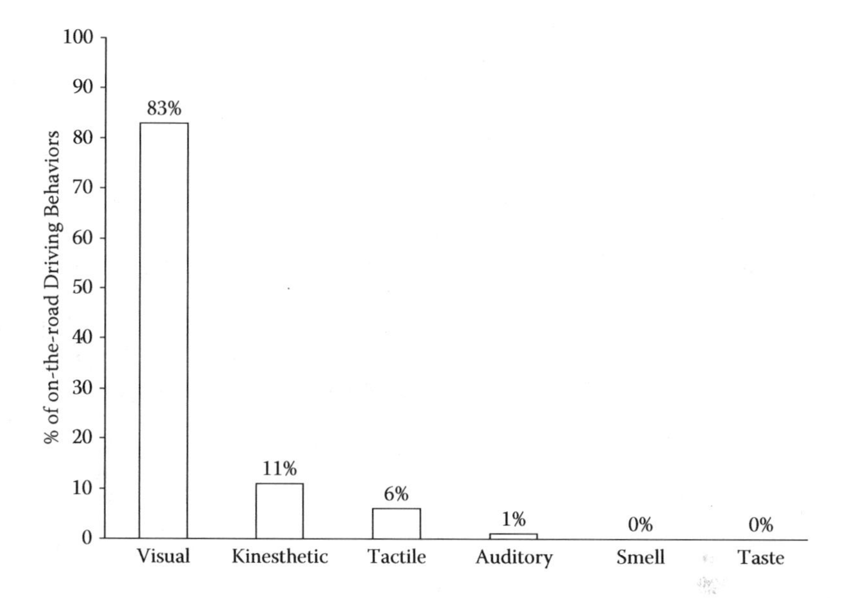

FIGURE 10.1 Sivak (1996) evaluated 89 of the most critical on-the-road behaviors (out of the 1500 identified by McKnight and Adams', 1970, task analysis of drivers) in terms of which sensory modality was required for their occurrence. Twenty-seven of these behaviors depended on input from more than one sense. (Note that the kinesthetic category presumably includes behaviors that rely on proprioceptive and/or vestibular inputs as well.)

that taps into the relatively underutilized auditory (or for that matter tactile) modality should result in relatively little cost, due to the putative separability of the attentional resources concerned (see also Wickens and Liu, 1988; see Spence and Driver, 1997, for a critical review of multiple resources theory).

10.2.1 Talking on the Mobile Phone

The majority of the evidence over the last 50 years, however, has shown that the performance of many auditory tasks actually has a noticeably detrimental effect on at least some aspects of driving performance. For example, the 24 male drivers in Brown et al.'s (1969) early study had to decide whether or not the car they were driving could be maneuvered through a range of openings that had been laid out on a test track (the gaps ranged from 3 inches narrower than the car they were driving to gaps that were 9 inches wider). At the same time, the drivers also had to verify the accuracy of a list of sentences read out over the car telephone. The drivers heard statements such as "A follows B" and then had to decide whether the letter pair that followed, such as "BA," was consistent with the preceding statement or not. Brown and his colleagues found that while performing this particular auditory task had relatively little effect on the more automatized aspects of driving (such as steering), it did have a markedly detrimental (or interfering) effect on their drivers' ability to decide whether or not they could steer through the narrow gaps on the test track. Brown et al. suggested that it was the switching of a driver's attention between the auditory and visual streams that impaired their perceptual and decision-making abilities (cf. Kahneman et al., 1973; Spence et al., 2001).

More recently, Horswill and McKenna (1999) conducted a laboratory-based study in which 121 drivers watched a series of car driving videos designed to test their risk-taking behaviors in close following and gap acceptance tasks while simultaneously performing an auditory monitoring task. In the close following task, the participants watched video footage of a car gradually approaching the back of another car on a motorway. They had to indicate when they reached the distance that they would normally travel behind another car, and to respond again when they felt uncomfortably close to the car in front. Meanwhile, in the gap acceptance task, the participants viewed video footage of a road junction from the point of view of a driver waiting to turn left. They had to indicate when they would have pulled out into the stream of traffic. The auditory task consisted of listening to a list of letters (read out at the rate of 1 per second) and responding whenever the target letter K appeared. Participants' performance on both of the simulated risk-taking tasks was impaired significantly whenever they monitored the auditory stream at the same time.

Horswill and McKenna's (1999) study has been criticized because of the artificial nature of the laboratory-based tasks used and because only relatively inexperienced undergraduate drivers were tested (who might be expected to make riskier decisions than more experienced drivers). Nevertheless, many other subsequent studies have now come to essentially the same conclusion, namely that people (i.e., car drivers) really do find it difficult to divide their attention simultaneously between their eyes and ears (e.g., see Horrey and Wickens, 2006; Spence and Read, 2003; Strayer and Johnston, 2001).

In fact, according to the results of a recent driving simulator study, using a mobile phone while driving may actually be just as dangerous as driving while under the influence of alcohol (Strayer et al., 2006). To put this result into some kind of perspective, an epidemiological study by Redelmeier and Tibshirani (1997) revealed that people who use a mobile phone while driving are actually four times more likely to be involved in an accident. Interestingly, Redelmeier and Tibshirani's analysis of the mobile phone records of nearly 700 drivers who had been involved in a motor vehicle accident over a 14-month period in the Toronto metropolitan area suggested that hands-free phones are no less dangerous than handheld phones (see also Horrey and Wickens, 2006, for similar findings). This result suggests that the difficulties experienced by drivers stem from the cognitive difficulties associated with trying to divide their attention between eye and ear, rather than manual limitations associated with holding, or manipulating, a phone. These findings are all the more troublesome given that approximately 8% of drivers use a cell phone at any given time during daylight hours (see Glassbrenner, 2005).

Well-controlled experimental studies have now started to break down the costs associated with using a mobile phone while driving into a number of relatively separable cognitive (attentional) components. For example, research by Spence and Read (2003) with the Leeds Advanced Driving Simulator has shown that part of the problem faced by drivers is that they find it difficult to divide their attention *spatially* between different locations (i.e., when trying to listen to the side while simultaneously attending visually to the front), something that has often been shown to cause people problems in laboratory studies of attention (see Driver and Spence, 2004, for a recent review). Meanwhile, Strayer and Johnston (2001) demonstrated that participants' simulated driving performance in the laboratory was significantly impaired when the experimenter engaged them in a cognitively demanding discussion of current affairs. In another study, Strayer et al. (2003) found that car drivers were both more likely to miss critical driving signals (such as the changing of the traffic lights or the car ahead braking), and to respond more slowly to the targets that they did detect, as well as having an increased risk of front-to-rear-end collisions, when talking on a mobile phone. Finally, Levy et al. (2006) recently demonstrated that if a driver had to make a speeded detection response to an imperative auditory or tactile signal then their braking responses were unavoidably slowed for a short time thereafter. These results highlight the central competition that goes on in the brain for access to the limited capacity central response selection bottleneck (known as the psychological refractory period; though see also Schumacher et al., 2001). A recent meta-analysis of 23 studies by Horrey and Wickens (2006) revealed that the costs associated with using a mobile phone while driving are primarily seen in reaction time tasks, with far smaller costs being associated with lane keeping or tracking tasks (just as reported in Brown et al.'s, 1969).

10.2.2 Talking to a Passenger

One might wonder whether using a mobile phone while driving is actually any more dangerous than talking to a passenger. The evidence now shows that talking to a passenger really does result in an increased accident risk. The most impressive data on this comes from a study by McEvoy et al. (2007b). They determined that the risk of a person having an accident while driving increased as a function of the number of passengers in the car. In particular, carrying two or more passengers increased the risk of a driver having an accident twofold when compared to when traveling alone (see also Briem and Hedman, 1995; McEvoy et al., 2007b; cf. Haigney and Westerman, 2001). However, while the evidence now shows that conversing with a passenger can be a risky business, it is worth noting that there are also some potentially important differences between the kinds of conversation that drivers have with their passengers versus with someone at the end of a mobile phone. It has, for example, been suggested that people may sometimes engage in more emotionally and/or intellectually demanding conversations when speaking to someone over the mobile phone than when conversing with their passengers. Furthermore, car passengers have been shown to pace (or regulate) their conversation with the driver in line with the current driving conditions. Thus, they tend to stop speaking when the driving conditions become more taxing (such as when driving on congested urban roads). No such conversational pacing is seen when drivers talk with someone on their mobile phones (see Crundall et al., 2005). These differences may therefore help to explain why talking on a mobile phone is more demanding (and hence more dangerous) than talking to a passenger.

10.2.3 Listening to the Radio

Finally, given that two-thirds of drivers listen to some form of in-car entertainment while at the wheel (Dibben and Williamson, 2007), it is worth considering what effect listening to the car radio or other in-car entertainment system might have on driving performance. Here, the evidence is rather more mixed (e.g., Brown, 1965; North and Hargreaves, 1999; Strayer and Johnston, 2001). For it seems that while listening to loud music can impair people's performance on certain driving-related tasks, such as those involving the detection of peripheral targets (Beh and Hirst, 1999), and that fast tempo music increases the number of virtual traffic violations people commit when playing a driving video game (Brodsky, 2002), listening to music can also facilitate driver responses under certain conditions due to the increased alertness it causes (e.g., Brown, 1965). Indeed, in-car entertainment may have a particularly important role to play in terms of relieving driver boredom, and hence reducing the risk of a driver falling asleep at the wheel (see Dibben and Williamson, 2007). To conclude, listening to an in-car entertainment system is typically less taxing (and hence less dangerous) than engaging in a conversation with a passenger or with someone on the mobile phone. Listening to music can also influence a driver's performance indirectly, in either a positive or negative manner, as a result of the effect it has on their arousal/mood (Ho and Spence, 2008).

10.3 INTERIM SUMMARY

The research published to date shows that drivers find it difficult to monitor two independent sources of information at the same time, such as when using the mobile phone while driving, when talking to passengers, or when listening to the car radio. All of these activities can result in a significant impairment on driving performance as shown by both laboratory- and simulator-based studies, as well as by epidemiological research. Taken together, the evidence therefore unequivocally shows that people really do find it difficult to divide their attention between different sources of information at the same time, even when that information is presented to separate sensory modalities. These findings argue against Wickens' (1980, 1992, 2002) MRT of human information processing. In fact, we now have a much better, and more detailed, understanding of the many different cognitive limitations that constrain an interface operator's ability to monitor (and to respond to) multiple sources of information at the same time. However, one important issue for the future will be to try and develop some kind of integrated theoretical framework for thinking about these various effects (indeed, this was one of the major strengths of MRT when it was originally formulated).

10.4 MULTISENSORY INFORMATION DISPLAYS

Given the commonly held belief that drivers' visual systems are overloaded (Rumar, 1990; Sivak, 1996; Wierwille et al., 1988), there is a growing interest in the use of auditory, tactile, and multisensory information displays, such as to provide route finding information (e.g., Liu, 2001; Van Erp and Van Veen, 2004). For example,

Liu presented advanced traveler information via an auditory, visual, or audiovisual (i.e., multimodal) display. The results showed that both younger (mean age of 22 years) and older drivers (mean age of 68 years) were able to drive more safely and experienced a lower subjective workload when using the auditory and multisensory display than when using a visual display instead. More recently, Van Erp and Van Veen (2004) investigated the relative merits of presenting navigational information (concerning whether a driver should turn left or right, and after what distance, e.g., 250, 150, or 50 m) using directional vibrotactile cues presented to the driver's thigh via eight tactors mounted in the car seat. The results of this driving simulator study showed that both the unimodal tactile and the combined visuotactile displays were more effective than a unimodal visual display. The participants responded significantly faster to the multisensory navigation messages than to the unimodal visual messages, with intermediate performance being reported in the unimodal tactile display conditions.

It is important to note that while the available research now shows that auditory, tactile, and multisensory information displays can sometimes be used to transmit information to car drivers more effectively than by the use of unimodal visual displays, human factors researchers must nevertheless still take care to avoid the potential dangers associated with drivers being distracted by the introduction of such displays (see Wiese and Lee, 2007). That is, by forcing the driver to monitor more than just the visual modality (i.e., by forcing them to divide their attention between vision, touch, and possibly also hearing), the driver may have less attentional resources to focus specifically on what they are seeing (Spence and Driver, 1997; Spence et al., 2001). It should, however, be noted in relation to this (quite legitimate) concern that Van Erp and Van Veen's (2004) participants actually reported lower subjective mental effort when using the tactile or visuotactile display than when trying to navigate using just the unimodal visual display (see also Liu, 2001).

10.5 WARNING SIGNALS: (RE)CAPTURING DRIVER ATTENTION

Many human factors researchers have now started to investigate how best to capture the attention of drivers who may be distracted by the proliferation of in-vehicle technologies that are now available, such as those described in the previous section (see also Ashley, 2001). This line of research is particularly important given that driver distraction (from some form of secondary task) is thought to contribute to up to one third of all serious car crashes (e.g., Klauer et al., 2006; McEvoy et al., 2007a). Researchers have therefore become increasingly interested in the potential benefits associated with the use of auditory, tactile, and/or multisensory warning signals to recapture a distracted driver's attention (e.g., Ferris et al., 2006; Graham, 1999; Ho and Spence, 2005a; Ho, Reed, et al., 2006; Ho, Tan, et al. 2006; Lee et al., 2006; Spence and Ho, in press; Suzuki and Jansson, 2003; Van Winsum et al., 1999).

Much of the research in this area has focused on the question of how to design warning signals that are both localizable and meaningful, while at the same time have the right level of perceived urgency (and, what's more, are presented at the right time; e.g., Edworthy and Hellier, 2006; Ho and Spence, 2005a, 2006; Spence and Ho, in press). The importance of developing novel multisensory warning signals

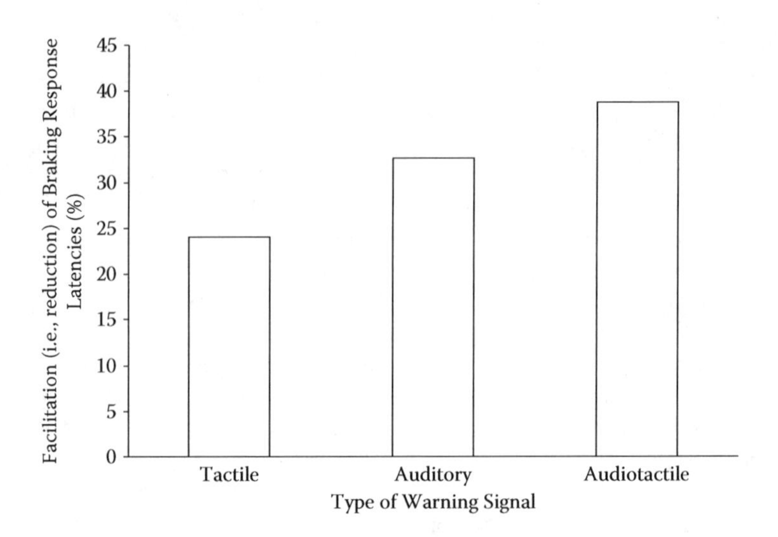

FIGURE 10.2 Graph highlighting the facilitation of braking latencies reported in simulator-based driving tasks associated with the use of tactile, auditory, and combined audiotactile (i.e., multisensory) warning signals when compared to when no warning signal was presented, as is common in many cars today (see Ho et al., 2006a, submitted, for details).

capable of getting a distracted driver's eyes back onto the road ahead (or wherever else they may be needed) was highlighted recently by the results of a 100-car naturalistic driving study in which it was shown that 78% of the crashes and 65% of all near-misses observed during the yearlong study involved the driver looking away from the forward roadway in the moments just prior to the incident (see Klauer et al., 2006). Promising recent developments in this area have come from the work of Ho, Reed et al. (2006) who have shown that presenting car drivers with a tactile cue to their stomach (to warn them about the braking of the car on the road ahead) led to a 400-ms reduction in driver braking response latencies in their driving simulator study. Spatial auditory cues (such as the sound of a car horn presented from the front) have been shown to reduce braking latencies by a further 100 ms or so (see Figure 10.2). Suetomi and Kido's (1997) recent suggestion that any device (or intervention) that could reduce driver braking latencies by 500 ms would lead to a 60% reduction in the incidence of front-to-rear-end collisions (currently the most common type of accident on the road, accounting for about 25% of all road traffic accidents) highlights the potential importance of Ho, Reed et al.'s findings in terms of improving road safety (and reducing car accidents or crashes). (Though note that the warning signals in Ho, Reed, et al.'s study were presented far more frequently than they would be in any real on-road implementation of this kind of system.)

10.5.1 Multisensory Warning Signals

Multisensory warning signals may be especially effective in capturing an interface operator's attention (Ho et al., 2007; Santangelo et al., 2007; Santangelo and Spence, 2007). For example, Ho et al. (2007) recently reported a driving simulator study in which multisensory (in this case audiotactile) warning signals were shown to be even more effective than unimodal warning signals in capturing a driver's attention (see Figure 10.2). Furthermore, laboratory-based research from Santangelo et al. (2007) has also shown that multisensory warning signals (or cues) are capable of capturing a person's attention no matter what else they happen to be doing at the same time. These results therefore suggest that stimulating multiple sensory channels may represent a particularly effective compensatory strategy to help overcome sensory overload and driver distraction (though see also Lee et al., 2006).

Laurienti et al. (2006) have also suggested that stimulating multiple sensory channels may represent a particularly effective compensatory strategy to help overcome the sensory decline (and consequent slowing of responses) experienced by the growing population of elderly drivers (see also Liu, 2001). In fact, it has been estimated that there will be more than a billion people over the age of 60 by 2020 (U.S. Senate Special Committee on Aging, 1985–1986). This is a particularly troublesome statistic given the rapidly increasing rate of crashes per mile driven that drivers are involved in once they reach the age of 55 years.

Multisensory warning signals have the advantage of combining the best features of each sensory modality. So, for example, while the effectiveness of tactile warning signals will not be compromised if the driver listens to loud background music (in contrast to auditory cues), auditory stimuli provide a much more effective means of presenting iconic (i.e., inherently meaningful) warning signals (such as the sound of a car horn; Graham, 1999; Ho and Spence, 2005a). Note also that tactile cues/information can be presented to car drivers not only via the seat or seat belt, but also potentially via the brake/accelerator pedals or even via the steering wheel itself for the case of a lane-departure warning system (Suzuki and Jansson, 2003). Interestingly, however, it seems that audiotactile warning signals only seem to capture a person's attention when they are presented from the same direction (i.e., both on the left or both on the right) but not when the tactile signal was presented to the center of the participant's stomach while the auditory signal was presented from the side (Ho et al., submitted; see also Selcon et al., 1995). With regard to the question of the timing of multisensory warning signals, it has been suggested that presenting the component unisensory signals slightly asynchronously (i.e., with the visual signal leading the auditory signal for the case of audiovisual warning signals) might actually be more effective than presenting them both simultaneously (see Chan and Chan, 2006; Spence and Driver, 1999). The reason for this being that slightly asynchronous warning signals might nevertheless end up in the brain centers thought to control overt attentional orienting (such as the superior colliculus) simultaneously, and hence prove more effective in triggering an orienting response (see Spence and Driver, 2004).

10.6 VIGILANCE: ALERTING THE SLEEPY DRIVER

Although the use of olfactory cues to warn a driver of impending danger is effectively precluded by the relatively slow transduction latencies for olfactory stimuli across the nasal epithelium (see Spence and Squire, 2003), it has nevertheless been suggested that olfactory cues could be used to provide a less unpleasant means of enhancing the alertness of sleepy drivers than, say, a loud sound. While the most responsible course of action would obviously be for a weary driver to take a break, for those who do decide to continue on their journeys, the presentation of alerting scents may help to enhance/maintain their alertness. In fact, the occasional presentation of olfactory cues can improve people's ability to detect occasionally presented targets in laboratory-based vigilance tasks by as much as 15% (see Gould and Martin, 2001; Ho and Spence, 2005b; Warm et al., 1991), and preliminary evidence now shows that olfactory cues (such as the smell of peppermint) can also facilitate driving performance in a simulator setting (e.g., Grayhem et al., 2005; Martin and Cooper, 2007; Raudenbush, 2005). Results such as these raise the possibility that the presentation of alerting odors, such as peppermint or citrus, may help to keep drowsy drivers vigilant (Baron and Kalsher, 1998; Bounds, 1996, *Rinspeed "Senso,"* n.d.; Schuler and Raudenbush, 2005). It should, however, be noted that odors are only minimally effective once people are asleep (e.g., Carskadon and Herz, 2004).

10.7 CONCLUSIONS

Although vision may be the dominant sense for driving, several of the other major human senses—including proprioception/kinesthesia/the vestibular system, touch, and audition—also provide potentially useful information to car drivers. What's more, the rapid development of modern technologies means that there is now a greater scope than ever before to present information via the auditory or tactile modalities, to recapture a distracted driver's attention using multisensory warning signals, and even possibly to use scent to enhance alertness/vigilance. It now seems inevitable that these multimodal (or multisensory) interfaces and warning signals for car drivers will become even more sophisticated (and prevalent) over the next few years (e.g., Ferris et al., 2006; Spence and Ho, in press). Finally, it is important to note that given that older drivers constitute the most rapidly growing section of the driving population (see Waller, 1991), human factors researchers will also need to focus their research efforts on designing multisensory interfaces that meet the sensory and cognitive needs, capabilities, and, of course, limitations of this population (see Liu, 2001).

REFERENCES

Ashley, S. (2001). Driving the info highway. *Scientific American, 285*(4), 44–50.

Baron, R. A., and Kalsher, M. J. (1998). Effects of a pleasant ambient fragrance on simulated driving performance: The sweet smell of…safety? *Environment and Behavior, 30,* 535–552.

Beh, H. C., and Hirst, R. (1999). Performance on driving-related tasks during music. *Ergonomics, 42,* 1087–1098.

Booher, H. R. (1978). Effects of visual and auditory impairment in driving performance. *Human Factors, 20,* 307–320.

Bounds, W. (1996, May 6). Sounds and scents to jolt drowsy drivers. *Wall Street Journal*, pp. B1, B5.

Briem, V., and Hedman, L. R. (1995). Behavioural effects of mobile telephone use during simulated driving. *Ergonomics, 38*, 2536–2562.

Brodsky, W. (2002). The effects of music tempo on simulated driving performance and vehicular control. *Transportation Research Part F, 4,* 219–241.

Brown, I. D. (1965). Effect of a car radio on driving in traffic. *Ergonomics, 8,* 475–479.

Brown, I. D., Tickner, A. H., and Simmonds, D. C. V. (1969). Interference between concurrent tasks of driving and telephoning. *Journal of Applied Psychology, 53,* 419–424.

Bryan, W. E. (1957). Research in vision and traffic safety. *Journal of the American Optometric Association, 29,* 169–172.

Carskadon, M. A., and Herz, R. S. (2004). Minimal olfactory perception during sleep: Why odor alarms will not work for humans. *Sleep, 27,* 402–405.

Chan, A. H. S., and Chan, K. W. L. (2006). Synchronous and asynchronous presentations of auditory and visual signals: Implications for control console design. *Applied Ergonomics, 37,* 131–140.

Crundall, D., Bains, M., Chapman, P., and Underwood, G. (2005). Regulating conversation during driving: A problem for mobile phones? *Transportation Research Part F: Traffic Psychology and Behaviour, 8,* 197–211.

Dibben, N., and Williamson, V. J. (2007). An exploratory survey of in-vehicle music listening. *Psychology of Music, 35,* 571–589.

Driver, J., and Spence, C. (2004). Crossmodal spatial attention: Evidence from human performance. In C. Spence and J. Driver (Eds.), *Crossmodal space and crossmodal attention* (pp. 179–220). Oxford, UK: Oxford University Press.

Edworthy, J., and Hellier, E. (2006). Complex nonverbal auditory signals and speech warnings. In M. S. Wogalter (Ed.), *Handbook of warnings* (pp. 199–220). Mahwah, NJ: Lawrence Erlbaum.

Ferris, T., Penfold, R., Hameed, S., and Sarter, N. (2006). The implications of crossmodal links in attention for the design of multimodal interfaces: A driving simulation study. *Proceedings of the Human Factors and Ergonomics Society 50th Annual Meeting,* 406–409.

Gibson, J. J., and Crooks, L. E. (1938). A theoretical field-analysis of automobile-driving. *American Journal of Psychology, 51,* 453–471.

Glassbrenner, D. (2005). Driver cell phone use in 2004—Overall results. In *Traffic safety facts: Research note* (DOT HS 809 847). Washington, DC: U.S. Department of Transportation.

Gould, A., and Martin, G. N. (2001). “A good odour to breathe?” The effect of pleasant ambient odour on human visual vigilance. *Applied Cognitive Psychology, 15,* 225–232.

Graham, R. (1999). Use of auditory icons as emergency warnings: Evaluation within a vehicle collision avoidance application. *Ergonomics, 42,* 1233–1248.

Grayhem, R., Esgro, W., Sears, T., and Raudenbush, B. (2005). *Effects of odor administration on driving performance, safety, alertness, and fatigue.* Poster presented at the 27th Annual Meeting of the Association for Chemoreception Sciences, Sarasota, FL.

Haigney, D., and Westerman, S. J. (2001). Mobile (cellular) phone use and driving: A critical review of research methodology. *Ergonomics, 44,* 132–143.

Ho, C., Reed, N. J., and Spence, C. (2006). Assessing the effectiveness of “intuitive” vibrotactile warning signals in preventing front-to-rear-end collisions in a driving simulator. *Accident Analysis and Prevention, 38,* 989–996.

Ho, C., Reed, N., and Spence, C. (2007). Multisensory in-car warning signals for collision avoidance. *Human Factors, 49,* 1107–1114.

Ho, C., Santangelo, V., and Spence,C. (2008, submitted). Multisensory warning signals: When spatial location matters. *Experimental Brain Research.*

Ho, C., and Spence, C. (2005a). Assessing the effectiveness of various auditory cues in capturing a driver’s visual attention. *Journal of Experimental Psychology: Applied, 11,* 157–174.

Ho, C., and Spence, C. (2005b). Olfactory facilitation of dual-task performance. *Neuroscience Letters, 389,* 35–40.

Ho, C., and Spence, C. (2006). Verbal interface design: Do verbal directional cues automatically orient visual spatial attention? *Computers in Human Behavior, 22,* 733–748.

Ho, C., and Spence, C. (2008). *The multisensory driver: Amplications for ergonomic car interface design.* Aldershot: Ashgate Publishing.

Ho, C., Tan, H. Z., and Spence, C. (2006). The differential effect of vibrotactile and auditory cues on visual spatial attention. *Ergonomics, 49,* 724–738.

Horrey, W. J., and Wickens, C. D. (2006). Examining the impact of cell phone conversations on driving using meta-analytic techniques. *Human Factors, 48,* 196–205.

Horswill, M. S., and McKenna, F. P. (1999). The effect of interference on dynamic risk-taking judgments. *British Journal of Psychology, 90,* 189–199.

Kahneman, D., Ben-Ishai, R., and Lotan, M. (1973). Relation of a test of attention to road accidents. *Journal of Applied Psychology, 58,* 113–115.

Kemeny, A., and Panerai, F. (2003). Evaluating perception in driving simulation experiments. *Trends in Cognitive Sciences, 7,* 31–37.

Klauer, S. G., Dingus, T. A., Neale, V. L., Sudweeks, J. D., and Ramsey, D. J. (2006). *The impact of driver inattention on near-crash/crash risk: An analysis using the 100-car naturalistic driving study data* (Technical Report No. DOT HS 810 594). Washington, DC: National Highway Traffic Safety Administration.

Laurienti, P. J., Burdette, J. H., Maldjian, J. A., and Wallace, M. T. (2006). Enhanced multisensory integration in older adults. *Neurobiology of Aging, 27,* 1155–1163.

Lee, J. D., McGehee, D. V., Brown, T. L., and Marshall, D. (2006). Effects of adaptive cruise control and alert modality on driver performance. *Transportation Research Record, 1980,* 49–56.

Levy, J., Pashler, H., and Boer, E. (2006). Central interference in driving: Is there any stopping the psychological refractory period? *Psychological Science, 17,* 228–235.

Liu, Y.-C. (2001). Comparative study of the effects of auditory, visual and multimodal displays on drivers' performance in advanced traveller information systems. *Ergonomics, 44,* 425–442.

Martin, G. N., and Cooper, J. A. (2007, March 21–23). *Odour effects on simulated driving performance: Adding zest to difficult journeys.* Poster presented at the British Psychology Society Annual Conference, York, UK.

Matthews, M. L., and Cousins, L. R. (1980). The influence of vehicle type on the estimation of velocity while driving. *Ergonomics, 23,* 1151–1160.

McEvoy, S. P., Stevenson, M. R., and Woodward, M. (2007a). The prevalence of, and factors associated with, serious crashes involving a distracting activity. *Accident Analysis and Prevention, 39,* 475–482.

McEvoy, S. P., Stevenson, M. R., and Woodward, M. (2007b). The contribution of passengers versus mobile phone use to motor vehicle crashes resulting in hospital attendance by the driver. *Accident Analysis and Prevention, 39,* 1170–1176.

McKnight, A. J., and Adams, B. B. (1970). *Driver education task analysis. Volume 1: Task description* (DOT HS 800 367). Washington, DC: Department of Transport.

McLane, R. C., and Wierwille, W. W. (1975). The influence of motion and audio cues on driver performance in an automobile simulator. *Human Factors, 17,* 488–501.

North, A. C., and Hargreaves, D. J. (1999). Music and driving game performance. *Scandinavian Journal of Psychology, 40,* 285–292.

Nugent, H. (2006, September 5). Man with no eyes was driving car. *Times Online.* Retrieved May 15, 2007, from http://www.timesonline.co.uk/tol/news/uk/article628374.ece.

Raudenbush, B. (2005). *Effects of odor administration on driving performance, safety, alertness, and fatigue.* Paper presented at the 27th Annual Meeting of the Association for Chemoreception Sciences, Sarasota, FL.

Redelmeier, D. A., and Tibshirani, R. J. (1997). Association between cellular-telephone calls and motor vehicle collisions. *New England Journal of Medicine, 336,* 453–458.

Rinspeed "Senso"—The car that senses the driver. (n.d.). Retrieved May 30, 2005, from http://www.rinspeed.com/pages/cars/senso/pre-senso.htm.

Rumar, K. (1990). The basic driver error: Late detection. *Ergonomics, 33,* 1281–1290.

Santangelo, V., Belardinelli, M. O., and Spence, C. (2007). The suppression of reflexive visual and auditory orienting when attention is otherwise engaged. *Journal of Experimental Psychology: Human Perception and Performance, 33,* 137–148.

Santangelo, V., and Spence, C. (2007). Multisensory cues capture spatial attention regardless of perceptual load. *Journal of Experimental Psychology: Human Perception and Performance, 33,* 1311–1321.

Schuler, A., and Raudenbush, B. (2005). *Effects of intermittent peppermint odor administration on alertness, mood, mobility, and sleep patterns.* Poster presented at the 27th Annual Meeting of the Association for Chemoreception Sciences, Sarasota, FL.

Schumacher, E. H., Seymour, T. L., Glass, J. M., Fencsik, D. E., Lauber, E. J., Kieras, D. E., et al. (2001). Virtually perfect time sharing in dual-task performance: Uncorking the central cognitive bottleneck. *Psychological Science, 12,* 101–108.

Selcon, S. J., Taylor, R. M., and McKenna, F. P. (1995). Integrating multiple information sources: Using redundancy in the design of warnings. *Ergonomics, 38,* 2362–2370.

Sivak, M. (1996). The information that drivers use: Is it indeed 90% visual? *Perception, 25,* 1081–1089.

Spence, C., and Driver, J. (1997). Cross-modal links in attention between audition, vision, and touch: Implications for interface design. *International Journal of Cognitive Ergonomics, 1,* 351–373.

Spence, C., and Driver, J. (1999). A new approach to the design of multimodal warning signals. In D. Harris (Ed.), *Engineering psychology and cognitive ergonomics, Vol. 4: Job design, product design and human–computer interaction* (pp. 455–461). Hampshire, UK: Ashgate.

Spence, C., and Driver, J. (Eds.). (2004). *Crossmodal space and crossmodal attention.* Oxford, UK: Oxford University Press.

Spence, C., and Ho, C. (in press). Multisensory warning signals for event perception and safe driving. *Theoretical Issues in Ergonomics Science.*

Spence, C., Nicholls, M. E. R., and Driver, J. (2001). The cost of expecting events in the wrong sensory modality. *Perception and Psychophysics, 63,* 330–336.

Spence, C., and Read, L. (2003). Speech shadowing while driving: On the difficulty of splitting attention between eye and ear. *Psychological Science, 14,* 251–256.

Spence, C., and Squire, S. B. (2003). Multisensory integration: Maintaining the perception of synchrony. *Current Biology, 13,* R519–R521.

Strayer, D. L., Drews, F. A., and Crouch, D. J. (2006). A comparison of the cell phone driver and the drunk driver. *Human Factors, 48,* 381–391.

Strayer, D. L., Drews, F. A., and Johnston, W. A. (2003). Cell phone-induced failures of visual attention during simulated driving. *Journal of Experimental Psychology: Applied, 9,* 23–32.

Strayer, D. L., and Johnston, W. A. (2001). Driven to distraction: Dual-task studies of simulated driving and conversing on a cellular telephone. *Psychological Science, 12,* 462–466.

Suetomi, T., and Kido, K. (1997). *Driver behavior under a collision warning system–A driving simulator study.* Society of Automotive Engineers Technical Publication, 970279, *1242,* 75–81.

Suzuki, K., and Jansson, H. (2003). An analysis of driver's steering behaviour during auditory or haptic warnings for the designing of lane departure warning system. *JSAE Review, 24,* 65–70.

Van Erp, J. B. F., and Van Veen, H. A. H. C. (2004). Vibrotactile in-vehicle navigation system. *Transportation Research Part F, 7,* 247–256.

Van Winsum, W., Martens, M., and Herland, L. (1999). *The effects of speech versus tactile driver support messages on workload, driver behaviour and user acceptance* (TNO-Report No. TM-99-C043). Soesterberg, The Netherlands: TNO Human Factors Research Institute.

Waller, P. F. (1991). The older driver. *Human Factors, 33,* 499–505.

Warm, J. S., Dember, W. N., and Parasuraman, R. (1991). Effects of olfactory stimulation on performance and stress in a visual sustained attention task. *Journal of the Society of Cosmetic Chemists, 42,* 199–210.

Wickens, C. D. (1980). The structure of attentional resources. In R. S. Nickerson (Ed.), *Attention and performance* (Vol. 8, pp. 239–257). Hillsdale, NJ: Erlbaum.

Wickens, C. D. (1992). *Engineering psychology and human performance* (2nd ed.). New York: HarperCollins.

Wickens, C. D. (2002). Multiple resources and performance prediction. *Theoretical Issues in Ergonomics Science, 3,* 159–177.

Wickens, C. D., and Liu, Y. (1988). Codes and modalities in multiple resources: A success and a qualification. *Human Factors, 30,* 599–616.

Wierwille, W. W., Antin, J. F., Dingus, T. A., and Hulse, M. C. (1988). Visual attention demand of an in-car navigation system. In A. G. Gale, M. H. Freeman, C. M. Haslegrave, P. Smith, and S. P. Taylor (Eds.), *Vision in vehicles—II* (pp. 307–316). Amsterdam: Elsevier.

Wiese, E. E., and Lee, J. D. (2007). Attention grounding: A new approach to in-vehicle information system implementation. *Theoretical Issues in Ergonomics Science, 8,* 255–275.

11 Interventions to Reduce Road Trauma

Narelle Haworth

CONTENTS

Reflection ... 201
11.1 Introduction ... 202
 11.1.1 The Haddon Matrix ... 202
11.2 Driver Licensing, Training, and Management ... 203
 11.2.1 Graduated Licensing ... 204
 11.2.2 Driver Training and Education ... 205
 11.2.3 Driver Testing ... 206
 11.2.4 Older Drivers ... 206
 11.2.5 Motorcycle Licensing ... 207
 11.2.6 Commercial Vehicle Driver Licensing ... 208
 11.2.7 Driver Management ... 208
11.3 Organizational Interventions ... 210
 11.3.1 Organizational Interventions in Heavy Vehicle Fleets ... 211
 11.3.2 Organizational Interventions in Small Fleets ... 212
11.4 Vehicle Changes ... 212
11.5 Road Environment and Traffic Engineering Changes ... 213
 11.5.1 Improving the Safety of Intersections ... 213
 11.5.2 Run-Off-Road Crashes ... 214
 11.5.3 Treatment of Hazardous Locations ... 214
 11.5.4 Speed Limit Reductions ... 215
 11.5.5 Road Environment Safety and Special Road-User Groups ... 215
11.6 Conclusions ... 216
References ... 216

REFLECTION

This chapter is written from the perspective of someone who started her research career as an experimental cognitive psychologist but spent the past 20 years in road safety. There is an interesting parallel here with the structure of the book as a whole.

It starts with detailed descriptions of the perceptual and cognitive issues in driving and then ends with a chapter that effectively says "well, this is what we can do to fix the problems." In this chapter I have tried to convey the multidisciplinary nature of interventions to reduce road trauma and to stress the point that the nature of the problem does not dictate the nature of the solution. Behavioral problems are not always best dealt with by behavioral interventions. This chapter attempts to show that road trauma prevention must combine interventions that prevent crashes and also those that reduce injury severity. I hope that this chapter fosters the concept of a "successful crash" as one in which the road and vehicle system succeeds in protecting the occupants from serious harm.

11.1 INTRODUCTION

The earlier chapters of this book have described the range of visual and cognitive demands that drivers face in safely navigating the traffic environment. This chapter discusses the interventions that have been developed to mitigate or overcome these challenges, and thus reduce crashes and subsequent injury to drivers and other road users.

The chapter commences with a description of interventions relating to the licensing, training, and management of individual drivers, ranging from graduated licensing for young novice drivers to measures designed to maintain the safety of older drivers. Specialist approaches to licensing and training of motorcyclists and heavy vehicle drivers are also detailed. The focus then moves to organizational interventions in fleets of light and heavy vehicles. Vehicle safety improvements are then grouped into those that help prevent crashes and those that help reduce injury in crashes. The chapter finishes with a description of road environment and traffic engineering countermeasures, both in terms of general design and treatment of particularly hazardous locations.

11.1.1 The Haddon Matrix

The Haddon matrix (Haddon, 1972) was developed to apply the basic principles of public health to road safety and has often been used to classify road crash countermeasures. Table 11.1 presents the simplest form of this matrix, where the rows represent the phase in which the countermeasure has its effect (precrash, crash, and postcrash). These countermeasures are often referred to as primary, secondary, and tertiary approaches to injury prevention, respectively. The columns represent the factors (human, vehicle and equipment, and environment) that the countermeasures operate on. Specific types of countermeasures can be placed into particular cells. Haddon (1972) recommended varying the matrix to deal with specific problems, such as dividing environment into physical and cultural components (or institutional factors, as in Murray, Newnam, Watson, Davey, and Schonfeld, 2003). Some recent writers have added a third dimension to the matrix to incorporate purpose of journey (Faulks and Irwin, 2002) or to facilitate its usefulness in selecting among available countermeasures (Runyan, 1998).

TABLE 11.1
Using the Haddon Matrix to Classify Countermeasures According to Phases (Precrash, Crash, and Postcrash) and Factors (Human, Vehicle and Equipment, and Environment)

	Human	Vehicle and Equipment	Environment (Physical and Social)
Precrash	Licensing Education and training Enforcement	Improvements to vehicle design to avoid crashes Vehicle maintenance	Road design and maintenance Government and company road safety policies Driver selection and management policies Vehicle selection policies
Crash	Helmets Protective clothing Enforcement	Vehicle design to protect occupants in a crash	Treatment of roadside hazards
Postcrash		Safer removal of helmets Automatic collision notification	Emergency medical systems Rehabilitation Crash investigation processes Crash data systems

Haddon criticized the focus on precrash to the exclusion of crash and postcrash countermeasures, claiming that "the unnecessary continuation of the slaughter and economic loss from the world's highways is largely, though by no means totally, the result" (Haddon, 1972). Yet many precrash countermeasures are relevant to the challenges of visual and cognitive performance in driving. This chapter will focus on countermeasures that seek to ameliorate the effects of human cognitive and perceptual errors, rather than those primarily designed to deal with deliberate risk taking or illegal behaviors. A comprehensive review of the whole range of countermeasures can be found in a number of works including Elvik and Vaa (2004).

11.2 DRIVER LICENSING, TRAINING, AND MANAGEMENT

Reductions in the number and severity of crashes can be achieved by reducing the amount of the activity being undertaken (exposure reduction) or by ensuring that the activity is undertaken more safely (risk reduction). Many licensing measures (for example, graduated licensing) combine these effects. Special measures (such as riding age restrictions and training requirements) may apply to particular subgroups known to be at the greatest risk (such as young, novice riders with little or no experience), whereas other measures are targeted toward the general population (such as maximum blood alcohol levels). Most driver management interventions focus on risk reduction. An ideal system of driver licensing, training, and management is one that is cost-effective and utilizes those measures shown to be most effective in reducing road trauma without unduly compromising the mobility and independence of the community.

11.2.1 Graduated Licensing

Young drivers are among the most vulnerable road users, particularly during their first month but also during the first 6 to 12 months of unsupervised driving (Mayhew, Simpson, and Pak, 2003; McCartt, Shabanova, and Leaf, 2003; Williams, 1999). Both youth and inexperience contribute to the high crash risk of young drivers. The contribution of youth is generally found to be greater in jurisdictions with low minimum licensing ages (Williams, 2006) and inexperience is more important where minimum licensing ages are higher (Drummond and Yeo, 1992; Maycock, Lockwood, and Lester, 1991; Mayhew, Simpson, and des Groseilliers, 1999).

Graduated licensing systems (GLS) have been introduced to reduce young driver fatalities and serious injuries. These systems require the individual to progress through a number of successive stages of licensing, each with requirements and restrictions particular to that stage, before progressing to a full license. Specific objectives include (National Highway Traffic Safety Administration [NHTSA], 1998):

- Expanding the learning process to maximize driving experience and maturity of the driver before an unrestricted license is issued.
- Reducing exposure to risk by requiring novices to build up important experience in low-risk situations (e.g., undersupervision).
- Improving driver proficiency by encouraging practice and by having multilevel testing, requiring well-developed basic skills before moving on to more advanced skills, and by delaying retesting after failures.
- Providing greater motivation for safe driving by rewarding good driving (progressively lifting restrictions) and imposing penalties for violations.

The implementation of restrictions for beginning drivers, in addition to other general road safety measures that apply to all age groups, has proven successful in reducing crashes in many jurisdictions (see Hartling, Wiebe, Russell, Petruk, Spinola, and Klassen, 2004; Senserrick and Whelan, 2003). The Cochrane review of graduated licensing for young drivers (Hartling et al., 2004) concluded that the existing evidence shows that it is effective in reducing the crash rates of young drivers but the magnitude of the effect is unclear. It also concluded that the relative contributions of the different provisions within GLS programs were uncertain. Reviews of evaluations of GLS initiatives (Senserrick and Haworth, 2005; Williams, 2007) found that the following measures showed clear associations with crash reductions:

- Increasing the minimum learner period.
- Introducing nighttime driving restrictions for provisional drivers.
- Introducing peer passenger restrictions for provisional drivers.

Education initiatives that encourage early licensure and extensive professional instruction in the absence of sufficient private supervised driving experience have been linked to increased crash risk. Senserrick and Whelan note that the effectiveness of any individual component is dependent on other components that comprise the model.

Although higher levels of compliance with GLS restrictions generate greater benefits, many authors have demonstrated benefits despite significant levels of noncompliance (Begg, Langley, Reeder, and Chalmers, 1995; Mayhew, Simpson, Ferguson, and Williams, 1998). Parents appear to play a larger role in enforcing GLS laws than police and efforts to increase parental management of newly licensed young drivers have shown promise (Simons-Morton and Ouimet, 2006). In some jurisdictions, the display of L (learner) or P (provisional) plates is required to facilitate police enforcement.

Recent discussions have concluded that comprehensive programs of interventions are needed to bring about a reduction in young driver crashes, combining graduated licensing approaches with insight training, insurer discounts, and involvement of parents and police to both model and enforce desired behaviors (Shope, 2006; Williams, 2006).

11.2.2 Driver Training and Education

Driver training refers to a specific instructional program or set of procedures that relates to car control (Christie, 2001; Horneman, 1993, Siegrist, 1999). Clear examples are vehicle-handling skills programs that teach the driver to control a vehicle in traffic. In contrast, driver education refers to the more contemplative and value-based instruction of knowledge and attitudes relating to safe driving behavior. It generally covers a broader range of topics than training and is carried out over a longer period. Driver training, therefore, can be viewed as a specific component of the broader field of driver education.

The traditional approach to training young novice drivers has focused on vehicle handling and control skills, with the aim of passing a practical test for a driving license, and usually includes some teaching of road and traffic laws. It tends to overlook higher-order cognitive skills (Herregods, Nowé, Bekiaris, Baten, and Knoll, 2001) and the motivational orientations behind driving, making it less likely that optimal safe driving practices will be adopted regardless of the level of congruity between driving skills and task demands of the young driver (Peräaho, Keskinen, and Hatakka, 2003).

In contrast, training from an insight approach involves raising awareness or improving insight into factors that contribute to road trauma. From this perspective, it can be argued that it is not the amount or level of skill a driver possesses that is important, but rather when and to what extent that skill is implemented to achieve and maintain safe driving (Dols, Pardo, Falkmer, Uneken, and Verwey, 2001; Peräaho et al., 2003).

While there is general agreement in the academic community that traditional skills-focused training is counterproductive for novices, there is still some uncertainty about whether insight training is effective in reducing crash involvement (see review in Senserrick and Whelan, 2003), although it has not found a counterproductive effect. From a theoretical viewpoint, the insight approach offers the most promise in developing effective training programs for the at-risk novice driver.

As noted earlier, a range of higher-order cognitive–perceptual skills has been identified as important for safe driving, including information processing, hazard perception situational awareness, attentional control, time sharing, and self-calibration.

The most widely researched of these skills in relation to driver training, and perhaps the most promising, is hazard perception (Elander, West, and French, 1993). Hazard perception is the ability to perceive and identify specific hazards in the driving environment (McKenna and Crick, 1994) and involves scanning the road environment, evaluating other drivers' location in the traffic environment, and predicting objects and other drivers' behavior (Ferguson, 2003). Hazard perception ability has been found to be associated with crash rates (Quimby, Maycock, Carter, Dixon, and Wall, 1986) and with driver experience (McKenna and Crick, 1994; Whelan, Groeger, Senserrick, and Triggs, 2002). Training programs and products to improve hazard perception have been developed (e.g., Fisher, Laurie, Glaser, Connerney, Pollatsek, Duffy, and Brock, 2002; Regan, Triggs, and Godley, 2000) and shown to be effective in simulator trials but have not been evaluated on road. Most of the emphasis has been on the development of hazard perception tests for use in licensing to test hazard perception abilities and also encourage learning in this area. Clear evidence of the effects of this approach has not yet emerged.

11.2.3 Driver Testing

Driver testing should improve safety by identifying those who lack the required competence to drive and not permitting them to enter the system. However, most candidates who fail a test simply undergo more practice and then take the test again so that in the end few drivers are screened out of the system. In this sense, the main purpose of driver testing is to encourage learner drivers to undergo sufficient training and practice (see review by Goldenbeld, Baughan, and Hatakka, 1999). Performance on theoretical and practical licensing tests has not been found to predict later crash involvement (Baughan, 2000; Maag, Laberge-Nadeau, Desjardins, Morin, and Messier, 2001; Maycock, 2002) and so while testing is an accepted part of driver licensing, it has not proven to be an effective safety measure in and of itself.

11.2.4 Older Drivers

Licensing for older drivers has become a topic of considerable research and political interest, given the aging of the driving population in developed countries. Older drivers have high fatal and serious crash rates per distance traveled that are comparable to, or greater than, young drivers. There has been considerable debate regarding the relative contributions of the higher prevalence of sensory and cognitive impairments in older drivers (compounded by increasing complexity in the driving and in-vehicle environments), increased frailty, and low driving distances to this increased crash rate (Langford, 2003a). The data suggests that crash risk (rather than injury risk) is increased only for specific subgroups, such as those suffering from dementia, epilepsy, or insulin-treated diabetes, rather than all older drivers (Langford, 2003a). Most older drivers compensate for the usually gradual decline in their driving abilities by avoiding driving in situations they find difficult (e.g., darkness, wet roads, heavy traffic), driving more slowly, and seeking longer time gaps for merging at intersections.

Accordingly, the issue of driving cessation and the appropriateness of tests designed to identify unsafe drivers is much debated in the literature. Removing the licenses of older drivers may not improve their safety if it increases their injuries as pedestrians, where injury risk is very high (Langford, 2003b). Across a range of jurisdictions, licensing practices range from no intervention based on age to a combination of regular on-road and medical tests to renew the license once drivers reach a certain age. There does not seem to be any clear-cut evidence that stricter licensing requirements reduce the crash risk of older drivers (Langford, 2003b).

11.2.5 Motorcycle Licensing

Licensing has been the major countermeasure approach to reduce the high level of involvement of motorcyclists in serious crashes in many jurisdictions. A separate license (or license endorsement) is generally required to ride a motorcycle and additional training requirements or restrictions are common (Haworth and Mulvihill, 2005). Considerable research into the effects of applying limits to the engine size or power of motorcycles ridden by novices and of taking rider education and training has found little benefit (Mayhew and Simpson, 2001; TOI, 2003). The safety benefits of other restrictions such as zero blood alcohol concentration (BAC), late-night riding restrictions, supervision, and no passengers have not been examined specifically for motorcyclists (reviewed in Haworth and Mulvihill, 2005).

Although older riders have a lower crash rate than younger riders, the huge growth in the numbers of older riders has increased the need for effective measures to improve the safety of this group. The design of most motorcycle licensing systems assumes that most license applicants are young and do not have a car license and includes exemptions for the expected small minority of riders who are older or already hold a car license. Yet now the exemptions apply to many novices and so the graduated nature of the licensing system has been eroded. Similarly, training programs were also designed for young nondrivers and may not be as effective for the new generation of riders.

The return to riding by a large number of previously inactive (although licensed) riders has also challenged licensing systems. The licensing practice that allows motorcycle licenses to remain current at no additional cost to people who hold car licenses facilitates this situation. Implementing a system in which there is an active requirement to maintain the currency of a motorcycle license could act to ensure that those individuals wishing to return to riding have to regain a minimum level of skill or competence before doing so. Promotion of refresher courses for license holders returning to riding may be of benefit to improve skills and reinforce to potential riders that their skills may not be up to date. The crash involvement of older riders could also be decreased by general motorcycle safety measures that would benefit riders of all ages, such as reductions in impaired driving and other unsafe road user behaviors by car drivers, reductions in both speeding and general travel speeds, and improvements in roadside safety to prevent or reduce injury in a crash.

11.2.6 Commercial Vehicle Driver Licensing

Most jurisdictions have implemented special licensing requirements for commercial vehicle drivers in recognition of the greater severity of crashes involving these vehicles. The requirements often involve medical examinations, higher minimum age and/or holding a car license for a minimum period, training programs, and monitoring of driving hours. Some countries such as the United States require testing of commercial drivers for drugs and alcohol (McCartt, Campbell, Keppler, and Lantz, 2007). In general, licensing requirements become more stringent as the size of the vehicles increases or as the cargo becomes more hazardous. There is little research that clearly measures the safety benefit of commercial driver licensing. The considerable amount of research examining driving hours and their relation to driver fatigue and stimulant use suggests that the formulation of driving hours legislation does not match what is known about fatigue and that driving in excess of the prescribed limits is common in Australia and North America (Dawson, Feyer, Gander, Hartley, Haworth, and Williamson, 2001). There has been a move toward implementation of fatigue management programs and educating drivers and schedulers about fatigue, and health and lifestyle management, but there has been little evaluation of the effect of these programs on their ultimate goal of reducing crash occurrence.

11.2.7 Driver Management

Once drivers have been licensed, there is a range of policy and legislative approaches to encourage them to drive in a way that minimizes their risk of involvement in a crash. Traditionally, these approaches have focused on police enforcement of traffic laws and punishment for infringing the laws, with the potential for license removal or imprisonment for the most severe offenses. Although this "stick" approach has been widespread, there has been considerable interest in developing effective "carrots" to encourage good behavior.

11.2.7.1 Enforcement Programs

Police enforcement of traffic laws has been a prime focus of efforts to reduce crashes related to speeding, drunk driving, nonuse of seat belts, and unlicensed driving. Enforcement aims to discourage people from undertaking these behaviors (general deterrence) and to discourage offenders from continuing their illegal behavior (specific deterrence; Ross, 1982). Countries differ in the extent to which they rely on general and specific deterrence (see a comparison of the United States and Australia in Williams and Haworth, 2007), and within jurisdictions, different approaches may be used for drunk driving compared to speeding, for example.

For both drunk driving and speeding, the effectiveness of police enforcement as a measure to reduce road trauma depends on three elements: the level of the legal limit, the perceived chance of getting caught when exceeding the limit, and the perceived severity of the sanctions. The role of driver perceptions in deterrence means that mass media advertising related to enforcement can play a large role in boosting its effectiveness (Delhomme et al., 1999; Elliott, 1993).

According to a meta-analysis carried out by Elvik and Vaa (2004), reducing the existing BAC limit for all drivers in a country leads to a reduction of 8% in fatal crashes and a reduction of 4% in injury crashes. Random breath testing leads to larger crash reductions than testing only where there is suspicion (Henstridge, Homel, and Mackay, 1997); however, a U.S. review concluded that sobriety checkpoints (which are a form of selective testing) may be as effective as random breath testing (Elder, Shults, Sleet, Nichols, Zaza, and Thompson, 2002). There is evidence that severity of sanctions may not be as important as the certainty (and perhaps swiftness) of sanctions (Elliott, 2003; Legge and Park, 1994; Ross, 1985).

The level of the legal limit is also important for speed enforcement: If the speed limit is too high for the level of safety at that location, then the location will remain dangerous even if enforcement is able to eradicate speeding. The enforcement threshold is often set above the posted speed limit and this constrains speed reductions because the behavior of drivers will relate to the de facto speed limit, rather than the posted limit.

Many jurisdictions have introduced electronic speed enforcement technology—speed cameras—to increase the chance of getting caught. A review of evaluation studies has shown reductions in fatalities of 17%–71% and reductions in injuries of 12%–65% (Pilkington and Kinra, 2005). Speed cameras at fixed locations have been shown to result in substantial reductions in speeds and crashes at those locations, while mobile speed camera programs have demonstrated crash reductions generalized beyond the camera sites (Cameron and Delaney, 2007).

11.2.7.2 Incentives and Rewards

Incentives are offered to encourage the desired behavior and rewards are provided to reinforce the behavior once it has occurred. The evidence suggests that incentives are more effective than rewards, which can sometimes have counterproductive effects. Insurance companies reward drivers who do not make claims by reducing premiums, but there is no clear evidence that this promotes safer driving. It may instead discourage drivers from submitting a claim in the case of a crash, by increasing the effective threshold for claiming on the policy. Rewards are built into some graduated licensing systems in terms of the requirement for a clean record in order to progress to the next level. However, Hurst (1980) commented that these rewards are "so hopelessly delayed that it is hard to see how they could reinforce specific safe driving behaviours."

The potential for effective rewards is questioned in the literature. Both Hurst (1980) and Warren (1982) agree that it is hard to use rewards to teach a driver *not* to do something (like speeding) and that the authorities cannot administer socially meaningful rewards when they are viewed as the enemy. Warren, though, disputes Hurst's conclusion that reward systems should be abandoned, claiming instead that "a necessary precondition for the introduction of meaningful reward systems would be the identification of a delivery mechanism which would be viewed by the public as having a vested interest in the promotion of safety (rather than an excessive preoccupation with punishment)" (Warren, 1982). He suggests that the Department of Health might be such an agency because it is responsible for dealing with the injuries.

One of the issues that needs to be considered in designing an effective incentive or reward system is balancing the costs of the rewards with the benefits of the crash reductions. Although road crashes are a major health problem, the likelihood of an individual being involved in a crash is relatively small and the decrease in this likelihood by engaging in a desired behavior is even smaller. Thus, the size of the reward that could be offered to each driver might be so small as to have little effect on behavior.

Both of these concerns were addressed in the DriveRight campaign conducted in Victoria, Australia. The Transport Accident Commission, which is the sole supplier of insurance to injured road users, mounted a campaign in which motorists were encouraged to place campaign stickers on the rear of their cars. Drivers of police and emergency services and roadside assistance vehicles reported the registration numbers of vehicles that were observed driving safely and prizes were awarded to a random sample of the owners of these vehicles. Thus the prizes could be sufficient to be considered worthwhile by drivers. There was no evaluation of the effect of this campaign on traffic infringements or crashes, but it did manage to avoid the pitfalls of many other reward campaigns.

11.3 ORGANIZATIONAL INTERVENTIONS

Many road safety countermeasures can be delivered as organizational interventions. There are two main reasons why organizational interventions have been introduced. First, road crashes are the most common form of work-related death in many countries (National Institute for Occupational Safety and Health [NIOSH], 2003; National Occupational Health and Safety Commission, 1998) and fleet or company drivers have a higher crash risk than drivers of privately registered vehicles (Bibbings, 1997; Broughton, Baughan, Pearce, Smith, and Buckle, 2003; Lynn and Lockwood, 1998; Newnam, Watson, and Murray, 2002). Thus, improving the safety of work-related driving has the potential to improve both road safety and work safety. Second, improving the safety of vehicles purchased by fleets is one of the most powerful tools to increase the uptake of new safety features into the wider vehicle population (Haworth, Tingvall, and Kowadlo, 2000).

Organizational interventions can be grouped into those that relate to drivers (selection, induction, training, and management of safe driving), vehicles (selection, maintenance, replacement cycles, and rules for use), and company practices (vehicle usage and incident data collection, incident management, allocation of tasks, scheduling of driving, and other work tasks).

Yet barriers often exist at the organizational level that impede the introduction of effective interventions, or promote or maintain ineffective interventions (such as traditional driver training programs that have not been tailored to the needs of the organization). The barriers can be categorized as follows: a denial that the organization needs to improve (we don't have a problem); a belief that current actions are sufficient (we have got it under control); uncertainty about what interventions are required (we don't know what to do about it); a belief that interventions are not possible or will not succeed (we can't do anything about it); or a belief that the responsibility for road safety lies outside the company (it isn't our problem).

The extent to which effective organizational interventions are introduced varies among companies with those organizations that have a greater commitment to occupational health and safety across their operations generally being more active in work-related road safety as well. Interestingly, it is the policy of the parent company rather than the road safety climate of the country in which the organization is operating that appears to be more influential. Koppel, Charlton, and Fildes (2007) demonstrated that vehicle safety was not a more important consideration among fleet purchasers in Sweden (with its strong road safety policy) than in Spain. This supports the view that organizational interventions are strongly reliant on the commitment of management at the highest levels and are difficult to introduce and sustain at the grassroots level. Organizational safety culture has been identified as crucial to the adoption of organizational interventions (BOMEL Limited, 2004).

11.3.1 Organizational Interventions in Heavy Vehicle Fleets

Organizational responsibility for fleet safety is a more salient issue in heavy vehicle fleets than for light passenger and light commercial fleets because driving and transport of goods is their primary business. In typical heavy vehicle crashes, most injury occurs to the general public, not to the heavy vehicle occupants (Haworth and Symmons, 2003). For this reason, government regulation plays a more prominent role for heavy vehicle fleets than light vehicle fleets. Thus many practices in heavy vehicles reflect regulatory requirements, rather than the policies of individual fleets.

As driving and transport is the core business for many companies with heavy vehicle fleets, vehicle management tends to take priority over driver management in many instances. As vehicles are very expensive and commonly take priority, companies often are not aware of the value in their employees or do not anticipate the cost to the company if their employees are injured. For many operators (and particularly those with older trucks), the motivation for maintenance is to keep the trucks on the road, rather than safety.

The lack of government incentives for buying safer trucks in many countries means that economic factors (in terms of the company, rather than the community) appear to be the most important determinants of heavy vehicle selection. The selection of the type of vehicle depends largely on the nature of the task rather than safety, although it is influenced by legislative constraints on the type or dimensions of vehicles.

Given that many heavy vehicles are purchased as used vehicles and may not have the current safety features, retrofitting is a more important issue for heavy vehicle than light vehicle fleet safety. There is potential for retrofitting integrated lap/sash belts in heavy vehicle driving positions, improvements to visibility, and better underrun protection, but the extent to which this occurs is not well known.

A review of best practice for heavy vehicle fleets (Haworth and Greig, 2007) recommended the following organizational interventions:

- Vehicle management
- Speed limiting trucks to improve safety and lower fuel consumption
- Fit underrun protection

- Purchase trucks with safer cabs
- Purchase bigger trucks to minimize distance traveled (if this is plausible)
- Retrofit better seat belts on older trucks
- Driver management
- Require seat-belt wearing
- Inform drivers of the safety behaviors expected
- Monitor vehicle speeds
- Provide feedback to drivers about vehicle speeds

11.3.2 Organizational Interventions in Small Fleets

Much of the research into improving organizational road safety has examined large fleets. Small fleets generally do not have well-developed policies, procedures, and guidance information. There is a perception that most small fleets do not have the resources to be Occupational Health and Safety (OHS)-led or be active in fleet management organizations (Murray et al., 2003) and these fleets may fall outside many existing governmental health and safety frameworks. Australian research has found that the majority of fleets are small and vehicles in small fleets are older on average than vehicles in large fleets (Symmons and Haworth, 2005). Thus, small fleets may not be benefiting from improvements to vehicle safety as quickly as larger fleets.

11.4 VEHICLE CHANGES

Vehicle-based countermeasures have traditionally been divided into precrash (active safety) and crash (passive safety or crashworthiness) categories. Until recently, most of the emphasis has been on passive safety, but now more attention is being paid to the development of active safety measures (including advanced driver-assistance systems) and particularly the integration of active and passive safety measures to create an overall safety system (Schöneburg and Breitling, 2005).

The precrash vehicle safety measures comprise those related to: communication (including vision and lighting), vehicle control (generally speed, steering, and braking), and monitoring the driver for impairment (alcohol or fatigue). Improvements to communication have generally provided real or potential safety benefits, although some have had limited adoption (e.g., headlights that track steering, infrared lighting, or other systems to detect pedestrians in low-light levels). Vehicle control improvements have sometimes been introduced by vehicle manufacturers and become popular without necessarily having significant safety benefits (e.g., antilock braking systems) but enhanced stability control programs appear to be more successful (Lie, Tingvall, Krafft, and Kullgren, 2006). Recent research into intelligent speed adaptation suggests that it may reduce crashes and injury severity to a considerable extent if issues of driver acceptance can be overcome (Jamson, Carsten, Chorlton, and Fowkes, 2006). Alcohol interlocks prevent drunk-driving crashes when fitted to the vehicles of convicted drink drivers (Beck, Rauch, and Baker, 1997; Bjerre, 2002; Morse and Elliot, 1992; Weinrath, 1997), but these effects largely disappear once the interlock is removed (Beck et al., 1997; Frank, Raub, Lucke, and Wark, 2002; Tippetts and Voas, 1998; Voas, Marques, Tippetts, and Beirness, 2000). More widespread acceptance will rely on the development of both noninvasive and accurate devices. Compulsory vehicle inspection has been a common precrash vehicle safety measure, but evaluations suggest that it is not cost-beneficial as a measure to prevent crashes although it may have some secondary benefits in terms of encouraging purchases of newer vehicles and reduced emissions (e.g., Fridstrøm and Bjørnskav, 1989; TFB-VTI Research, 1991; Ylvinger, 1998).

Crash countermeasures have two aims: to increase the ability of the vehicle to protect its occupants (improved crashworthiness) and to decrease the injury caused to other parties in the crash (reduced aggressivity). The major contributors to aggressivity are body style (a light truck or van is more aggressive than a passenger car), mass, geometry, and stiffness (Austin, 2005). The most effective measures to increase crashworthiness include increased mass, vehicles of similar mass and geometry, better design of vehicles to absorb impact, seat belts, air bags, seat-belt interlocks/reminder systems, and better design of seat belts to prevent whiplash. It has been estimated that fatalities would reduce by one third if each car was replaced with the safest in its class (Krafft, 1998.).

However, it is not enough to understand the characteristics that make vehicles safer; there is a need to promote the purchase of safer vehicles, particularly by more vulnerable road-user groups, such as novice and older drivers, and by fleet buyers (Koppel, Charlton, and Fildes, 2007). Part of this effort should include persuading people to buy passenger cars rather than light trucks (including SUVs) or vans, which generally have increased likelihood of rollover and poorer passive safety.

11.5 ROAD ENVIRONMENT AND TRAFFIC ENGINEERING CHANGES

Road environment and traffic engineering changes can bring about both crash prevention and injury prevention outcomes. Traditionally, road environment design focused on crash prevention by measures such as appropriate radii for curves, standards for superelevation, and minimum sight distances. In recent years, there has been more emphasis on preventing or minimizing injury by safer designs of roadsides and development of more effective barrier systems. This has been a consequence of the adoption of road safety philosophies such as Vision Zero and Sustainable Safety that require that the road infrastructure be capable of preventing serious injury.

11.5.1 Improving the Safety of Intersections

About 40% of all injury crashes reported to the police occur at intersections (Elvik and Vaa, 2004), with this percentage being higher in cities and towns. Traffic signals and roundabouts are two approaches to improving traffic safety and reducing delays. Control of traffic by signals reduces the number of crashes by about 15% at T-junctions and 30% at crossroads (Elvik and Vaa, 2004), with similar benefits for injury and property damage crashes. Crashes involving cross-traffic are reduced, although there is an increase in rear-end collisions. Yet, traffic signals rely on driver compliance and serious crashes can result from red-light running. For this reason, there has been a move to replace some signalized intersections with roundabouts (as well as installing roundabouts at previously uncontrolled intersections). Roundabouts force drivers to slow down and also change the geometry of many crashes, together reducing the frequency of high-speed, side-impact crashes. Thus, while replacing traffic signals with roundabouts reduces crashes by about 10% overall, it almost eliminates fatal crashes.

Despite some concerns about the safety of roundabouts for vulnerable road users, the evidence suggests that the improvement in pedestrian safety is similar to that of vehicle occupants, although the improvement for cyclists is somewhat less (Elvik and Vaa, 2004).

11.5.2 Run-Off-Road Crashes

Run-off-road crashes can have very serious outcomes if vehicles roll over or hit rigid objects. Collisions with poles and trees are particularly severe and comprise about 30%–40% of fatal crashes in many countries. Treatments to prevent run-off-road crashes include improvements to roadway alignment, delineation, and road surface friction. Installation of guardrails, removal or protection of roadside hazards, and flattening of side slopes can reduce the severity of injury to occupants of an out-of-control vehicle.

11.5.3 Treatment of Hazardous Locations

A common approach to maximizing the benefits from limited resources available for improving road environment safety is to focus on treatment of hazardous locations (also termed accident black spots) that have a history of crash involvement. Evaluations of programs of treating hazardous locations have demonstrated crash reductions and the benefits of these crash reductions have generally exceeded the costs of implementation (Bureau of Transport and Communications Economics, 1995; Newstead and Corben, 2001).

To maximize the returns from investment in accident black-spot programs, rigorous and systematic procedures need to be applied to identify black-spot locations. Then, the crash problems at individual locations need to be clearly analyzed to ensure that the most appropriate treatment type is chosen and that the chosen treatments comprehensively address the predominant crash types (Duarte and Corben, 1998). Not all accident black-spot treatments are equally effective. Newstead and Corben (2001) concluded that new roundabouts, fully controlled right-turn phases, channelization, splitter islands at intersections, pavement resealing along routes, edgeline marking, and shoulder sealing resulted in statistically significant reliable reductions in casualty crash frequencies and costs to the community. Corben and Deery (1998) found that road surface improvements (including skid-resistant pavements and shoulder sealing) significantly reduced the incidence and cost of single-vehicle crashes into fixed objects, and improvements to road and roadside geometry (including horizontal geometry) significantly reduced the incidence of these crashes.

11.5.4 Speed Limit Reductions

There is overwhelming international evidence that lower speeds result in fewer collisions, and lesser severity in the crashes that do occur. Andersson and Nilsson (1997) developed a model to relate the increase in severity to the increase in speed that was based on studies of the effects of speed limit changes in Sweden. The model states that the probability of a fatal crash is related to the fourth power of the speed. This means that a 10% reduction of mean speed results in a reduction in the number of fatalities of approximately 40%.

Research undertaken in the United States after the raising of the interstate speed limits (cited in Finch, Kompfner, Lockwood, and Maycock, 1994) found that an increase in mean speed of 2–4 mph (approximately 3–6 km/h) was associated with a 19%–34% increase in the number of fatalities. This roughly translates into an 8%–9% increase in fatalities on U.S. interstate highways for every 1 mph change in mean speed. Later work confirmed that the relationship derived by Finch et al. (1994) holds for the general case, that is, every 1 km/h reduction in speed across the network leads to a 3% drop in crashes (Taylor, Lynam, and Baruya, 2000). However, greater crash reductions per 1 km/h reduction in speed are achieved on residential and town center roads, and lower reductions are achieved on higher quality suburban and rural roads.

Australian research has generated new evidence on the increases in crash risk with increasing travel speed. A study in metropolitan Adelaide reported that traveling at 5 km/h over the speed limit (in 60 km/h zones) doubles the risk of an injury crash, the same effect as a BAC of 0.05 (Kloeden, McLean, Moore, and Ponte, 1997). For pedestrian crashes, McLean, Anderson, Farmer, Lee, and Brooks (1994) reported a strong relationship between impact speed and injury severity.

Speed limit reductions typically result in a reduction in mean speed that is considerably less than the actual speed limit reduction (Haworth, Ungers, Corben, and Vulcan, 2001); however, even small reductions in mean speeds can have significant effects on fatal and serious injury crashes as demonstrated earlier. Analyses of speed profiles following the reduction in the general urban speed limit from 60 km/h to 50 km/h in Australia have shown that the speed reductions are greatest for the fastest vehicles (Haworth et al., 2001). Thus, the crash savings may be at least partly a function of a change in behavior of those who would be most likely to crash (or have the most severe crashes).

11.5.5 Road Environment Safety and Special Road-User Groups

Recent research has examined ways of adapting roadway design to better cope with the needs of older road users, including redesign of intersections and clearer signage. Similarly, guidelines have been produced for the design and maintenance of roads that will result in infrastructure that is safer for motorcyclists (Association des Constructeurs Européens des Motocycles [ACEM], 2005; Institute of Highway Incorporated Engineers [IHIE], 2005). The benefits of these measures have not yet been clearly quantified.

11.6 CONCLUSIONS

The visual and cognitive demands of the traffic environment can result in human error and contribute to crashes. Many analyses have shown that human factors are present in most crashes (e.g., Hanowski, Olsen, Hickman, and Dingus, 2006). Yet the history of road safety interventions has shown that efforts to change driver behaviors that are proximal to the crash have been less successful than efforts to change behaviors that are less immediate. For example, exhortations to "drive safely" have had much less effect than widespread police enforcement leading to drivers deciding to drive more slowly.

This is not to say that changing behavior is unimportant or that vehicle and roadway improvements can succeed without behavioral changes. Seat belts were a vehicle engineering solution to reduce injury severity in crashes that occurred because of human error, but they only provide a benefit if vehicle occupants buckle up. Although significant reductions in injury severity are possible as a result of safer vehicle technology, these gains will only be fully realized if private individuals and organizations are aware of these benefits and change their behavior in terms of the levels of safety that they demand when purchasing vehicles. This will encourage vehicle manufacturers to market safer vehicles. Widespread implementation of improvements in road design require not only engineering competence but an acceptance by governments and road builders and managers that they have a responsibility for providing a safe system of infrastructure that protects against the consequences of human error. Thus, further reductions in road trauma require an emphasis on cognitive factors in the legislature, the office, and the marketplace as well as behind the wheel.

REFERENCES

Andersson, G., and Nilsson, G. (1997) *Speed management in Sweden.* Linkoping: Swedish National Road and Transport Institute VTI.

Association des Constructeurs Européens des Motocycles (ACEM). (2005). *Guidelines for PTW-safer road design in Europe.* Brussels: Association des Constructeurs Européens des Motocycles.

Austin, R. (2005). *Vehicle aggressiveness in real world crashes* (Paper number 05-0248; DOT HS 809 825). Paper presented at the 19th Conference on the Enhanced Safety of Vehicles, Washington DC.

Baughan, C.J. (2000, June). *Review of the practical driver test.* Proceedings of the DTLR Novice Driver Conference, Transport Research Laboratory, UK.

Beck, K.H., Rauch, W.J., and Baker, E.A. (1997). *The effects of alcohol ignition interlock license restrictions on multiple offenders: A randomized trial in Maryland.* Arlington, VA: Insurance Institute for Highway Safety.

Begg, D.J., Langley, J.D., Reeder, A.I., and Chalmers, D. (1995). The New Zealand graduated licensing system: Teenagers' attitudes toward and experiences with this car driver licensing system. *Injury Prevention, 1,* 177–181.

Bibbings, R. (1997). Occupational road risk: Toward a management approach. *Journal of the Institution of Occupational Safety and Health, 1*(1), 61–75.

Bjerre, B. (2002). A preliminary evaluation of the Swedish ignition interlock program and recommended further steps. Proceedings of the 16th *International Conference on Alcohol, Drugs and Traffic Safety* [CD-ROM], Montreal, Canada.

BOMEL Limited. (2004). *Safety culture and work-related road accidents* (Research Report No. 51). London: Department for Transport

Broughton, J., Baughan, C., Pearce, L., Smith, L., and Buckle, G. (2003). *Work-related road accidents* (TRL Report No. 582). Crowthorne, UK: Transport Research Laboratory.

Bureau of Transport and Communications Economics. (1995). *Evaluation of the black spot program* (Report 90). Canberra, Australia: Australian Government Publishing Service.

Cameron, M., and Delaney, A. (2007). *Development of strategies for best practice for speed enforcement in Western Australia: Final report* (Report No.270). Melbourne, Australia: Monash University Accident Research Centre. Available at http://www.monash.edu.au/muarc/reports/ muarc270.pdf.

Christie, R. (2001). *The effectiveness of driver training as a road safety measure* (RACV Literature Report No. 01/03). Melbourne, Victoria, Australia: Royal Automobile Club of Victoria.

Corben, B.F., and Deery, H.A. (1998, May). *Road engineering countermeasures: An evaluation of their effectiveness and in-depth investigations of unsuccessful treatments.* Paper presented to the 9th Road Engineering Association of Asia and Australasia, Wellington, New Zealand.

Dawson, D., Feyer, A., Gander, P., Hartley, L., Haworth, N., and Williamson, A. (2001). *Fatigue Expert Group: Options for regulatory approach to fatigue in drivers of heavy vehicles in Australia and New Zealand.* Canberra, Australia: Australian Transport Safety Bureau.

Delhomme, P., Vaa, T., Meyer, T., Harland, G., Goldenbeld, C., Järmark, S., Christie, N., and Vlasta, R. (1999). *Evaluated road safety media campaigns: An overview of 265 evaluated campaigns and some meta-analysis on accidents* (Contract No. RO-97-SC.2235). Arcueil, France: GADGET project, INRETS. Available at http://www.kfv.or.at/gadget/wp4/index.htm.

Dols, J.F., Pardo, J., Falkmer, T., Uneken, E., and Verwey, W. (2001, August 14–17). *The trainer project: A new simulator-based driver training curriculum.* Proceedings of the 2001 International Driving Symposium on Human Factors in Driver Assessment, Training and Vehicle Design, Aspen, CO.

Drummond, A., and Yeo, E. (1992). *The risk of crash involvement as a function of driver age* (MUARC Report No. 42). Melbourne, Australia: Monash University Accident Research Centre.

Duarte, A., and Corben, B. (1998). *Improvement to black spot treatment strategy* (Report No.132). Melbourne, Australia: Monash University Accident Research Centre.

Elander, J., West, R., and French, D. (1993). Behavioural correlates of individual differences in road traffic crash risk: An examination of methods and findings. *Psychological Bulletin, 113,* 279–294.

Elder, R.W., Shults, R.A., Sleet, D.A., Nichols, J.L., Zaza, S., and Thompson, R.S. (2002). Effectiveness of sobriety checkpoints for reducing alcohol-involved crashes. *Traffic Injury Prevention, 3,* 266–274.

Elliott, B. (1993). *Road safety mass media campaigns: A meta analysis* (CR 118). Canberra, Australia: Federal Office of Road Safety.

Elliott, B. (2003, September 24–26). *Deterrence theory revisited.* Paper presented at the Australasian Road Safety Research, Policing and Education Conference, Sydney, Australia.

Elvik, R., and Vaa, T. (2004). *The handbook of road safety measures.* Amsterdam: Elsevier.

Faulks, I.J., and Irwin, J.D. (2002). Can Haddon's matrix be extended to better account for work-related road use? In I.J. Faulks (Ed.), *STAYSAFE 57: Work-related road safety. Proceedings of a seminar held at Sydney, Thursday 8 February 2001. Report of the Joint Standing Committee on Road Safety.* Sydney, Australia: Parliament of New South Wales.

Ferguson, S.A. (2003). Other high-risk factors for young drivers—how graduated licensing does, doesn't, or could address them. *Journal of Safety Research, 34,* 71–77.

Finch, D.J., Kompfner, P., Lockwood, C.R., and Maycock, G. (1994). *Speed, speed limits and accidents* (TRL Project Report 58). Crowthorne, UK: Transport Research Laboratory.

Fisher, D.L., Laurie, N.E., Glaser, R., Connerney, K., Pollatsek, A., Duffy, S., and Brock, J. (2002). The use of a fixed base driving simulator to evaluate the effects of experience and PC based risk awareness training on drivers' decisions. *Human Factors, 44,* 287–302.

Frank, J.E., Raub, R., Lucke, R.E., and Wark, R.I. (2002). Illinois ignition interlock evaluations. *Proceedings of the 16th International Conference on Alcohol, Drugs and Traffic Safety, Montreal, Canada* [CD-ROM].

Fridstrøm, L., and Bjørnskav, T. (1989). The determinants of road traffic accidents in Norway—an aggregate modelling approach (TØI report 0039/1989). Norway: Institute of Transport Economics. (English summary in *Nordic Road and Transport Research, 1989*(2), 21–23.)

Goldenbeld, C., Baughan, C.J., and Hatakka, M. (1999). Driver testing. In S. Siegrist (Ed.), *Driver training, testing and licensing— towards theory-based management of young drivers' injury risk in road traffic: Results of EU-Project GADGET, Work Package 3* (BFU Report 40). Berne, Switzerland: Beratungsstelle fur Unfaliverhutung.

Haddon, W., Jr. (1972). A logical framework for categorizing highway safety phenomena and activity. *The Journal of Trauma, 12,* 193–207.

Hanowski, R.J., Olsen, R.L., Hickman, J.S., and Dingus, T.A. (2006). *The 100-car naturalistic driving study: A descriptive analysis of light vehicle–heavy vehicle interactions from the light vehicle driver's perspective, data analysis results* (FMCSA-RRR-06-004). Washington, DC: National Highway Traffic Safety Administration.

Hartling, L., Wiebe, N., Russell, K., Petruk, J., Spinola, C., and Klassen, T.P. (2004). Graduated driver licensing for reducing motor vehicle crashes among young drivers. *The Cochrane Database of Systematic Reviews*, Issue 2, Art. No.: CD003300.

Haworth, N., and Greig, K. (2007). *Improving fleet safety—Current approaches and best practice guidelines.* Sydney, Australia: Austroads.

Haworth, N., and Mulvihill, C. (2005). *Review of motorcycle licensing and training* (Report No. 240). Melbourne, Australia: Monash University Accident Research Centre.

Haworth, N., and Symmons, M. (2003). *Review of truck safety. Stage 2—Update of crash statistics* (Report No. 205). Melbourne, Australia: Monash University Accident Research Centre.

Haworth, N., Tingvall, C., and Kowadlo, N. (2000). *Review of best practice road safety initiatives in the corporate and/or business environment.* Melbourne, Australia: Monash University Accident Research Centre.

Haworth, N., Ungers, B., Corben, B., and Vulcan, P. (2001, November 18–20). *A 50 km/h default urban speed limit for Australia?* Paper presented at 2001 Road Safety Research, Policing and Education Conference, Melbourne, Australia.

Henstridge, J., Homel, R., and Mackay, P. (1997). *The long-term effects of random breath testing in four Australian states: A time series analysis* (CR 162). Canberra, Australia: Federal Office of Road Safety.

Herregods, D., Nowé, H., Bekiaris, A., Baten, G., and Knoll, C. (2001, August 14–17). *The trainer project: Matching training curricula to drivers' real needs using multimedia tools.* Proceedings of the 2001 International Driving Symposium on Human Factors in Driver Assessment, Training and Vehicle Design, Aspen, CO.

Horneman, C. (1993). *Driver education and training: A review of the literature* (RTA Research Note RN6/93). Rosebery, New South Wales, Australia: Roads and Traffic Authority, Road Safety Bureau.

Hurst, P.M. (1980). Can anyone reward safe driving? *Accident Analysis and Prevention, 2,* 217–220.

Institute of Highway Incorporated Engineers (IHIE). (2005). *IHIE guidelines for motorcycling: Improving safety through engineering and integration.* UK: Institute of Highway Incorporated Engineers.

Jamson, S., Carsten, O., Chorlton, K., and Fowkes, M. (2006). *Intelligent speed adaptation. Literature review and scoping study*. Leeds, UK: University of Leeds and MIRA Ltd and TfL.

Kloeden, C.N., McLean, A.J., Moore, V.M., and Ponte, G. (1997). *Travelling speed and the risk of crash involvement* (CR172). Canberra, Australia: Federal Office of Road Safety.

Koppel, S., Charlton, J., and Fildes, B. (2007). How important is vehicle safety in the new vehicle purchase/lease process for fleet vehicles? *Traffic Injury Prevention, 8,* 130–136.

Krafft, M. (1998). *Non-fatal injuries to car occupants. Injury assessment and analysis of impacts causing short and long term consequences with special reference to neck injuries*. Thesis, Karolinska Institute, Stockholm, Sweden.

Langford, J. (2003a). Older drivers and the greying of Australasia. In J. Langford and B. Fildes (Eds.), *Australasian road safety handbook* (Vol. 1; AP–R234/03; pp. 16–24). Sydney: Austroads.

Langford, J. (2003b). Licensing options for managing older driver safety. In J. Langford and B. Fildes (Eds.), *Australasian road safety handbook* (Vol. 1; AP–R234/03; pp. 25–29). Sydney: Austroads.

Legge, J.S., and Park, J. (1994). Policies to reduce alcohol-impaired driving: Evaluating elements of deterrence. *Social Science Quarterly, 75*(3), 594–606.

Lie, A., Tingvall, C., Krafft, M., and Kullgren, A. (2006). The effectiveness of electronic stability control (ESC) in reducing real life crashes and injuries. *Traffic Injury Prevention, 7,* 38–43.

Lynn, P., and Lockwood, C.R. (1998). *The accident liability of company car drivers* (TRL Report 317). Crowthorne, Berkshire, UK: Transport Research Laboratory.

Maag, U., Laberge-Nadeau, C., Desjardins, D., Morin, I., and Messier, S. (2001, June 10–13). *Three-year injury crash records and test performance of new Quebec drivers.* Proceedings of the Canadian Multidisciplinary Road Safety Conference XII, London, Ontario, Canada.

Maycock, G. (2002). *Novice driver accidents and the driving test* (TRL Research Report 527). Crowthorne, UK: Transport Research Laboratory.

Maycock, G., Lockwood, C.R., and Lester, J.F. (1991). *The accident liability of car drivers* (TRL Research Report 315). Crowthorne, UK: Transport Research Laboratory.

Mayhew, D.R., and Simpson, H.M. (2001). *Graduated licensing for motorcyclists*. Ottawa, Canada: Traffic Injury Research Foundation.

Mayhew, D.R., Simpson, H.M., and des Groseilliers, M. (1999). *Impact of the graduated driver licensing program in Nova Scotia*. Ottawa, Canada: Traffic Injury Research Foundation.

Mayhew, D.R., Simpson, H.M., Ferguson, S.A., and Williams, A.F. (1998). Graduated licensing in Nova Scotia: A survey of teenagers and parents. *Journal of Traffic Medicine, 26,* 37–44.

Mayhew, D.R., Simpson, H.M., and Pak, A. (2003). Changes in collision rates among novice drivers during the first months of driving. *Accident Analysis and Prevention, 35,* 683–691.

McCartt, A.T., Campbell, S.F., Sr., Keppler, S.A., and Lantz, B.M. (2007). *The domain of truck and bus safety research: Enforcement and compliance* (TRB Circular Number E-C117, pp. 41–57). Washington, DC: Transportation Research Board.

McCartt, A.T., Shabanova, V.I., and Leaf, W.A. (2003). Driving experience crashes and traffic citations of teenage beginning drivers. *Accident Analysis and Prevention, 35,* 311–320.

McKenna, F., and Crick, J.L. (1994). *Hazard perception in drivers: A methodology for testing and training* (TRL Contract Report No. CR3131). Crowthorne, UK: Transport Research Laboratory.

McLean, A.J., Anderson, R.W.G., Farmer, M.J.B., Lee, B.H., and Brooks, C.G. (1994). *Vehicle travel speeds and the incidence of fatal pedestrian collisions. Volume 1.* (CR 146). Canberra, Australia: Federal Office of Road Safety.

Morse, B.J., and Elliott, D.S. (1992). Effects of ignition interlock devices on DUI recidivism: Findings from a longitudinal study in Hamilton County, OH. *Crime and Delinquency, 38*(2), 131–157.

Murray, W., Newnam, S., Watson, B., Davey, J., and Schonfeld, C. (2003). *Evaluating and improving fleet safety in Australia. Road Safety Research Grant Report.* Canberra, Australia: Australian Transport Safety Bureau.

National Highway Traffic Safety Administration (NHTSA). (1998). *Saving teenage lives: The case for graduated licensing.* Available at http://www.nhtsa.dot.gov/people/injury/new-driver/SaveTeens/Index.html.

National Institute for Occupational Safety and Health (NIOSH). (2003). *Work-related roadway crashes: Challenges and opportunities for prevention* (Publication No. 2003-119). Available at http://www.cdc.gov/niosh/docs/2003-119/.

National Occupational Health and Safety Commission. (1998). *Work-related traumatic fatalities in Australia, 1989 to 1992. Summary report.* Sydney, Australia: National Occupational Health and Safety Commission.

Newnam, S., Watson, B., and Murray, W. (2002). A comparison of the factors influencing the safety of work-related drivers in work and personal vehicles. *Proceedings Road Safety Research, Policing and Education Conference, 4-5 November, Adelaide, Australia,* pp. 488–494 [CD-ROM]. Adelaide, South Australia: Causal Productions.

Newstead, S.V., and Corben, B.F. (2001). *Evaluation of the 1992-1996 Transport Accident Commission funded accident black spot treatment program in Victoria* (Report No. 182). Melbourne, Australia: Monash University Accident Research Centre.

Peräaho, M., Keskinen, E., and Hatakka, M. (2003). *Driver competence in a hierarchical perspective; Implications for driver education.* Unpublished manuscript, University of Turku, Traffic Research.

Pilkington, P., and Kinra, S. (2005). Effectiveness of speed cameras in preventing road traffic collisions and related casualties: Systematic review. *British Medical Journal, 330,* 331–334.

Quimby, A.R., Maycock, G., Carter, I.D., Dixon, R., and Wall, J.G. (1986). *Perceptual abilities of accident involved drivers* (TRL Report RR27). Crowthorne, UK: Transport Research Laboratory.

Regan, M.A., Triggs, T.J., and Godley, S.T. (2000). Simulator-based evaluation of the DriveSmart novice driver CD-ROM training product. *Road Safety: Research, Policing and Education Conference: Handbook and proceedings* (pp. 315–320). Brisbane, Australia.

Ross, H.L. (1982). *Deterring the drinking driver.* Lexington, MA: Lexington Books.

Ross, H.L. (1985). Deterring drunken driving: an analysis of current efforts. *Journal of Studies on Alcohol, 1*(Suppl.), 122–128.

Runyan, C. (1998). Using the Haddon matrix: Introducing the third dimension. *Injury Prevention, 4,* 302–307.

Schöneburg, R., and Breitling, T. (2005, June). *Enhancement of active and passive safety by future PRE-SAFE systems.* Proceedings of the 19th International Technical Conference on the Enhanced Safety of Vehicles (ESV Conference), Washington, DC.

Senserrick, T., and Haworth, N. (2005). *Review of literature regarding national and international young driver training, licensing and regulatory systems* (Report No. 239). Melbourne, Australia: Monash University Accident Research Centre.

Senserrick, T., and Whelan, M. (2003). *Graduated driver licensing: Effectiveness of systems and individual components* (Report No. 209). Melbourne, Australia: Monash University Accident Research Centre.

Shope, J.T. (2006). Influences on youthful driving behaviour and their potential for guiding interventions to reduce crashes. *Injury Prevention, 12*(Suppl. 1), i9–i14.

Siegrist, S. (Ed). (1999). *Driver training, testing and licensing—towards theory based management of young drivers' injury risk in road traffic. Results of EU project GADGET, Work package 3*. Berne, Switzerland: Schweizererische Beratungsstelle fur Unfaliverhutung (BFU).

Simons-Morton, B.G., and Ouimet, M. (2006) Parent involvement in novice teen driving: A review of the literature. *Injury Prevention, 12*(Suppl. 1), i30–i37.

Symmons, M., and Haworth, N. (2005). *Safety attitudes and behaviours in work-related driving. Stage 1: Analysis of crash data*. Melbourne, Australia: Monash University Accident Research Centre.

Taylor, M., Lynam, D., and Baruya, A. (2000). *The effects of drivers' speed on the frequency of road accidents* (TRL Report 421). Crowthorne, UK: Transport Research Laboratory.

TFB-VTI Research. (1991). *Calculations—Road, traffic and vehicle engineering measures*. Linköping, Sweden: Swedish Transport Research Board and Swedish Road and Traffic Research Institute.

Tippetts, A.S., and Voas, R.B. (1998). The effectiveness of the West Virginia interlock program. *Journal of Traffic Medicine, 26*(1-2), 19–24.

TOI. (2003). *Motorcycle safety—a literature review and meta-analysis of countermeasures to prevent accidents and reduce injury*. (English summary). Available at http://www.vv.se/templates/page3____19023.aspx.

Voas, R.B., Marques, P.R., Tippetts, A.S., and Beirness, D.J. (2000). Circumventing the alcohol safety interlock: The effect of the availability of a noninterlock vehicle. *Proceedings of the 15th International Conference on Alcohol, Drugs and Traffic Safety, Stockholm, Sweden* [CD-ROM].

Warren, R.A. (1982). Rewards for unsafe driving? A rejoinder to P.M. Hurst. *Accident Analysis and Prevention, 14,* 169–172.

Whelan, M., Groeger, J.A., Senserrick, T.M., and Triggs, T.J. (2002). Alternative methods of measuring hazard perception: Sensitivity to driving experience. *RS2002 Road Safety: Research, Policing and Education: Conference Proceedings* [CD-ROM]. Adelaide, South Australia: Causal Productions.

Weinrath, M. (1997). The ignition interlock program for drunk drivers: A multivariate test. *Crime and Delinquency, 43*(1), 42–59.

Williams, A.F. (1999). Graduated licensing comes to the United States. *Injury Prevention, 5,* 133–135.

Williams, A.F. (2006). Young driver risk factors: Successful and unsuccessful approaches for dealing with them and an agenda for the future. *Injury Prevention, 12,* 4–8.

Williams, A.F. (2007). Contribution of the components of graduated licensing to crash reductions. *Journal of Safety Research, 38,* 177–184.

Williams, A.F., and Haworth, N.L. (2007). *Overcoming barriers to creating a well-functioning safety culture: A comparison of Australia and the United States*. In *Improving traffic safety culture in the United States: The journey forward* (pp. 77–91). Washington, DC: AAA Foundation for Traffic Safety.

Ylvinger, S. (1998). The operation of Swedish motor-vehicle inspections: Efficiency and some problems concerning regulation. *Transportation, 25,* 23–36.

12 On Not Getting Hit
The Science of Avoiding Collisions and the Failures Involved in That Endeavor

Peter A. Hancock

CONTENTS

Reflections 224
12.1 Introduction 224
12.2 The Minnesota Experiments 226
12.3 Behavioural Accident Avoidance Science: Understanding Response in Collision Incipient Conditions 227
12.3.1 Abstract 227
12.3.2 Introduction 228
12.3.2.1 Accident Information 229
12.3.2.2 Accident Evaluation 229
12.3.2.3 Investigative Rationale 231
12.3.3 Experimental Method 233
12.3.3.1 Experimental Facility 234
12.3.3.2 Scenario Description 234
12.3.3.3 Experimental Participants 236
12.3.3.4 Experimental Procedure 238
12.3.3.5 Experimental Design 239
12.3.4 Experimental Results 239
12.3.4.1 Intersection Scenario Results 240
12.3.4.2 Hill Scenario Results 242
12.3.5 Discussion 247
12.3.6 Summary and Practical Recommendations 248
12.3.7 Acknowledgments 249
References 249
12.4 Afterword 251
References 252

REFLECTIONS

There are many reasons why one might want to know about driver response capacity. Frequently, manufacturers want to sell their cars and component manufacturers want to sell technologies that appear in the vehicle. These companies certainly want to know what connotes safety in vehicle operation and when they might be creating circumstances that lead to unsafe behavior. Similarly, transport regulators and civil engineers are required to create safe and efficient highway systems that encourage the smooth and uninterrupted flow of traffic. They want to know about driver behavior because it affects the way that the architecture of the infrastructure is framed and developed. These are reasonable and understandable goals given the individual mandates of the institutions involved. But what of the behavioral scientist; what should be the primary motivation behind such research? For the manufacturer, the primary motivation is profit, for the regulator it is social facilitation, but for the behavioral scientist it should be collision prevention. This turns out to be a difficult enterprise because avoiding collisions is only partially related to the primary metric of transport, which is the transition of people, goods, and services from origin to destination.

12.1 INTRODUCTION

I make no apology for first referencing the classic work of Gibson and Crooks (1938), which I would argue is the most important single paper ever written on the behavioral aspects of driving. In this work, Gibson and Crooks referenced the goal of transport as the transition from origin to destination noted above. They indicated how the mappings of the vehicle controls themselves were intimately linked to this goal. Thus, for example, the accelerator is designed to facilitate speed of transition by initiating and sustaining the forward velocity of the vehicle. Let us imagine for a second, that we are out on Bonneville Salt Flats and have been cleared for a run. Almost by definition there are no obstacles to our progress, nor are there other vehicles in our path. The sole purpose of the exercise is to transition as quickly as possible between two measured points. The accelerator here is our prime control. True, we need to use the steering wheel for some lateral control, but the real need is for speed! As we know, these are the types of settings for establishing world land speed records. During such attempts there are no efforts to introduce other traffic. There is no absolute land speed record that incorporates obstacle avoidance. These are the pure circumstances in which virtually the whole focus is on the speed of transition.

Unlike such pristine, record-breaking circumstances, normal everyday driving (and even other forms of race driving) does not (at least at present) permit these untrammeled passages of progress. Indeed, in urban driving it appears that obstacles and especially other vehicles are constantly in one's way. It is here that the braking control plays a major role. Braking allows one to decelerate and thus avoid collision with objects and vehicles in the longitudinal *field of travel* in front of one's own vehicle. But note now, especially, that pure braking in this fashion runs counter to the central goal of transportation! In fact, the more one uses the brake, the less one is achieving the central goal of being in the vehicle in the first place (although I readily admit there are certainly individuals who drive for pure pleasure—so-called Sunday drivers whom I always find myself driving behind). Thus the accelerator and the brake are antagonistic not merely at the kinetic level but at the goal level also. Again, this situation is seen in its most pristine form in train operations. The presence of the railway tracks mean that control is always longitudinal and never lateral (see Branton 1979; Oborne, Branton, Leal, Shipley, and Stewart 1993). However, this frustrating situation for the driver is especially evident on the urban freeways of major cities in which the crowded lanes of the highway mean that the only progress is a sequence of stop–start events. In light of the frustration this builds up, it is unsurprising that major incidents of road rage seem to occur more frequently when these conditions pertain.

Again, the above examples really draw on relatively unusual conditions of world speed records and total freeway congestion. For many of us, we most often exercise the third control option of using the steering wheel. This permits lateral control and thus facilitates the compromise between the goals of the accelerator and the brake by permitting the continuation of passage without the need to slow or even halt progress. But this degree of freedom comes with a price. It mandates the need for almost constant driver attention to the roadway, especially in conditions in which obstacles are liable to appear in one's path of progress. In driving in general, the vehicle has a fairly high degree of dynamic stability but if uncorrected still wanders off of a straight-line course through variations in the roadway or circumstances such as side winds. Here, the driver must make constant, small adjustments to heading to keep the vehicle on track. However, in more complex maneuvers such as passing a line of parked vehicles, the driver must now select a path of progress that avoids all other obstacles of both a static and dynamic nature. Gibson and Crooks (1938) labeled this envelope of acceptable paths the *field of safe travel*. This represents both the longitudinal and lateral compromises between goal achievement and collision avoidance.

There are some rather obvious sequelae to these observations. If, as a driver, you fail at the task of obstacle avoidance and hit something, you will most probably fail in your primary goal of getting from origin to destination. However, if you focus your effort exclusively on obstacle avoidance it is doubtful whether there is much point in using the vehicle. Since powered transportation is only augmented locomotion, it implies that in order to be effective, driving must be faster than walking. We know that this is often not true in the downtown area of major cities and hence many residents choose not to have a vehicle. However, for the majority of us, we possess a vehicle because we could not reach our goal destination effectively without one. The other important thing to note is that collision is such a *relatively* rare event (that is, rare for any one individual driver), that we tend to underestimate its importance. Unfortunately, when major collisions do occur they are often life-changing and even life-ending events. So, while manufacturers and regulators continue with their business of overall transport efficiency, it is the prime mandate of the behavioral scientist to understand the processes of collision and to seek ways of ensuring that they do not happen. Paradoxically, we know very little of this phenomenon.

Despite the previous assertion, we do know a considerable amount about collision events themselves. For example, epidemiologists work to collect terrabytes of data concerning the spatial and temporal distribution of such collisions. Accident analysts perform an almost endless number of investigations seeking post hoc explanations of specific events. Mechanical engineers have engaged in decades of laudable activity, searching for ways to dissipate kinetic energy during collisions so that drivers do not pay the ultimate price for failure in the collision-avoidance task. In the latter effort, we have seen many significant and praiseworthy gains in terms of crumple zones, air-bag technology, and the like. But what about avoiding the collision in the first place? What do we know about those vital few seconds just before a collision and, more importantly, what do we know about those vital few seconds before the successful avoidance of an imminent collision? That is, what do we know about the behavioral response during the process of incipient collision itself? The answer is lamentably little. In some sense this is understandable. It is very hard for a scientist or researcher to gain access to these "moments of terror." We cannot have people exposed to the dangers of actual collision purely for research purposes and memory-based accounts are almost inevitably impoverished and biased, especially in the case of severe collisions. Indeed, for very severe collisions, the individuals involved may have no memory of the impact at all. What then can we do to access these vital moments of behavior to understand them better and to facilitate their training and transfer to promote consistent and successful resolution?

12.2 THE MINNESOTA EXPERIMENTS

This was the puzzle that faced us at the University of Minnesota where we had the good fortune to have the support of the Minnesota Department of Transportation and the university's Center for Transportation Studies (CTS) under the direction of Richard Braun, and especially the CTS's Intelligent Transportation Studies Institute under the direction of Dennis Foderberg. For a considerable period of time before these opportunities arose, I had been concerned with collisions, especially those involved with left turns (Caird and Hancock 1994; Hancock, Wulf, Thom, and Fassnacht 1990). Although some of this work, which I designed, had been done on actual roadways (Rahimi, Briggs, and Thom 1990), the work at Minnesota was primarily conducted in simulation facilities (Manser, Hancock, Kinney, and Diaz 1997). One continuing problem with respect to these studies was the behavior of other vehicles in the environment. Most often these were programmed to follow a set path at a set velocity or a set distance from a preceding or following vehicle. In short, these were "dumb" vehicles that did not respond to anything in their environment. This contrasted sharply with actual events in which it was evident that vehicles involved in multivehicle collisions respond to each other in complex ways. This is especially evident in marine accidents where the size and inertia of the vehicles involved often reveal the dynamics of collision on an extended time scale (Perrow 1984). As a result of these observations, and especially because I was lucky enough to possess two functioning, full-vehicle simulators, I had the idea to link the two simulators together and to explore two drivers in a collision-likely situation. There were many technical barriers to this achievement that individuals more capable with computers than I, including Jim Klinge, Peter Easterlund, and Erik Arthur, fought to solve in terms of software and hardware. In and of themselves, these issues provided nontrivial barriers and required innovations in simulation capacity to solve. Further, I had to devise an experimental protocol that allowed me to observe incipient collision behavior without letting the individuals involved know that a collision was possible. The following section, which is reprinted from the journal publication that emerged from these experiments, provides the account of that research program. At the end of this reproduced work, I provide a summarizing commentary on the work itself and some possible future directions for such efforts.

12.3 BEHAVIOURAL ACCIDENT AVOIDANCE SCIENCE: UNDERSTANDING RESPONSE IN COLLISION INCIPIENT CONDITIONS*

12.3.1 Abstract

Road traffic accidents are the single greatest cause of fatality in the workplace and the primary cause of all accidental death in the U.S. for individuals up to the age of seventy-eight. However, behavioural analysis of response in the final seconds and milliseconds before collision has been a most difficult proposition since the quantitative recording of such events has largely been beyond cost feasibility for road transportation. Here, a new and innovative research strategy is reported that permits just such a form of investigation to be conducted in a safe and effective manner. Specifically, a linked simulation environment has been constructed in which drivers are physically located in two adjacent, full-vehicle simulators acting within a shared single virtual driving world. As reported here for the first time, this innovative technology creates situations that provide avoidance responses paralleling those observed in real-world conditions. Within this shared virtual world forty-six participants (25 female, 21 male) were tested who met in two ambiguous traffic situations: an intersection and a hill scenario. At the intersection the two drivers approached each other at an angle of one-hundred thirty-five degrees and buildings placed at the intersection blocked the view of both drivers from early detection of the opposing vehicle. The second condition represented a 'wrong' way conflict. Each driver proceeded along a three-lane highway from opposite directions. A hill impeded the oncoming view of each driver who only saw the conflicting vehicle briefly as it crested the brow of the hill. Driver avoidance responses of steering wheel, brake, and accelerator activation were recorded to the nearest millisecond. Qualitative results were obtained through a postexperience questionnaire in which participants were asked about their driving habits, simulator experience, and their particular response to the experimental events which they had encountered. The results indicated that: 1) situations have been created which provided avoidance responses as they have been recorded in real-world circumstances, 2) the recorded avoidance responses depended directly upon viewing times, and 3) the very short viewing times in this experiment resulted in a single avoidance action, largely represented by a random choice of swerve to either right or left. The present results lead us to posit that in order to be able to design accident avoidance mechanisms that respond appropriately in the diverse situations

* Reprinted from Hancock, P., and De Ridder, S. (2003). Behavioural accident avoidance science: Understanding response in collision incipient conditions, *Ergonomics, 46*(12), 1111–1135. Article reprinted with permission of Taylor & Francis Ltd. (http://www.tandf.co.uk).

encountered, there is a need to pay particular attention to mutual viewing times for drivers. The general implications for a behavioral science of collision avoidance are evaluated in light of the present findings.

12.3.2 Introduction

The greatest single cause of fatality in the workplace is road traffic accidents. This startling fact is masked by two fundamental but obscuring issues. First, the workplace is traditionally considered to be a static location and so accidents which occur in vehicles in diverse locations are often excluded from the figures concerning workplace injury. Second, transportation accidents are themselves considered a single epidemiological category and so the traffic injuries associated with work are included in the general count of all road traffic crashes. The result of this form of categorization is that vehicle injuries are frequently overlooked or even excluded in the examination of the hazards of working life. Ergonomists work very hard to improve workplace safety and while we especially respect the achievements of allied researchers involved in traffic safety, we believe that a fruitful marriage can be made between ergonomic knowledge and the problems posed by traffic accidents. It is this overarching theme that motivates our work.

The present traffic safety community labours against a particularly insidious problem, which is that road traffic accidents are often considered by the public as somehow predestined. This popular fatalism is especially evident after high-profile accidents. For example, although the fund established in the name of Princess Diana provides millions of pounds to support efforts in areas as safety critical as land-mine decommissioning, it directs no substantive funds toward road accident reduction, the cause of her death. In his book debunking various conspiracy theories, Gregory (1999) expresses this attitude clearly in noting "in the shock of Diana's death, many had sought to impose a kind of romantic unity on her senseless end, speculating on a marriage which would lend an air of classical tragedy to what was *a thoroughly ordinary death in an avoidable car crash*" (p. 125, italics added). Indeed, such fatalism is reflected also in the fact that the vast majority of safety resources which to date have been directed to the accident question have focused overwhelmingly on crash survival. We are second to none in our admiration of those who have made crucial advances in air bag, crush zone, and restraint technology. They have assuredly saved many thousands of lives. However, it is almost as if collision were a given and the primary safety mandate is the protection of those already involved in such untoward events.

We believe this emphasis needs to be changed and that a formal science of *behavioural accident avoidance* should be established which draws heavily upon the armory of knowledge and tools possessed by those in Ergonomics research. We claim no unique precedent in this establishment and indeed point to the fast growing technical developments of collision-warning and collision-avoidance technologies of Intelligent Transportation Systems (ITS) as evidence of such burgeoning concern. The human-centered approach is clearly one in which Ergonomists provide the lead. Thus, when the vehicle is the workstation, there is a crucial role for those in both physical and cognitive ergonomics in the battle against this silent but most deadly of occupational hazards. Further, we see this marriage as one that benefits both traffic safety and ergonomics since the fundamental issues of human error and response limitation are a strong mutual concern of each (see Hancock 2003). The field of behavioral accident avoidance has only recently become open to empirical investigation through technical innovations in linked simulation and it is this approach we have helped pioneer to produce the results first reported in this work. As a first step, we address the larger picture of accident occurrence as found in major epidemiological accident databases.

12.3.2.1 Accident Information

While the number of motor vehicle collisions relative to the number of vehicles on the road has diminished, the increase in the absolute number of collisions and thus the total number of people killed and injured indicates the persistent and destructive global impact that motor-vehicle accidents have. In 1998, in the United States alone, there were over 6.3 million police-reported traffic crashes. Over 37,000 people lost their lives and 4.3 million people were injured. More than four million collisions involved property damage only and it is reasonable to assume that there were many more collisions of lesser severity that went unreported to any database. Our efforts here are initially most relevant to multiple vehicle collisions and in 1998, there were 16,184 such fatalities. Of these 46.3% (7,489) occurred with vehicles approaching at an angle, 32.4% (5243) occurred in a head-on configuration, 11.7% (1,896) were rear-end collisions, and 3.7% (599), were side-on collisions. This national pattern is also reflected in crash statistics for the state of Minnesota. In 1998, Minnesota reported 92,926 traffic crashes in which 650 people lost their lives and 45,115 were injured. In crashes of known configuration, 81.7% (51,820) involved multiple vehicles in which both were in motion.

One level of clarification of these findings can be found by examining reported vehicle tracks prior to collision. These data are derived from diagrams in police reports and are presented in Table 12.1. In examining the adjusted crash figures, we find that the top three categories each involve multiple vehicle configurations. These include: rear-end collisions, left-turns against oncoming traffic and right-angle crashes (and see Hancock et al. 1988; Hancock et al. 1991; Caird and Hancock 2002). Each of these is particularly relevant to the form of investigation considered in the present experimental procedure. Thus, crash data confirm that intervehicle collision is a crucial concern and one that addresses the majority of crashes including fatality and major injury (see also Treat 1980). These collective findings confirm the societal damage, including occupational injury and death, resulting from road traffic accidents. Further, such data show the relevance of our particular concern for injury and fatality reduction. In this sense, the epidemiological data serve to focus and direct our efforts.

12.3.2.2 Accident Evaluation

Accidents are examined by many different disciplines at many different levels. We have illustrated this in Figure 12.1 with a Cartesian coordinate system using the axes of space and time of progressively increasing magnitude. For example, the epidemiological perspective we have initially employed examines accident patterns on a very large scale. Typically, databases are generated at the State and Federal level

TABLE 12.1
Crash Involvement Illustrated by Police Diagrams

Maneuver	Reported	Percentage	Adjusted	Percentage
Rear end	20,143	21.7	20,143	21.7
Right angle	17,363	18.7	8,682	9.3
Ran off road, right	6,703	7.2	6,703	7.2
Sideswipe passing	5,370	5.8	5,370	5.8
Ran off road, left	4,918	5.3	4,918	5.3
Left turn, oncoming traffic	4,537	4.9	13,218	14.2
Head on	2,516	2.7	2,516	2.7
Sideswipe opposing	1,381	1.5	1,381	1.5
Right turn, cross traffic	510	0.5	510	0.5
Other/unknown	29,485	31.7	29,485	31.7

Source: Data from Minnesota Accident Facts 1998.

Note: In the original reported data, as given in the first column, the "right angle" category is the second largest. This is reported, however, as a significant error. Traffic engineers have measured the true number of right-angle accidents to be half the number the police reports. Crashes that are coded as "right angle" are often "left turn into oncoming traffic." The adjusted numbers take this into account. The large number in the category "unknown" accounts for the fact that in many cases the diagram is left blank. (See also Minnesota Department of Public Safety, 1998, Table 1.23.)

and are compiled yearly, thus integrating information over large spatial and temporal ranges. As we have shown, such information helps us to frame National policy and show general areas in which to focus more specific research, e.g., the problems experienced by very young and older drivers as shown by the classic 'bath-tub' curve (Dewar 2002). At the other end of the scale we have mechanical engineers involved with crash severity mitigation technologies such as 'crush zone,' 'air bags,' and similar developments. The window on the accident process for these engineers is framed in terms of milliseconds and centimeters since this is the 'scale' of their phenomena of interest. In the growth of any one area of research, scientists endeavour to expand their range of concern. For example, traffic engineers have traditionally constructed models of traffic flow to better help design and manage roadways. Often, such models focused upon freeway flow with 'node' points for every mile in the model. Today, such researchers are refining their spatial and temporal scales, advocating the addition of arterials and local streets and digitizing at the scale of yards while also significantly increasing the temporal frequency of their sampling. Thus, in the search for causation it is often the case that scientists appeal for explanation to other levels of spatio-temporal levels of analysis than their own.

To truly understand crash causation one has to integrate information from all levels of analysis. However, we suggest that it is most important to feature information from the behavioural level. We have to comprehend events over the ranges of metres and seconds, since these are the scales of immediate human perception (James 1890; Hancock and Chignell 1995). Until recently, quantitative information concerning

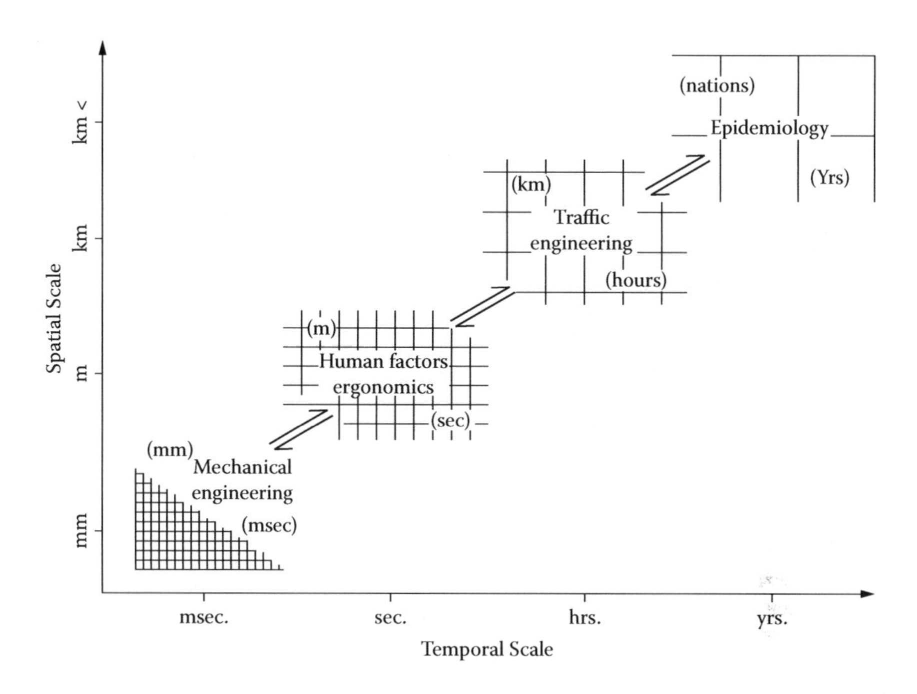

FIGURE 12.1 Spatio-temporal representation of the scales of action involved in accident research. At the largest scale, epidemiology identifies trends on a national basis at an annual rate. At the lowest extreme, crash mitigation technologies developed by mechanical engineers deal with millimeters and milliseconds. The present behavioral level analysis permits the investigation of accidents the human scale of seconds and meters.

behavioural response in accident events has been most difficult to collect since we cannot intentionally expose any individual to that level of danger. Subjective accounts of crashes are beset by the severe problems associated with recall memory. While some forms of reconstruction can inform us as to precollision physical manoeuvres, almost no technique can elucidate the human perceptual, cognitive, and motor responses that occur in the last fateful seconds before impact. Thus we affirm that the present experimental innovation provides a new window on the accident process that we hope to exploit to provide new information on such crucial events in transportation and indeed other realms beyond.

12.3.2.3 Investigative Rationale

In view of the above observations, there should be relatively few experimental research reports on driver performance in incipient crash circumstances and indeed this is the case. Beyond the vehicle trajectories and subjective report, it is immensely difficult to assemble this portrait of momentary driver response (Hancock and Scallen 1999). Most existing research has concentrated on who gets into dangerous or crash-likely situations (Hakinnen 1979; Summala 1987, 1996; Rothengatter 1997; Trimpop and Kirkcaldy 1997; Berthelon et al. 1998). However, evaluating and comprehending quantitative aspects of behavioural response in the vital milliseconds before collision has rarely been reported. Such research that does exist concentrates mainly on obstacle avoidance manoeuvrering where the obstacle put in the field of travel is controlled in some preset fashion (e.g., Barrett et al. 1968; Malaterre et al. 1988; Lerner et al. 1995). Whenever another vehicle was present it was controlled by the experimenter (Malaterre et al. 1988; Lechner and Malaterre 1991). Such approaches render very important data, however, they are limited in that they cannot ascertain and evaluate the reciprocal action between drivers who mutually adapt to the incipient demands.

There are other forms of investigation, which could inform us as to behaviour in collision-likely conditions. These can be divided into three basic categories. The first category focuses on time-to-contact, time-to-passage, and curve negotiation (see for example, Manser and Hancock 1996). The questions here concern the nature of the information drivers use to determine 'safe' behaviour with respect to the constraints of the roadway and the actions of other drivers (Groeger 1999; Caird and Hancock 1994; Manser and Hancock 1996; Sidaway et al. 1996). The relationship to collision is an implicit one with the often unstated but pervasive expectation that poor time-to-contact performance will be correlated and/or causally linked with collision involvement. This is especially the case when drivers in the real-world are required to judge motion-in-depth, such as in the case for oncoming vehicles at left-turns. Evidence for such a correlational relationship is sparse and a causal relation in the real world has still to be demonstrated. Nevertheless, the obvious fact that time-to-contact estimates and collisions are both intimately involved with navigating around complex, changing environments cannot be denied and it is upon this general basis at least, that such research is hopeful of adding to crash comprehension (see Hancock and Manser 1997). While time-to-contact research is providing an important theoretical foundation it does not represent the whole picture of collision avoidance.

The second relevant field of research that is directed to determining crash causation is epidemiology. As we noted, epidemiology seeks to understand what exogenous factors contribute to crash involvement, such as age, gender, etc. Endogenous factors such as cognitive and or visual impairments, attitudes or risk-taking behaviour, reaction time, field dependence, and close-following behaviour are often inferred from epidemiological information (Babarik 1968; Heyes and Ashworth 1972; Elander et al. 1993; Shinar 1993; Summala 1996). While much understanding has been gleaned from this form of investigation (see Evans 1991), many causal mechanisms have yet to be clarified. It has been suggested and there is some evidence that variations in attention are related causally to accident involvement (Kahneman et al. 1973). However, as might be suspected, providing online evaluation of momentary attention as crashes occur imposes exceptionally difficult methodological challenges, although such challenges are being taken up. The third contributory field concerns traditional traffic engineering. This includes elements of the driving environment such as road characteristics, control devices, and traffic flow and how these factors 'cause' possible hazardous situations (Rajalin et al. 1997; Steyvers and de Waard 1997). The confluence of this collective evidence provides a general framework for behavioural accident avoidance, however, it does not inform us as to the exact behavioural response just prior to the collision or more importantly, inform us as to what characteristics of response permit successful avoidance.

In this work, we are trying to determine what reaction patterns occur when drivers encounter an accident-likely situation and more importantly, successfully avoid collision. The determination of what constitutes a near-accident situation is largely up to the driver and may be construed as the point at which other road users enter their *'safe field of travel'* (Gibson and Crooks 1938). Drivers generally adapt to changes in the traffic system, whether these changes occur in the vehicle, in the road environment, in the weather and road surface conditions, or in their own skills or state. Such reactions occur in accordance with their motivations (Summala 1987, 1997; Summala and Mikkola 1994). One of the few experimental evaluations of such response is the report of Rizzo et al. (1997). These authors developed a graphic tool for analyzing driver performance and possible errors that may lead to crashes. Their participants were a group of older, licenced drivers, who were cognitively impaired due to mild or moderate Alzheimer's disease. They report the advantage of using a high-fidelity simulator in combination with this experimental evaluation tool as a new way of looking at accidents and individual differences in driver behaviour. Another relevant study relating to the issue of individual differences in driver response is that of Babarik (1968) where it is argued that people getting into (multiple) rear-end accidents are not necessarily slower drivers than others, but actually faster. Drivers who are faster to react to somebody else braking in front of them change the ratio of the cars to intervehicle space and make it harder for following drivers to avoid them. Thus slow reaction may be an advantage in this common driving manoeuvre.

Our hypothesis of multiple-vehicle accidents is a specific one. We view the sequence of events as a form of Markov process in which the avoidance actions of each driver are necessarily linked together and act to negate each other. Thus our hypothesis is amenable to modelling through a closed-loop feedback architecture. A critical feature of the model is that the timing of the respective avoidance actions fall within the respective response times of the two involved drivers. Thus, while each driver seeks specifically to avoid the other, their sequential responses act to nullify their mutual goal of mutual avoidance. The fact that these 'conditions' in which the respective responses become 'locked' together are rare, is reflected in the relatively infrequency of collisions in general as set against the opportunity of their occurrence. Below, we examine our dynamic systems-based theory in a specific situation but we are especially aware that our conception can well address other collision configurations and indeed collision etiology in circumstances well beyond transportation alone.

12.3.3 Experimental Method

In order to answer the question of how drivers perform in an accident-likely situation, a simulated environment was constructed in which two drivers meet each other in the same virtual world in a situation that has a strong potential for a collision. Driver performance is assessed by velocity control, braking, as well as steering response. We chose the respective scenarios in this study based upon accident statistics for the State of Minnesota and the whole United States. (In countries which drive on the left side of the road, clearly, these selections would be different.) In the U.S., the three most common accidents situations are the angled, head-on, and rear-end collision.

For Minnesota, the situation is somewhat different, since the accident statistics are differently grouped. However, when we sum left-turn oncoming traffic, right-turn cross traffic, and right-angle collisions together, we end up with a percentage of over 24%, which is comparable to the numbers reported for the whole USA. Simply providing possible crash scenarios does not necessarily mean that the crash will end up in that same category. We cannot predict driver performance to that detail. This means that we need to provide scenarios that will include as many as possible of the prominent categories of accidents: angle (right/left and turning), head-on, and rear end. For this particular study we choose two major crash types, the head-on collision and intersection collisions.

12.3.3.1 Experimental Facility

In order to accomplish the task of investigating collision-likely conditions, we used the dual simulation facility at the Human Factors Research Laboratory at the University of Minnesota that is shown diagrammatically in Figure 12.2.

This configuration is represented by two adjacent, full-vehicle simulators, which share a common, virtual world. The vehicles 'appear' to one another in a shared virtual world and thus the drivers can interact with each other. In comparable forms of simulation, the alternate vehicles either follow prescribed, preset paths and essentially do not interact with the human driver at all, or they follow some form of avoidance algorithm generated in the software, which represents a programmer's view of avoidance behaviour not normal dynamic response. It is only in our shared environment that live drivers mutually interact with one another.

One of the vehicles (a 1990 Honda Accord) was located in front of a flat screen display that was 260 cm from the driver's eye point. An Electrahome three-lens projector projected a 225 by 165 cm field of view composed of a 1024 by 768-pixel display. Sound feedback was provided through a Sony Stereo receiver with home theatre speakers and a base shaker system that gave a representation of road and vehicle noise as calibrated to the momentary speed of the vehicle. A second vehicle (a 1990 Acura Integra) was located in a wraparound simulator, whose dimensions were 549 cm at maximum and 492 cm diameter at the floor. The eye point of the driver was located 240 cm from the screen. Sound feedback was provided by a satellite-subwoofer speaker system in the vehicle trunk and high-powered subwoofers under the driver's seat.

12.3.3.2 Scenario Description

In order to explore driver behaviour enacted in collision-likely conditions, the first requirement is to generate such conditions. This presents a number of conceptual and methodological challenges. In order that the findings from such simulation research be valuable in understanding real-world collisions, the development of the scenarios has to be as realistic as possible. That is, the drivers cannot be in the position of 'expecting' either a collision or a near-collision event. Further, in order to understand the unconstrained behaviour of drivers, it is not possible to then constrain their behaviour in terms of free control of the vehicle. Therefore, one of the first problems to be faced is how to coordinate the actions of the two drivers without their being aware of the on-coming event. We achieve this objective through use of traditional

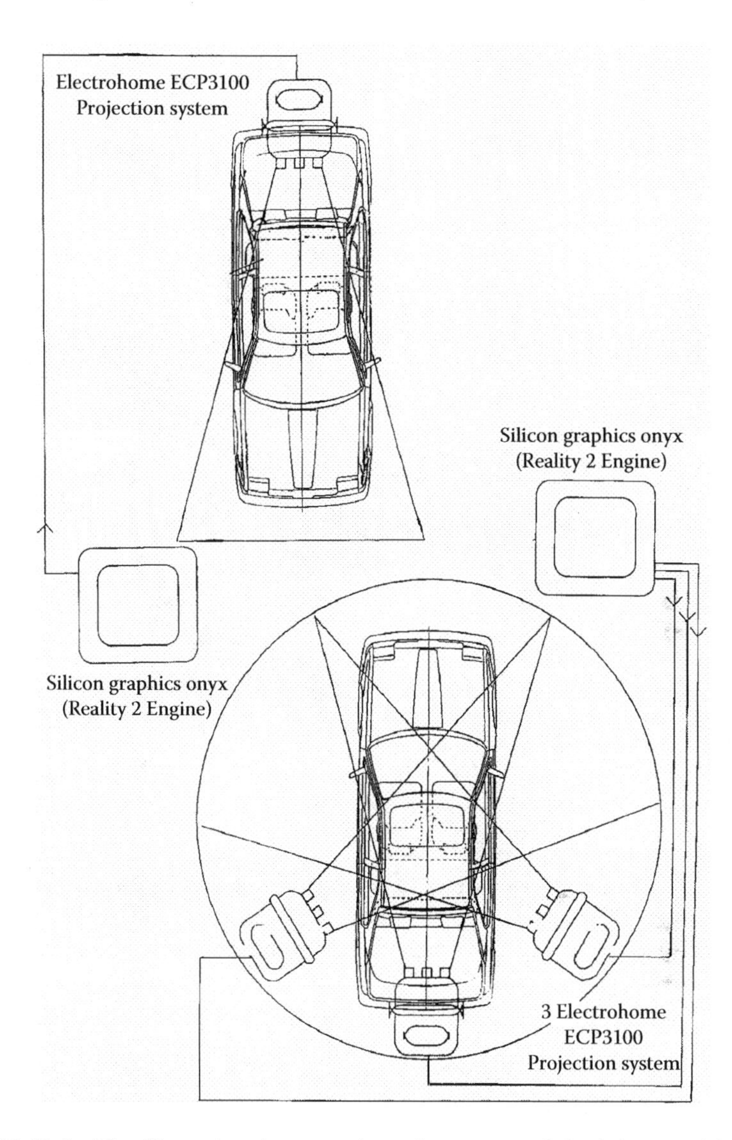

FIGURE 12.2 The illustration shows a schematic representation of the two side-by-side simulators in the Human Factors Research Laboratory of the University of Minnesota. The lower, wraparound facility provided a panoramic front field of view, while the single field of view simulator is shown above. The two facilities were linked to, and coordinated by, a single central computer that created the shared virtual world and synchronized actions within the world.

traffic control devices by having the drivers stopped at a traditional stop-light. When both drivers are in position, we let them proceed into one of the two scenarios (see Figure 12.3). As a result, we developed two scenarios that sought to answer these concerns and these are illustrated in Figure 12.4 and Figure 12.5.

FIGURE 12.3 In order to bring drivers into an accident-likely circumstance without interfering with their natural driving, we used traffic control devices. Here, a driver is waiting at a red light in the wraparound simulation facility and when each driver is in position, we change the light to green, which then triggers the collision-likely situations as described in the text.

The first scenario involved an unregulated, off-angle intersection. Both drivers approached the intersection and their mutual sight distance and therefore time prior to conflict could be controlled through the imposition of obstructive buildings positioned on the two corners of the intersection. This is a realistic circumstance for collision, although in many countries, sight distances at intersections are regulated to avoid this form of crash. In the second scenario, two drivers were placed on a unidirectional, three-lane highway and told to proceed in a safe manner obeying the traffic central laws. The drivers proceeded toward each other while their mutual progress was obscured by a hill whose dimensions and characteristics were manipulated in software, in order to influence sight distance and thus time for avoidance. This general condition is the equivalent in the real world to a 'wrong-way' incursion along a one-way thoroughfare. Thus the circumstance was unusual but not unrealistic.

12.3.3.3 Experimental Participants

Forty-six participants (25 female, 21 male) were recruited from staff and students of the University of Minnesota. All participants included in the analysis currently held a Minnesota driver's licence; they had normal or corrected to normal vision, and were between the ages of 18 and 80 years. Specifically, the mean age was 22.14 years of age (standard deviation 4.07 year). All drivers completed a driving questionnaire concerning their driving experience and driving habits and were debriefed as to the

FIGURE 12.4 The illustration shows the "hill" scenario from a high-up, side-on perspective. The respective drivers approach each other from the two ends of the roadway. By instruction set as to desired speed and change in the curvature (apparent steepness) of the hill, the experimenter can manipulate crucial independent variables such as mutual sight distance and, therefore, time-to-contact. This is accomplished without prejudicing the situation by warning the driver of a potential impending collision. We suggest here that any such prior warning negates the value of data collected when the driver is "on guard." Our method provides a way of circumventing this problem.

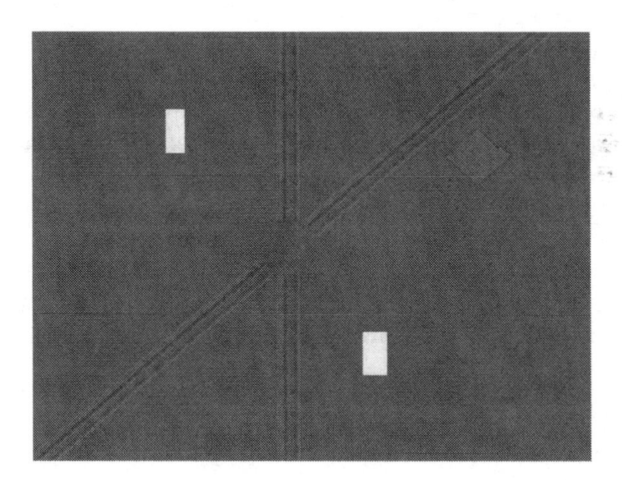

FIGURE 12.5 God's-eye view of the intersection collision scenario. As the curvature of the hill provides the control of certain independent variables in that scenario, so the positioning of the buildings accomplishes the same function in the intersection. Two caveats are important. First, in the real world, many roadway design and driver regulation manuals would prevent the minimal sight distance we have used in this experiment. Second, we experienced much greater difficulties in generating conflicts in this situation where the cars approached at an angle, compared to the head-on situation of the hill scenario.

nature of the experimental procedure and their reactions to the procedure following completion. The rules and regulations of the permission of the Human Subjects Committee were adhered to at all times.

12.3.3.4 Experimental Procedure

Participants came into the Human Factors Research Laboratory in pairs. Unbeknownst to each other, these two participants drove in the same simulated environment together. If, however, one of the two participants did not show up, one of the experimenters would drive the flat-screen simulator and act as an unresponsive driver, meaning the experimenter drove at a constant speed of 45 mph (72.4 kph) and was totally inactive when an accident likely situation occurred. These cases are referred to as a 'single-case' and were subject to separate analyzes. Participants were randomly assigned to either the wraparound simulator, or the flat-screen simulator. In the single-case trial, however, the participant always drove in the wraparound simulator.

All participants were given practice that lasted five minutes, or until they felt comfortable driving the vehicle in our simulated environment. At the end of practice all participants were asked via a standardized checklist if they felt comfortable enough to proceed with the next stage of the experiment. The experimenter then stepped out of sight and both participants were presented with four subsequent scenarios. Participants were asked to accelerate up to 45 mph (72.4 kph), in the lane that they were positioned in, at the start of the trial. During the trial they were informed that it was their task to drive at a safe and comfortable speed and obey any traffic laws that may apply. In this way driving behaviour is structured as it occurs in the real world, but not constrained unrealistically. All scenarios started with a red traffic light displayed on the screen. Participants were instructed to start driving when it turned green. The first and third scenarios consisted of a straight two-way road with buildings on either side. Other vehicles appeared both in the driver's own lane and the oncoming lane, but no accident-likely situations occurred. In these two scenarios that each lasted about two minutes, the two cars were not coupled.

The two cars were coupled into the same simulated environment in the second and the fourth scenario. After confrontation in the second scenario, both participants drove for another minute and were then uncoupled to drive the third trial. This trial again lasted two minutes where other traffic again was present but no accident-likely situations occurred. Participants were then coupled again and the fourth scenario was displayed. Following confrontation, participants would drive for another minute until the experimenter reappeared and told them the experiment had ended. Immediately after the experiment, participants were asked to fill in a questionnaire that consisted of questions about themselves and their driving habits, a survey on accidents the participants were involved in the past, and their remembrance of perceptions and actions before and during any accident they had been involved in. Questions about the feeling of control of the simulator and car, and questions to gain information on the remembrance of perceptions and actions of the participants during the trials and possible accidents, were also asked. After completing the questionnaire participants were debriefed as to the purpose of the study. The experimenter finally ensured that all participants left the experiment feeling relaxed and comfortable.

12.3.3.5 Experimental Design

In only one trial scenario (the intersection) the participants have a different viewpoint of the simulated world approaching from different directions to the 'target location.' The 'target location' is where the two cars are in an accident-likely situation and where avoidance strategies were measured. The intersection scenario is a case in which the two participants are both positioned in front of a stoplight and start driving at the same time when the traffic light turns green. In this way we can ensure, as far as possible, that the participants are coupled in a timely manner and thus give the greatest probability of conflict. After 200 metres both cars approached the intersection where the view from the other car is blocked by a building standing at the corner of the intersection. The two drivers cannot see each other and because there are no stop signs positioned at the intersection this is an accident-likely situation. The second coupled trial scenario involves the hill. Both cars started driving through a rural environment and were positioned on the middle lane of a three-lane, one-way road. They each start at a stoplight at the base of the hill. Both participants presumably 'assume' that no traffic will face them, but they are driving in the same lane on the same road approaching each other head on. They are not able to see however, because of the intervening hill. At the crest, or a little beyond (the 'target location'), the two cars meet and it is here that avoidance strategies are measured. To examine avoidance strategies, we examined three responses: swerving, acceleration, and braking. We look upon braking and acceleration as active responses whereas releasing the accelerator is a more passive, waiting response. We recorded the 20 seconds before, during, and after the point of closest approach. Even if drivers did not collide, they often swerve off the road seconds after the avoidance manoeuvre, as they do not appear to be able to stabilize due to, for example, distraction or shock.

12.3.4 Experimental Results

For the purposes of analysis, the results from the two scenarios were examined individually. In the intersection scenario, we evaluated the reactions of 13 pairs of drivers compared to the hill scenario in which we examined responses from 16 driver pairs. Decisions to exclude data for specific pairs from analysis were based on a number of factors. The first factor, consisting of four cases, involved the intersection scenario and was represented by a significant discrepancy in velocity between the two participants (>30 kilometres per hour [18.6 mph] at point of first sight). This led to situations where only one of the two participants briefly saw another vehicle passing the intersection far away in the distance and in these cases, neither of the two drivers engaged in any avoidance behaviour. These velocity discrepancies are evidence of just how difficult it is to create collision-likely conditions when no direct control can be exerted over driver response. The second exclusion of three cases involved the hill scenario and was justified by the fact that one of the two participants decided to drive in a lane other than the middle one by changing lanes prior to encountering the conflict situation. Again this represents an individual driving decision which our protocol permitted but which essentially negated the sought-after avoidance response. In one hill trial, the speed difference between the two vehicles meant that the cars encountered each other near the base of one side of the hill. This led to a situation with greatly extended viewing times and therefore was incompatible with all other recorded trials. However, from this trial, information was individually very useful and we employed this particular result as illustrative of a multiple response avoidance event that is the basis of a following investigation. We discuss this particular trial later in greater detail.

For the analyzed trials, point of first sight and point of closest approach were calculated using the following procedure. First, we determined the distance between the two vehicles throughout the whole trial by using the following coordinate equation:

$$d = \sqrt{\begin{matrix}((x\,coordinate\,of\,car\,\#1) - (x\,coordinate\,of\,car\,\#2)2) + \\ ((y\,coordinate\,of\,car\,\#1) - (y\,coordinate\,of\,car\,\#2)2)\end{matrix}}.$$

Once the distance between the cars for every data point was determined the respective points at which the two drivers are able to see each other for the first time are specified. These were calculated as 56 metres (61.2 yards) for the hill scenario and 209 metres (228.6 yards) for the intersection scenario. The point of closest approach is specified as the location where the minimal value of d is recorded. The following results are discussed in terms of, first, the intersection trials and then the hill trials.

12.3.4.1 Intersection Scenario Results

The mean age of the eight males and eighteen female drivers in this scenario was 21.4 years. All had valid drivers' licences that had been in their possession for an average of five years and they drove an average of 600 miles (965 kilometers) per month. Each participant was asked to answer a debriefing questionnaire designed to elicit responses concerning their driving habits, their perception of the simulator and the simulator controls, their perception of the trial conditions, and their perception of their own behaviour and performance. The questionnaire was composed of a combination of Likert-type, forced choice, and open-ended questions. Of their own on-road driving, they reported using city streets and highways more often than rural roads and almost never following a car too closely but almost always knowingly driving faster than the posted speed limit. They only periodically drove faster than the weather, traffic, or road conditions allowed. Eight participants had been involved in a self-reported accident. In general participants reported normal driving behaviours and felt comfortable in the simulated environment. They felt in control of the steering, accelerator, and brake, and drove at a speed that felt safe and comfortable. Twenty-five out of the forty-six participants felt their vision of traffic was obscured during part of the experience with most comments related to the intersection situation. This was reasonable given that our intended manipulation of sight distance in the intersection was specifically through the use of buildings to obstruct such sight distance. Characteristics of the participants based on the results of the questionnaires specific for these trials can be found in Table 12.2.

TABLE 12.2
Characteristics of the Participants in the Intersection Trial Based on the Questionnaire

Question	Min	Max	Mean	SD
Age	18	31	21.4	4.04
Year first acquired driver's license	1984	1998	1994.4	3.26
Number of kilometers per month	0	3218	978.9	943.4
Number of accidents involved in	0	3	.52	.87
Question in Likert-Type Scale (1= always, 3 = sometimes, 5 = never)			**Mean**	**SD**
How often do you drive?			1.9	0.93
How much of your driving occurs on city streets?			2.4	0.96
How much of your driving occurs on rural/country roads?			3.5	0.81
How much of your driving occurs on highways?			2.4	0.81
How often do you knowingly follow a car in front of you too closely?			3.7	0.84
How often do you knowingly drive faster than the posted speed limit?			2.1	1.01
How often do you knowingly drive faster than weather, traffic or road conditions allow?			3.6	1.02
Question in Likert-Type Scale (1 = always, 3 = mostly, 5 = none)			**Mean**	**SD**
I felt nauseous			4.2	1.20
I felt in control of the steering			2.8	1.10
I felt in control of the accelerator			2.3	1.04
I felt in control of the brake			2.5	1.24
I felt in control of the car			2.3	0.72
I drove at a speed that was comfortable			1.8	0.88
I drove at a speed that was safe			2.3	1.02

In respect of the quantitative results for the intersection trials, the first outcome was that the intersection scenario evoked considerably fewer active avoidance manoeuvres compared with the hill scenario. Only nine participants felt it likely at some point in the trial they were getting into an accident and only two drivers reported having experienced an accident. Speed differences between drivers had an overwhelming influence here since any significant difference meant that no conflict occurred. The closest point of approach had a wide range (5.47–44.33 metres), resulting a mean of 19.4 metres (21.2 yards) and a standard deviation of 14.22 metres (15.5 yards). Given the longitudinal difference for an accident (i.e., instant colocation of the two virtual vehicles) was only 4.5 metres (4.92 yards) and the comparable lateral distance was 2.0 metres (2.19 yards), it is evident that few actual collisions occurred. Although accident-likely situations in this particular scenario were thus infrequent, it is interesting that only three participants chose to register no response reaction at all as they approached the intersection. An overview of the response behaviours that participants manifested can be found in Table 12.3. As is evident, the strongest response pattern is one of conservatism in the uncertain situation as represented by the reduction of speed. However, this is a relatively passive and cautious response consisting of an

TABLE 12.3
Avoidance Manoeuvres for the Intersection Trial

Avoidance Manoeuvre	Number Occurred	Percentage
Brake	6	23.07
No brake	20	76.9
On accelerator	4	15.39
Off gas	22	84.61
Brake plus off gas	5	19.23
No brake plus on gas	3	11.53
Sped up	1	3.85

'Off Acceleration' reaction. Positive brake activation was itself relatively rare. Few drivers exhibited any form of aggressive response, although there was one participant who sped up in order to 'beat' the other driver to the intersection. In keeping with our hypothesis, drivers who respond with different strategies, e.g., cautious versus aggressive, do not meet in this present scenario since they start at a common distance from the intersection. However, those with common response strategies do tend to encounter each other. Although this might, in general, be considered a limitation of the present intersection scenario, examining collision-likely conditions between drivers of difference response type can be accomplished in this configuration by staggering start distance. However, since the hill scenario answers this particular concern and produced significantly more conflicts, it is to these results we now turn.

12.3.4 Hill Scenario Results

Thirty-two drivers, with a mean age of 22 years, participated in the sixteen trials. They drove 600 miles per month on average and they had possessed a valid Minnesota driver's licence for approximately six years. They classed their own driving as 'normal' and reported driving on city streets and highways 'almost always' as to 'almost never' on rural roads, which is a reasonable pattern given our local Metropolitan sample. The drivers reported almost never following a car too closely, almost always driving faster than the posted speed limit, but never faster than the road or weather conditions would allow. Fifteen participants reported having been involved in an accident and filled in our special questionnaire on these accidents. In relation to simulator control, participants felt in control of the steering, the gas, the brake, and the car, in general drove at a speed that was safe and comfortable.

Twenty-three participants reported that they felt their vision was obscured at some point in the trial. When asked more specifically about the obstruction, all of these individuals referred to problem of not being able to see over the hill. While this accords with our experimental design to control mutual sight distance, it suggests that participants were aware of the problem of the configuration of this road. Five participants reported having lost their attention at some point in the trial. When asked directly they again referred to the road configuration as the reason for this. Twenty-nine participants reported in retrospect that they felt they were getting into

TABLE 12.4
Characteristics of the Participants in the Hill Trial Based on the Questionnaire

Question	Min	Max	Mean	SD
Age	18	34	22.8	4.9
Year first acquired driver's licence	1981	1997	1992.5	4.4
Number of kilometers per month	0	3218	973.7	805.8
Number of accidents involved in	0	5	.87	1.18
Question in Likert-Type Scale (1 = always, 3 = sometimes, 5 = never)			**Mean**	**SD**
How often do you drive?			2.03	0.97
How much of your driving occurs on city streets?			2.38	0.94
How much of your driving occurs on rural/country roads?			3.50	0.80
How much of your driving occurs on highways?			2.20	0.79
How often do you knowingly follow a car in front of you too closely?			3.60	1.00
How often do you knowingly drive faster than the posted speed limit?			2.20	0.95
How often do you knowingly drive faster than weather, traffic, or road conditions allow?			3.50	1.10
Question in Likert-Type Scale (1 = always, 3 = mostly, 5 = never)			**Mean**	**SD**
I felt nauseous			4.34	1.12
I felt in control of the steering			2.60	1.02
I felt in control of the gas			2.03	0.78
I felt in control of the brake			2.25	1.04
I felt in control of the car			2.25	0.72
I drove at a speed that was comfortable			1.70	0.88
I drove at a speed that was safe			2.25	1.05

an accident. They referred to a fear of another car at the other side of the hill but this was after the event had occurred. Characteristics of the drivers based on the questionnaire results are presented in Table 12.4.

All driver pairs experienced an accident-likely event in this scenario. The closest distance between the two cars ranged from 2.91 metres (3.18 yards) to 0.374 metres (0.4 yards). This means that all drivers needed to perform a control maneuver to avoid colliding with the car that entered their forward safe field of travel (Gibson and Crooks 1938). Twelve participants reported a crash in this situation. The distance between the cars is measured from the midpoint of each car model. When the cars are positioned head on toward each other, the minimum distance without being in collision is 4.5 metres (4.92 yards). A smaller distance is required when the cars are passing each other, at which point the minimum distance is only 2 metres (2.2 yards). If, at the point of closest approach, the distance between two cars does not exceed 2 metres (2.2 yards) they have collided. In 8 of the 16 pairs this was the case and a collision did occur. Two participants reported a collision that in fact, according to the quantitative data for point of closest approach represented a very near miss. Ten participants correctly identified collision, and four reported not to have collided while

TABLE 12.5
Avoidance Manoeuvres for the Hill Condition

Avoidance Manoeuvre	Number of Occurrences	Percentage
Swerve left	17	53.1
Swerve right	14	43.8
No swerve	1	3.1
Brake	9	28.1
No brake	23	71.9
Off gas	29	90.6
Not off gas	3	9.4
Swerve plus brake	8	25.0
Swerve plus no brake	23	71.9
No swerve plus brake	1	3.1
No swerve/no brake	0	0

in fact they did. All participants performed at least one avoidance manoeuvre and these are detailed in Table 12.5. A representation of one of these individual avoidance manoeuvres is illustrated graphically in Figure 12.6.

As was evident in the intersection situation, the predominant response on the hill is also a passive, off-the-gas response. In most cases, this is not accompanied by a braking response, rather this seems to be a 'wait and see' strategy as to how the situation will develop. As for the actual avoidance manoeuvre itself, it is overwhelmingly a change in direction, that is, lateral control of the vehicle, rather than braking which represents longitudinal control. We are very aware that our scenario promotes this form of response and indeed a valuable future contribution will be to distinguish how and in what manner the configuration of the roadway and the approaching vehicle trajectory dictates the predominant form of response. In the present circumstance, the lateral avoidance manoeuvre is certainly consistent with Gibson and Crooks' (1938) 'field of safe travel' conception, however, it is important to note that given that each vehicle travels in the centre lane, the option to go either right or left is not specified by the 'field of safe travel' proposal. As we discuss below, the response of the individuals in this experiment is informative as to our own specific hypothesis.

Of the 21 participants who reported what direction they swerved in, only two accurately identified their own response. Given that this was a relatively benign simulation with no legal ramifications, the misidentification rate strongly illustrates the problem of memorial recall of these forms of emergency event. An important observation is that participants did not react in any systematic fashion. Right and left swerves occurred almost equally and these did not seem to be directly contingent upon any preemptive action on behalf of the other conflicting driver. Why this is the case is at present not clear. In point of fact, some drivers report having been taught to swerve to the right in such a condition, a most useful strategy. Therefore we performed a post hoc calculation ascertaining that the mean mutual viewing time for each pair was small (approximately a 1.2 seconds mean). Given so limited a viewing

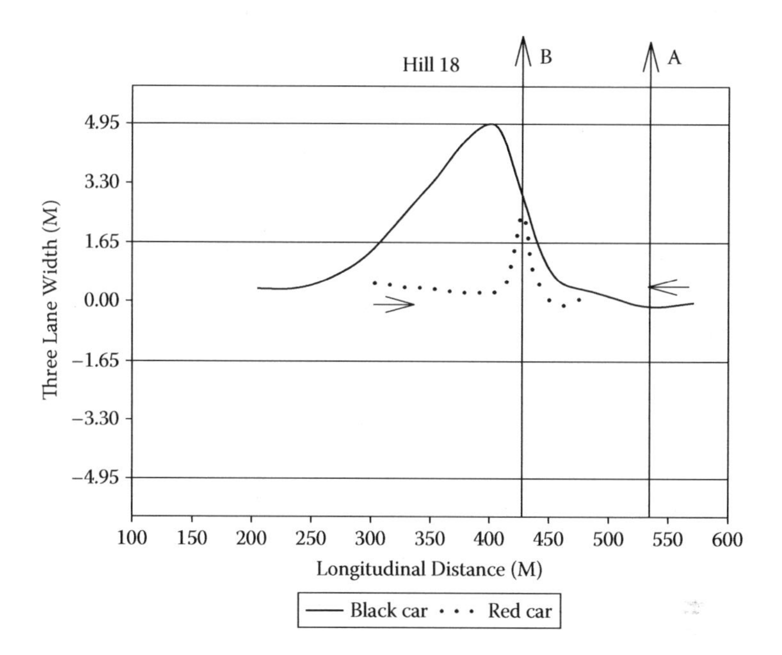

FIGURE 12.6 This graph shows a "window" of the original data stream. Once the point of closest approach is measured between the cars, 10 seconds before and 10 seconds after this point are plotted in order to examine driver performance in the few essential seconds before and after an accident-likely situation. In this particular trial (Hill 18) participants meet each other after the actual crest of the hill due to different velocities, which lengthens viewing times and also results in the difference between the length of the plotted lines. (A = crest of the hill, B = point of closest approach.)

time, it is evident that response patterns are essentially single reactions rather than avoidance strategies per se and thus the swerve right strategy would serve drivers well in such conditions. More evidence for the restriction to a single response lies in correlations between the time of first possible sight and the onset of the first avoidance action for each driver being very high (0.998 and 0.996, respectively) as well as the fact that braking occurred only infrequently (71.8% did not use the brake pedal at all). In essence, this was a 'see and avoid' situation, which did not permit enough time for multiple, linked avoidance responses to occur. Interestingly, however, the correlation between the reaction times of both drivers even in this brief interval is high (0.95). This supports the contention that the behaviour of the two drivers is still 'interlocked' in some fashion even for these brief mutual, viewing times. The results presented in Figure 12.7 as well as Table 12.6 confirm these observations. Of course we recognize the general problem of time restriction here, i.e., the drivers only have a certain 'window' of time in which to respond anyway. As a consequence of these findings, we are proceeding with subsequent experiments that open up the window of possible response by permitting longer viewing times.

Evidence that more extended viewing times may result in more interactive patterns of response come from the data for one pair of drivers (where the trial was designated 'Hill 18,' see Figure 12.6). Due to the large speed difference between the

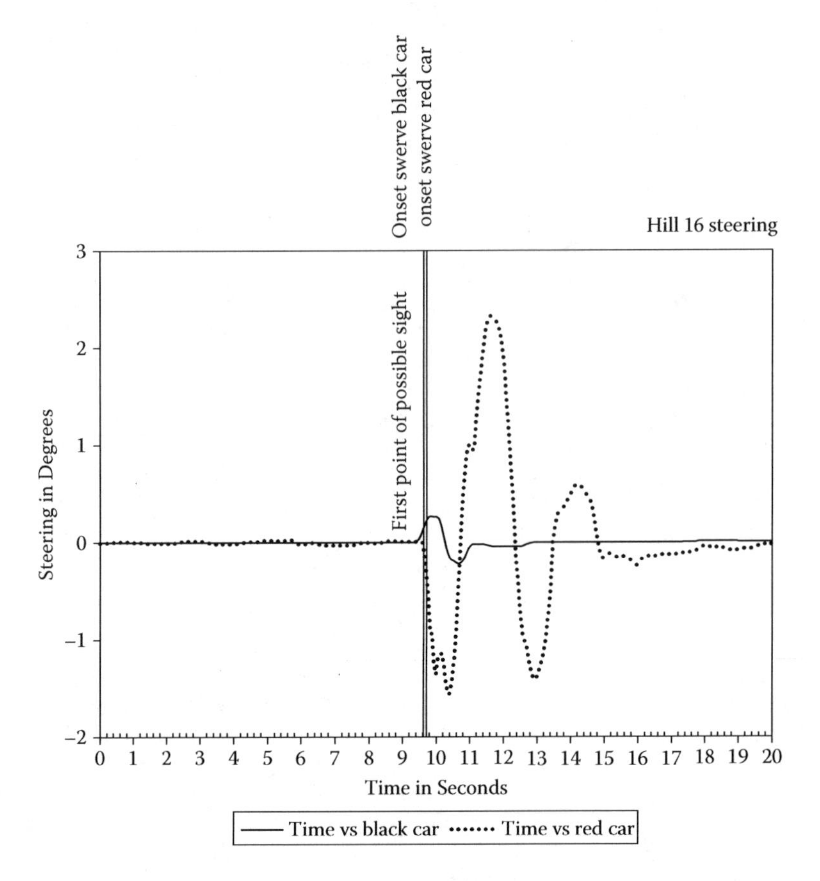

FIGURE 12.7 Illustration of the atypical joint kinematic traces of the opposing vehicles. This measure shows vehicle lateral control through change in steering direction.

two vehicles (one had crested the hill as the other began the ascent) these drivers had a much longer mutual viewing time, in the order of several seconds. This gave the opportunity to examine interaction for a greater period of time. In this case a mutual interaction did occur and although we have the evidence in the kinematic traces for the trial as illustrated in Figure 12.6, it is perhaps best expressed by the subjective report of one of the drivers:

TABLE 12.6
Driving and Avoidance Profile of Both Cars in the Hill Trials

Measurement	Car 1	SD	Car 2	SD
Mean speed in kilometers per hour	56.7	12.2	59.3	8.3
Mean onset of swerve in seconds	8.78	0.63	9.13	0.61
Mean speed at onset of swerve	50.5	17.4	55.9	11.9
Mean total reaction time	1.82	2.36	1.48	1.61

> When a car emerged over the top of the hill, in the lane I was in, I steered to the right, then left when the car facing me followed my direction. The car appeared to follow me when I tried to avoid it by steering right.

There is perhaps no better evidence as yet to date for the linked avoidance response hypothesis, in which the intended avoidance actions of each individual cancel each other out to result in unwanted collision.

12.3.5 Discussion

It is our hope that, using the tools and methods of Ergonomics, we have opened a new window on the accident process by examining avoidance response at a behavioural level. In terms of the present results, we have found that when there is a relatively ambiguous driving situation in which drivers identify cues that suggest possible problems, the primary response is one of caution, expressed as an 'off the accelerator' action. In effect, this action, by reducing velocity, serves to increase the global time-to-contact and thus time to reach the general problem area. As evident in the formulation of Gibson and others (see Gibson 1966, 1979; Hancock et al. 1995), this action response may itself allow time for the situation to disambiguate itself and for the appropriate response to become evident. Given the relative infrequency of accidents compared to the number of opportunities for their occurrence, it is evident that this response is overwhelmingly effective and it is only in very rare or unusual circumstances that such ambiguity persists. In both of the scenarios we have investigated, the preferred acute avoidance response is one of lateral control (i.e., swerving the vehicle), as compared to our original expectation of much greater use of braking. In part, this is, of course, a response to the configuration we have exposed our participants to. However, it remains a surprising finding given the supposed greater efficacy of both brake and steering response in mitigating high-momentum impact.

In the present experimental research we have shown that realistic avoidance behaviours can be created and replicated in the interactive simulation environment. As such the first, and in essence, the major contribution of this work is that a new technique is now available for the investigation and amelioration of all vehicle collisions. This conclusion is buttressed by both the objectively recorded driver responses and their concomitant subjective report of the validity of the experiences they encountered during the different scenarios. In addition, we have also addressed and provided one innovative solution to the highly intractable problem of behaviour shaping. In many experiments in the behavioural research laboratory, the experimenter 'frames' the participant's response through instruction sets and testing protocols. In the present work, we sought specifically to overcome this form of self-fulfillment. To do so, we created purpose-specific conditions in which through the simplest of instructions 'drive safely and follow the rules of the road' we have managed to bring drivers into a surprise conflict situations which only they can resolve. Together, with these successes of methodology we have also created an interactive simulation environment in which the time-lag problem across two facilities has been sufficiently controlled to permit essentially coincident driving. Thus the present work has exhibited technical as well as investigative success.

Having described these successes, it is equally important to indicate current shortfalls that provide areas in which substantive improvement is possible. In the case of the intersection scenario, the result of permitting each driver complete freedom is that the velocity differential between vehicles often negated the occurrence of conflict. This itself is evidence that providing participants freedom of action will often 'compromise' an experimental procedure to the point where the experimenter's purpose is obviated. To remedy the intersection situation, we are in the process of developing dynamic software manipulations, which, without the knowledge of either driver or any change in the perceptual environment, can momentarily change the relative positions of the respective vehicles to increase the probability of a conflict, although drivers' avoidance responses will not be affected in any way.

With respect to the hill scenario, a major problem in the present experiment was mutual viewing time. With the hill curvature we have chosen, in combination with the speed selected by the drivers, the viewing time on average was sufficiently small that only a single avoidance action could be taken. In our continuing experiments we are providing longer viewing times by changing hill curvature and through the use of simulated levels of fog. However, our basic thesis concerning interaction between drivers received most encouraging support from the hill trial in which viewing time was extended by the great speed differential. In addition to these useful advances, we have had to develop some new approaches to examining the contingent dual kinematic traces, an illustration of which is shown in Figure 12.6.

With respect to the specific findings of the present experiment in the two different scenarios, the hill trial showed unequivocally that the reaction times of the drivers permitted only a single avoidance manoeuvre. Overwhelmingly these manoeuvres consisted of a swerve in a single direction. In respect of the limitation of viewing time, this is not surprising. Further, the swerving tactic may well have been encouraged by the presence of an open lane on either side of the oncoming vehicle. We expect that specific avoidance patterns (i.e., swerving, braking, or swerving and braking) will be contingent on the characteristics of the roadway in a manner consistent with Gibson and Crooks' (1938) notion of the field of safe travel. Our present methodology that we have developed permits the first true test of this proposition over sixty years since its postulation.

12.3.6 Summary and Practical Recommendations

Our first simple and practical recommendation relates to head-on collisions. Our information confirms that in the process of driver education, young drivers should be taught to 'swerve to the kerb side' in the case of incipient, head-on collision. It is clear that in any multivehicle collision the opportunity for avoidance and propensity for damage and injury is contingent on the actions of both drivers. Thus while one driver might make a significant avoidance response, collision may still not be avoided if the other driver makes no response, or worse makes a response that cancels out that of the other. Through the recommendation of the kerb-side swerve strategy, we will maximize the chance of collision avoidance even if both drivers can make only a minor manoeuvre. Parenthetically, this will require different directional response contingent upon whether one is in a country that drives on the left or the right side of the roadway. Thus, in head-on conflicts, swerve to the kerb.

In the conception, fabrication, and installation of computer-assisted collision-warning and collision-avoidance systems, currently envisaged under many ITS programmes, the optimal design configuration is one that reinforces and supports the natural driver avoidance response. While it is clear that the specific situation will prove a primary influence on what tactic it is best to adopt, it is clear from the present results that a system which complements the anticipatory process and assists in vehicle slowing when approaching ambiguous situations does serve the process of support for human-centered, rather than technology-centered, avoidance activities. Thus, in potentially ambiguous situations, assistive devices should focus on prediction and prevention rather than instantaneous amelioration as current technologies are envisaged.

Our final recommendation is one that occurs in the vast majority of experimental papers and that is a call for further research. However, here we wish to articulate such a need in a little more detail. In the present, we have sought to help open a new window on the accident process. However, this is only a start. What is clearly required as a next step in the process is a programmatic and sustained effort on behalf of many researchers in order to take advantage of the opportunity which dynamic, interactive simulation presents. In the very first statements of the present work we established the clear societal importance of this effort for both occupational concerns and general injury. However, also evident was that the sheer number of researchers in behavioural accident avoidance research is too small for the task. Therefore, by the present work, we appeal to fellow Ergonomists to take up this challenge. Interactive simulation can certainly address traffic collisions, however, judiciously developed, such a technology can also inform many other areas of human–human–machine interaction. If this capability can help in the battle to save life and reduce injury, we shall have earned our salt.

12.3.7 Acknowledgments

We are very grateful for the help of Neil Ganey and Derek Diaz in the final preparation of this manuscript. The experimental situation would not have been possible without the technical skills of Jim Klinge and Peter Easterlund and we gratefully acknowledge their assistance. An earlier version of the present work was awarded the Liberty Mutual Prize for Ergonomics and Occupational Safety, given by the International Ergonomics Association (IEA). We are most grateful for the comments of the respective award reviewers and adjudicators on an earlier version of this paper. Finally, we would like to thank the editor and the unknown reviewers for their comments, which helped to finalize the work.

REFERENCES

Babarik, P. (1968). Automobile accidents and driver reaction pattern. *Journal of Applied Psychology, 52*, 49–54.

Barrett, G., Kobayashi, M., and Fox, B. (1968). Feasibility of studying driver reaction to sudden pedestrian emergencies in an automobile simulator. *Human Factors, 10*, 19–26.

Berthelon, C., Mestre, D., Pottier, A., and Pons, R. (1998). Is visual anticipation of collision during self-motion related to perceptual style? *Acta Psychologica, 98,* 1–16.

Caird, J.K., and Hancock, P.A. (1994). The perception of arrival time for different oncoming vehicles at an intersection. *Ecological Psychology, 6,* 83–109.

Caird, J.K., and Hancock, P.A. (2002). Left turn and gap acceptance crashes. In: R.E. Dewar and P. Olson (Eds.), *Human Factors in Traffic Safety*. Tucson, AZ: Lawyers and Judges Publishing, 613–652.

Dewar, R.E. (2002). Age difference: Drivers old and young. In: R.E. Dewar and P. Olson (Eds.), *Human Factors in Traffic Safety*. Tucson, AZ: Lawyers and Judges Publishing, 209–233.

Elander, J., West, R., and French, D. (1993). Behavioral correlates of individual differences in road-traffic crash risk: An examination of methods and findings. *Psychological Bulletin, 113,* 279–294.

Evans, L. (1991). *Traffic Safety and the Driver.* New York: Van Nostrand Reinhold.

Gale, A.G., Brown, I.D., Haselgrave, C.M., and Taylor, S.P. (Eds.). (1998). *Vision in Vehicles VI.* Amsterdam: North-Holland.

Gibson, J.J. (1966). *The Senses Considered as Perceptual Systems.* Boston: Houghton-Mifflin.

Gibson, J.J. (1979). *The Ecological Approach to Visual Perception.* Boston: Houghton-Mifflin.

Gibson, J.J., and Crooks, L.E. (1938). A theoretical field-analysis of automobile driving. *American Journal of Psychology, 51,* 453–471.

Gregory, M. (1999). *The Diana conspiracy exposed.* Milford, CT: Olmstead Press.

Groeger, J.A. (1999). Expectancy and control: Perceptual and cognitive aspects of the driving task. In: P.A. Hancock (Ed.), *Handbook of Visio-Spatial Cognition XVII: Human Performance and Ergonomics.* San Diego: Academic Press, 243–264.

Hakinnen, S. (1979). Traffic accidents and professional driver characteristics: A follow-up study. *Accident Analysis and Prevention, 11,* 7–18.

Hancock, P.A. (2003). The tale of a two-faced tiger: Identifying commonalties across accident types to reduce their frequency and impact by design. *Ergonomics in Design,* in press.

Hancock, P.A., Caird, J.K., and Johnson, S.B. (1991). The left-turn. In: Y. Queinnec and F. Doidlou (Eds.), *Designing got everyone. Proceedings of the International Ergonomics Association,* Vol. 2. London: Taylor & Francis, 1428–1430.

Hancock, P.A., and Chignell, M.H. (1995). On human factors. In: J.M. Flach, P.A. Hancock, J.K. Caird, and K. Vicente (Eds.), *Global Approaches to the Ecology of Human Machine Systems.* Hillsdale, NJ: Erlbaum, 14–51.

Hancock, P.A., Flach, J.M., Caird, J.K., and Vicente, K.J. (Eds.), (1995). *Local Applications in the Ecological Approaches to Human-Machine Systems.* Hillsdale, NJ: Erlbaum.

Hancock, P.A., and Manser, M.P. (1997). Time-to-contact: More than tau alone. *Ecological Psychology, 9,* 265–297.

Hancock, P.A., Rahimi, M., Wulf, G., and Briggs, R. (1988). Analyzing the behavior of left-turning drivers, In: *Proceedings of the International Technical Conference on Advanced Safety Vehicles.* Gothenburg, Sweden: ASV.

Hancock, P.A., and Scallen, S.F. (1999). The driving question. *Transportation Human Factors, 1,* 47–55.

Heyes, M.P., and Ashworth, R. (1972). Further research on car following models. *Transportation Research, 6,* 287–291.

James, W. (1890). *The Principles of Psychology.* New York: Holt.

Kahneman, D., Ben-ishai, R., and Lotan, M. (1973). Relation of attention to road accidents. *Journal of Applied Psychology, 58,* 95–108.

Lechner, D., and Malaterre, G. (1991). Emergency maneuver experimentation using a driving simulator. Presented at *Autotechnologies, 5th Conference and Exposition,* Monte Carlo, Monaco. Report No. SAE 910016.

Lerner, R., Huey, R., McGee, H., and Sullivan, A. (1995). *Older driver perception-reaction time for intersection sight distance and object detection*, FHWA Report FHWA-RD-93-168 (Washington, DC: Federal Highway Administration).

Malaterre, B., Ferrandez, F., Fleury, D., and Lechner, D. (1988). Decision making in emergency situations. *Ergonomics, 31*, 643–655.

Manser, M.P., and Hancock, P.A. (1996). The influence of approach angle on estimates of time-to-collision. *Ecological Psychology, 8*, 71–99.

Minnesota Department of Public Safety. (1998). *Minnesota motor vehicle 1998 crash facts.* Office of Communications, Minnesota Department of Public Safety, St. Paul, MN.

Rajalin, S., Hassel, S., and Summala, H. (1997). Close-following drivers on two-lane highways. *Accident Analysis and Prevention, 29*, 723–729.

Rizzo, M., Reinach, S., and McGehee, D. (1997). Dissection of car crashes in high fidelity driving simulations. In: K. Brookhuis, D. de Waard, and C. Weikert (Eds.), *Simulations and Traffic Psychology*. Groningen: Centre for Environmental and Traffic Safety, University of Groningen.

Rothengatter, T. (1997). Errors and violations as factors in accident causation. In: T. Rothengatter and C. Vaya (Eds.), *Traffic and Transport Psychology: Theory and Application.* Amsterdam: Pergamon, 59–64.

Shinar, D. (1993). Traffic safety and individual differences in drivers' attention and information processing capacity. *Alcohol, Drugs and Driving, 9,* 219–237.

Sidaway, B., Fairweather, M., Sekiya, H., and Mcnitt-Gray, J. (1996). Time-to-collision estimation in a simulated driving task. *Human Factors, 38,* 101–113.

Steyvers, F.J.J.M., and de Waard, D. (1997). Road-edge delineation in rural areas: Effects on driving behaviour. In: K. Brookhuis, D. de Waard, and C. Weikert (Eds.), *Simulations and Traffic Psychology.* Groningen: Centre for Environmental and Traffic Safety. University of Groningen, 163–178.

Summala, H. (1987). Young driver accidents: Risk taking or failure of skills? *Alcohol, Drugs and Driving, 3*, 79–91.

Summala, H. (1996). Accident risk and driver behavior. *Safety Science, 22,* 103–117.

Summala, H. (1997). Hierarchical model of behavioural adaptation and traffic accidents. In: T. Rothengatter and E. Vaya (Eds.), *Traffic and Transport Psychology: Theory and Application.* Amsterdam: Pergamon, 41–52.

Summala, H., and Mikkola, T. (1994). Fatal accidents among car and truck rivers: Effect of fatigue, age, and alcohol consumption. *Human Factors, 36,* 315–326.

Treat, J.R. (1980). A study of pre-crash factors involved in traffic accidents. *HSRI Research Review, 10,* 1–35.

Trimpop, R., and Kirkcaldy, B. (1997). Personality predictors of driving accidents. *Personality and Individual Differences, 23,* 147–152.

12.4 AFTERWORD

I should make very clear that I view the reported experiment as only the first step toward a science of behavioural accident avoidance. Unfortunately, as is the case with many such projects, time injected a number of changes, which meant I was unable to pursue this issue as I would have liked. I was recruited to Florida by the University of Central Florida and in leaving Minnesota had only limited access to driving simulation facilities. Although the effort to understand driver response capacity continues, there have been few experiments that have followed the present protocol, perhaps for technical reasons as linked simulation is difficult to achieve, and perhaps for financial reasons since funding for such work is still lamentably limited. Road traffic fatalities continue to be one of the major sources of early death in society and much is now being done to create technology-inspired, collision-avoidance systems. Like the palliative strategies of in-vehicle defense, such as seat belts and air bags, these are praiseworthy efforts. However, many collisions derive from human-centered phenomenon and virtually all collisions involve human concern. Thus approaching the issue from a human-centered, rather than an engineering/physics-based perspective promises to render even greater return in the prevention of these awful events. I hope others will take up the challenge laid down here to further articulate a science of behavioural collision avoidance. It is a both a scientific and a moral challenge that we should all embrace.

REFERENCES

Branton, P. (1979). Investigations into the skills of train-driving. *Ergonomics, 22,* 155–164.

Caird, J.K., and Hancock, P.A. (1994). The perception of arrival time of on-coming vehicles at an intersection. *Ecological Psychology*, *6*(2), 83–109.

Gibson, J.J., and Crooks, L.E. (1938). A theoretical field-analysis of automobile driving. *American Journal of Psychology, 51*, 453–471.

Hancock, P.A., Wulf, G., Thom, D., and Fassnacht, P. (1990). Driver workload during differing driving maneuvers. *Accident Analysis and Prevention*, *22*(3), 281–290.

Manser, M.P., Hancock, P.A., Kinney, C., and Diaz, J. (1997). Understanding driver behavior though application of advanced technological systems. *Transportation Research Record, 1573*, 57–62.

Oborne, D.J., Branton, R., Leal, F., Shipley, P., and Stewart, T. (Eds.). (1993). *Person-centered ergonomics: A Brantonian view of human factors*. London: Taylor & Francis.

Perrow, C. (1984). *Normal accidents: Living with high risk technologies*. Princeton, NJ: Princeton University Press.

Rahimi, M., Briggs, R.P., and Thom, D.R. (1990). A field evaluation of driver eye and head movement strategies toward environmental targets and distractors. *Applied Ergonomics*, *21*(4), 267–274.

Index

A

Acceptance, information acquisition, 176–181
Accidents, *see also* Collisions; Crashes
analysis, 132–139
distractions, 77
evaluation, 229–231
fundamentals, 141–142
information, 229
liability relationship, 89–93
risk, adverse weather conditions, 124–125
Accuracy, driving simulators, 14
Activation demands, distractions, 79
Adaptation
adverse weather conditions, 125–127, 140
lane width, 120, 140
roadway luminance, 41
visual demands and driving, 24–26
ADAS, *see* Advanced driver assistance systems (ADAS)
Adequacy, driving simulators, 14
Advanced driver assistance systems (ADAS), 82–83
Adverse weather conditions
accident risk, 124–125
behavioral adaptation, 125–127
fundamentals, 140–141
Advertisements, *see also* Signs
exploratory selection, 69–70
horizontal search window, 97–98
Aerial view training technique, 111
Affordance, information acquisition model, 7
Aging drivers, *see* Older drivers
Alarms, 177–181
Alcohol, *see also* Drugs
deliberate selection, 70
distractions, 79
enforcement programs, 208–209
interlocks, 212
mobile phone simulation, 190
older drivers, 91
road markings, 121
speeding effect, 215
younger drivers, 91
Alignment, roadway design, 123–124
Alzheimer's disease, 233, *see also* Dementia
Ambient (dorsal) vision, *see* Ambient-focal mechanisms
Ambient-focal mechanisms
ambient (dorsal) vision, 33–35
diagnostic signature, 35–37
estimation tasks, 53–54
eye allocation, 160–161
field of view, 42–44
focal (ventral) vision, 33–35
fundamentals, 32–34
heuristic value, 45–47
modes of visual processing, 33–35
preview distance/time, roadway, 44–45
properties, 34–35
roadway luminance, 40–41
theoretical framework, 32–37
two-level model, 35–36
visual acuity, 37–40
Ambient insufficiency hypothesis
heuristic value, ambient-focal framework, 47
visual requirements, 32
Analysis, collisions
accidents, 132–139, 229–231
design, 239
discussion, 247–248
evaluation of accidents, 229–231
experimental method and results, 233–247
facility, 234
fundamentals, 227–229, 233–234
hill scenario, 242–247
intersection scenario, 240–242
investigative rationale, 231–233
participants, 236, 238
procedure, 238
recommendations, 248–249
results, 239–247
scenario, 234–236, 240–247
summary, 248–249
Anticipation, mental model, 113
Applying makeup, 93
Attention, dividing and focusing, 113
Attention, testing, 16
Attentional efficiency, 76–77
Attentional selection
deliberate selection, 70–71
dimensions, 64–65
exploratory selection, 69–70
fundamentals, 64, 71
habitual selection, 68–69
modes, 65–71
reflection, 63–64
reflexive selection, 67–68

Attention systems, 81–83
Attentive selection, 64
Attribution of responsibility, 77–78
Auditory warnings, 68, *see also* Alarms
Australia
  commercial vehicle drivers, 208
  DriveRight campaign, 210
  driving simulators, 15, 15
  speed limit reductions, 215

B

Bandwidth, 5
Bathtub curve, 230
Behavioral accident avoidance, 228, *see also* Collisions, avoidance
Behavioral adaptation, 125–127, 140, *see also* Adaptation
Behavioral analysis and experiment, collision response
  accidents, 229–231
  design, 239
  discussion, 247–248
  evaluation of accidents, 229–231
  experimental method and results, 233–247
  facility, 234
  fundamentals, 227–229, 233–234
  hill scenario, 242–247
  intersection scenario, 240–242
  investigative rationale, 231–233
  participants, 236, 238
  procedure, 238
  recommendations, 248–249
  results, 239–247
  scenario, 234–236, 240–247
  summary, 248–249
Behavior prediction
  hazardous situations, 106–107
  preview distance/time, 44–45
  road users, 112–113
"Black box" recordings, 59
"Blindness"
  habitual selection, 69
  inattentional, thinking tasks, 84–85
Blinks and blinking, 81
Blur, 40
Boredom, 192
Brakes and braking
  collision avoidance, 224–225, 233
  driving simulators, 14
  estimation performance capacities, 56–57
  fast reaction time, 233
  hill scenario response, 242–247
  passive, off-the-gas response, 244
  warning signals, 194
Buildings, 122
Bus-shelter advertisements, 98, *see also* Advertisements
Button controls, adjusting, 91

C

Car controls, adjusting, 91
Car following, 97
Cell phones, *see also* Hands-free phones
  conversation pacing, 191
  distractions and accidents, 77
  enforcement, 77–78
  likelihood of collisions, 91, 93
  similar to alcohol, 190
  situation awareness, 101
  using, 189–191
  younger drivers, 93
Center for Transportation Studies (CTS), 226
Cessation of driving, 207
Channel bandwidth, 5
Channelization, 214
Chevron markings, 58, *see also* Herringbone markings; Markings
Citrus, 196
Civil twilight, 41
Cognitive demands, *see also* Decisions; Mental loads; Thinking tasks
  distractions, 77, 79
  eye movement recording, 20
  resources, conservation, 53
  thinking tasks, 83
Collisions, *see also* Accidents; Crashes
  auditory and vibrotactile warnings, 68
  likelihood estimations, 54–55
Collisions, avoidance
  accidents, 229–231
  design, 239
  discussion, 247–248
  evaluation of accidents, 229–231
  experimental method and results, 233–247
  facility, 234
  fundamentals, 224–229, 233–234
  hill scenario, 242–247
  intersection scenario, 240–242
  investigative rationale, 231–233
  Minnesota experiments, 226–227
  participants, 236, 238
  procedure, 238
  recommendations, 248–249
  reflections, 224
  results, 239–247
  scenario, 234–236, 240–247
  summary, 248–249
Comfort, 120, 140
Commercial vehicle drivers, 208

Compulsory vehicle inspection, 213
Concierge services, 153, *see also* In-vehicle technologies (IVIS)
Conflicts
  environment, 132–139, 141–142
  quantification, 134–136
Consciousness, distractions, 77–78
Conscious selection, 64
Conspicuity of a stimulus, 22, 158–159
Continuous memory task (CMT), 132
Continuous signal, 6–7
Continuous steering performance, 39
Controlled processing, 64–65
Controls, adjusting, 91
Conversation, *see also* Passengers; Talking
  mental load, 85
  pacing, 191
Cortico-centric dorsal-ventral stream mechanism, 34
Cost 331 study (European Cooperation in the Field of Scientific and Technical Research), 44
Countermeasures, 171–172, 175, *see also* Vehicle changes
Crashes, *see also* Accidents; Collisions; Hazards
  distractions and accidents, 77
  incentives and rewards, 209–210
  reflexive selection, 68
  run-off-road type, 214
Crossmodal information processing
  dividing attention, eye and ear, 188–192
  fundamentals, 188, 196
  interim summary, 192
  listening to radio, 192
  multisensory information displays, 192–193
  multisensory warning signals, 194–195
  reflection, 187
  sleepiness, 195–196
  talking on mobile phone, 189–191
  talking to a passenger, 191
  warning signals, 193–195
CTS, *see* Center for Transportation Studies (CTS)
Curves
  negotiation, 232
  road markings, 122
  speed, 140
  underestimating sharpness, 123–124
Cyclists, 96

D

DBI, *see* Driving Behavior Inventory (DBI)
DBQ, *see* Driver Behavior Questionnaire (DBQ)
Death rate, 53
Decisions, *see also* Cognitive demands; Thinking tasks
  distraction errors, 78
  perception models, 9
  societal forces, 60
Deliberate selection, 67, 70–71
Dementia, 206, *see also* Alzheimer's disease
Denmark, 60
Dependent variables (DV), 16, 23
Design, 239
Devices, 69–70, *see also* Technologies
Diabetes, 206
Diagnostic signature, 36–37
Diana, Princess, 228
Dimensions, attentional selection, 64–65
Distance
  estimation performance capacities, 55–56
  from vehicle in front, 14
Distractions
  accidents, 77
  activation distractions, 79
  anticipation distractions, 79
  attentional efficiency, 76–77
  attention systems, 81–83
  attribution of responsibility, 77–78
  cognitive distractions, 79
  cognitive effort, 83
  consciousness, 77–78
  drivers' expectations, 82–83
  errors, 76–79
  fundamentals, 76, 81, 86
  imposition of demands, 82
  inattentional blindness, 84–85
  information processing errors, 78–79
  internal distraction, 83
  measurements, mental loads, 83
  parameters, 80–81
  performance, 84–85
  reflection, 75–76
  scenario characterization, 80
  search, 80–81
  speed, 84–86
  thinking tasks, 83–86
  types of, 78–79
  vision, 84–85
  visual attention, 79–81
  visual distractions, 79
  visual search and performance effects, 84–85
  warning demands, 82
Dividing attention, eye and ear
  listening to radio, 192
  talking on mobile phone, 189–191
  talking to a passenger, 191
DOCTOR conflict quantification, 134–136
Driver Behavior Questionnaire (DBQ), 21
Driver comfort, 120, 140
DriveRight campaign, 210
Driver licensing, training, and management
  commercial vehicle driver licensing, 208

enforcement programs, 208–209
fundamentals, 203
graduated licensing, 204–205
incentives and rewards, 209–210
management, 208
motorcycle licensing, 207
older drivers, 206–207
testing, 206
training and education, 205–206
Driver management, 208–210
Drivers' expectations, 82–83
Driver's field of view, 42–44
Driver Skill Inventory (DSI), 21
Drivers' self-assessment, 20–21
Driver workload, 132, *see also* Cognitive demands; Decisions; Thinking tasks
Driving Behavior Inventory (DBI), 21
Driving on wrong side of road, 236
Drowsiness, *see also* Fatigue; Sleepiness
distractions, 79
electroencephalography, 24
likelihood of collisions, 93
olfactory cues, 196
Drugs, *see also* Alcohol
distractions, 79
driving errors, 77
Drunk driving enforcement programs, 208–209, *see also* Alcohol
DSI, *see* Driver Skill Inventory (DSI)
DV, *see* Dependent variables (DV)
Dynamic public lighting, 130

E

Eating, 91
Edgeline markings, 214
Education, 205–206, *see also* Graduated licensing; Training
Effort, 159
Electrocardiography, 24
Electroencephalography, 24
Elements, Endsley's model, 9–10
Emergency vehicles
distraction due to warning systems, 81–82
reflexive selection, 68
Empirical evidence, experiments
driver's field of view, 42–44
roadway luminance, 40–41
roadway preview distance/time, 44–45
visual acuity, 37–40
Endogenous factors, 232
Endogenous selection, 65
Endsley's model, 9, 101
Enforcement programs
driver management, 208–209
graduated licensing, 205
Environment
accident risk, 124–125
accidents, 132–139, 141–142
adverse weather conditions, 124–127, 140–141
alignment, road, 123–124
analysis, 132–139
behavioral adaptation, 125–127
conflicts, 132–139, 141–142
DOCTOR conflict quantification, 134–136
driver workload, 132
environment, 122–123
freeways, 129–130
fundamentals, 118–119, 139–140
gap acceptance, 132–139
implications, 136–139
lane width, 120
lateral clearance, 121
markings, road, 121–122
modeling, 136–139
pavement, 119–121
public lighting, 128–132, 141
reflection, 118
roadway design, 118–124, 139–140
roughness, road, 119–120
rural roads, 130–131
video observations, 133–134
Environmental distractions
attention systems, 81–83
drivers' expectations, 82–83
fundamentals, 81
imposition of demands, 82
warning demands, 82
Epidemiology, 232
Epilepsy, 206
Errors, distractions, 76–79
Estimations
collision likelihood, 54–55
fundamentals, 59–60
gap acceptance, 57–58
making, 52–54
movement perception, 58–59
performance capacities, 55–57
real-world application, 51–52
reflection, 51
European Cooperation in the Field of Scientific and Technical Research, *see* Cost 331 study
Evaluation, accident, 229–231
Events, 5–6
Exogenous selection, 65
Expectations, 82–83, 157
Experience and visual attention
accident liability relationship, 89–93
attention, dividing and focusing, 113
behavior prediction, other road users, 112
experienced drivers, 98–102

management, 113
mental model of situation, 113
novice drivers, 98–102, 112–113
reflection, 89
road users, predicting behavior, 112
and scanning, 93–98
training, 108–112
visual field, 102–107
visual search, 98–102, 108–112
Experienced drivers, *see also* Older drivers
anticipation, 113
cell phone use, 101–102
dangerous situations, 113
fixation, 104
hazardous situations, 96, 105
visual field, 102–107
Experiments, *see also* Collisions, avoidance; Testing
advantages, 16
driver's field of view, 42–44
roadway luminance, 40–41
roadway preview distance/time, 44–45
visual acuity, 37–40
Experiments, collision response behavioral analysis
design, 239
discussion, 247–248
facility, 234
fundamentals, 233–234
hill scenario, 242–247
intersection scenario, 240–242
participants, 236, 238
procedure, 238
recommendations, 248–249
results, 239–247
scenario, 234–236, 240–247
summary, 248–249
Exploratory selection, 66, 69–70
External validity, 16
Eye-freezing, 81, *see also* Fixation
Eyes, allocation, *see also* Visual attention
ambient-focal visual processing, 160–161
fundamentals, 152, 162–163
glance behavior and distribution, 155–156, 161–162
in-vehicle technologies, 152–156
modeling visual attention allocation, 156–160
reflection, 151–152
telematics, 154–156
visual demands, 154–156
Eyes, movement, *see also* Visual search
hazardous situations, 94, 97–98
horizontal search window, 97–98
novice drivers, 98–100
police pursuit video clips, 97
recording techniques, 17–20
training, 109–110
Eyes, pupils
measurements, 81, 83
mental load and speed, 85

F

Facility, 234
False alarms, 180
"Far" road ahead, 36, 45–46, *see also* Ambient-focal mechanisms
Fatality Analysis Reporting System (FARS), 53
Fatigue, *see also* Drowsiness; Sleepiness
deliberate selection, 70
distractions, 79
driving errors, 77
Field of safe travel
accident evaluation, 233
collision avoidance, 224, 225
information acquisition model, 7–8
Field of useful expansion, 54
Field of view
deliberate selection, 71
empirical evidence, experiments, 42–44
heuristic value, ambient-focal framework, 46
Field of vision, 14
Finland, 126
Fixation, *see also* Gazing
hazardous situations, 95–97
measurements, 80–81
variance locations, 98–100
Flashing lights
distraction due to warning systems, 81–82
reflexive selection, 68
Flowing array of stimulus energy, 55
Focal (ventral) vision, *see* Ambient-focal mechanisms
Fog, *see also* Weather conditions, adverse
automatic warning system, 127
behavioral adaptation, 125, 141
warning system, London, 126
Following a lead vehicle, 97
Foreseeing near future, 10, 112–113
France, 14, 15
Freeways, public lighting, 129–130
Freezing rain, *see* Weather conditions, adverse
French National Research Institute for Transport and Safety Research, 14

G

Gap acceptance
environment, 132–139
estimations, 57–58
modeling, 136–137
movement perception, 58

Gazing, *see also* Fixation
distractions, 79
mental load, 85
Gibson's model, 7–8
Glances and glance behavior
distractions, 93
distribution, 161–162
in-vehicle technologies, 155–156
visual demands, 154
GPS, *see* Navigation systems
Graduated licensing, 171, 204–205, 209
Guardrails, 214

H

Habitual selection
attentional selection modes, 68–69
fundamentals, 65–66
Haddon matrix, 202–203
Häkkinen classification, 11
Hands-free phones, *see also* Cell phones
mental load, 85
safety misconceptions, 78
simulations, 190
situation awareness, 101
Hazardous location treatment, 214
Hazards, *see also* Collisions; Crashes
avoidance, visual acuity reduction, 38
habitual selection, 68–69
management, 113
novice drivers, 171
perception test, 104
roadway luminance, 41
and scanning, 93–98
Heads-up displays, 69–70
Heavy vehicle fleets, 211–212
Herringbone markings, 122, *see also* Chevron markings; Markings
Heuristics, 45–47, *see also* Ambient-focal mechanisms
High-spatial-frequency information, reduction, 37
Hill scenario, 242–247
Holland, 15
Horizontal search window, 97–98
Housing, 122
Human Factors Research Laboratory, 234
Hypervigilance, 24

I

IAAV, *see* Integral Approach of the Analysis of Traffic Accidents (IAAV)
Ice-cream van example, 113
Implications, environment, 136–139
Imposition of demands, 82
Inattentional blindness
habitual, 69
thinking tasks, 84–85
Inattentional selection, 64
In-car distractions, 91, *see also* In-vehicle technologies (IVIS)
In-car entertainment, 192
Incentives and rewards, 209–210
Incomprehensible alarms, 180
Independent variables (IV), 16
Information, *see also* Ambient-focal mechanisms
acquisition, memory, 22
bits, 4
collision response analysis and experiment, 229
defined, 7
high-spatial-frequency information, reduction, 37
infovores, 66
models, information acquisition, 3–7
processing errors, 78–79
short-term context, 4–5
what *vs.* where, 33
Information acquisition
acceptance, 176–181
countermeasures, 171–172, 175
fundamentals, 168–169, 181–182
older driver crashes, 169–170
reflection, 167–168
reliance, 176–181
younger driver crashes, 170–171
Infovores, 66
Injury severity, 215
Institute for Road Safety and Research (SWOV), 134
Institut National de Recherche sur les Transports et leur Sécurité (INRETS), 14, 15
Instituto Nacional de Técnica Aerospacial (INTA), 14
Insulin-treated diabetes, 206
Integral Approach of the Analysis of Traffic Accidents (IAAV), 135
Intelligent transportation systems (ITS), 228
Intent, alarms, 178
Intentional selection, 64
Interim summary, 192
Internal distraction, 83
Internal validity, 16
International Traffic Conflicts Workshop, 134
Intersection safety, 213–214
Intersection scenario, 240–242
Interviews, drivers' self-assessment, 21
In-vehicle technologies (IVIS), *see also* In-car distractions
benefits, 152–154
cognitive demands, 82

concierge services, 153
costs, 152–154
distractions and accidents, 77
glance behavior, factors that impact, 155–156
modeling visual attention allocation, 156–160
navigation systems, 77–78, 152–153
novice drivers, 171
older drivers, 170
speech-based in-vehicle technologies, 85
visual demands, 82, 154–156
weather conditions, 127
Inventario de Situaciones Ansiógenas en el Tráfico, 21
Inventory lists, 21
Inventory of Situations Producing Anxiety in Traffic, 21
Investigative rationale, 231–233
IPods, 93, *see also* Technologies
ITS, *see* Intelligent transportation systems (ITS)
IV, *see* Independent variables (IV)
IVIS, *see* In-vehicle technologies (IVIS)

K

Knob controls, adjusting, 91

L

Lane keeping and lane excursions
field of view reduction, 42–43
visual acuity reduction, 39–40
visual attention, 80
Lanes
multiple, gap acceptance, 57
width, roadway design, 120, 140
Lateral clearance, 121
Lateral lane position, 44
Lateral speed, 39–40
Lead car following, 97
Learning ability, 16
Leeds Advanced Driving Simulator, 190
Left-hand turn maneuver, see Gap acceptance
Liability relationship, accidents, 89–93
Licensing
commercial vehicle drivers, 208
graduated, 171, 204–205
motorcycle licensing, 207
older drivers, 206–207
younger drivers, 204–205
Lighting, see Public lighting
Listening to radio, 192
Longitudinal markings, 58, 80, see also Markings
"Looking but not seeing" accidents
accidents, 77
distractions, 77, 79
heuristic value, ambient-focal framework, 47
older drivers, 32
perception impairment, 84
traffic accidents, 22
visual demands, 2
Loss, potential information, 5
Loss of vehicle control, 120
Low-spatial-frequency motion perception, 46
Luminance, see also Public lighting
reflexive selection, 67–68
of roadway, 40–41
Luoma classification, 12

M

Makeup application, 93
Making estimations, *see* Estimations
Management
drivers, 208–210
hazards, 113
Markings
driving simulators, 14
edgeline, 214
movement perception, 58
roadway design, 121–122
visual attention, 80
Measurements, 22–24, 83
Memory
hazardous situations, 94–95
performance measurements, 23–24
recording techniques, 21–22
testing, 16
Mental loads, *see also* Cognitive demands; Decisions; Thinking tasks
pupil dilation, 81, 83
speed, 84–85
visual search and performance effects, 84–85
Mental model development, 113
Mesopic range, vision, 41
Methodologies, information acquisition, 11–13
Middle-aged drivers, *see also* Older drivers; Younger drivers
likelihood of accidents, 89–90
preview distance/time, roadway, 44
Minnesota experiments, *see* Collisions, avoidance
Mirror use, 100
Mobile phones, *see* Cell phones; Hands-free phones
Models
Endsley's model, 9
environment, 136–139
Gibson's model, 7–8
information theory, 3–7
methodologies used, 11–13
models, information acquisition, 3–11

reflections, 2
sign detection theory, 8–9
visual attention allocation, 156–160
Modes, attentional selection
deliberate selection, 70–71
exploratory selection, 69–70
fundamentals, 65–67
habitual selection, 68–69
reflexive selection, 67–68
Modes, visual processing, 33–35
Moore model, 9
Motorcycles, 207
Mount Cotton driver training course, 42
Movement perception estimations, 58–59
MRT, *see* Multiple resources theory (MRT)
Multiple lanes, *see* Lanes
Multiple resources theory (MRT), 188
Multisensory information displays, 192–193
Multisensory warning signals, 194–195
Music, *see* Radio

N

NASA TLX workload index, 83
Navigation systems, *see also* In-vehicle technologies (IVIS)
alarms, 177–181
costs and benefits, 152–153
enforcement, 77–78
information acquisition, 172
multisensory displays, 192–193
Near-crashes, distractions and accidents, 77, *see also* Collisions, avoidance
"Near" road ahead, 36, 45–46, *see also* Ambient-focal mechanisms
Netherlands
automatic fog warning system, 127
dynamic public lighting, 130
road categories, 119
societal forces, 60
Neuropsychological systems, 24
Night driving
adaptation, 25, 26
ambient-focal mechanisms, 41
freeway lighting, 130
lighting and illumination, 128
likelihood of accidents, 90
older drivers, 170
road markings, 121, 140
roadway luminance, 41
slippery road condition sign, 126
North America, commercial vehicle drivers, 208
Novelty effect, 69–70, 156
Novice drivers, *see also* Younger drivers
adapting to changing traffic conditions, 92, 100–101
car purchasing, 213
cell phone use, 101–102
distractions, 92
eye movements, 98–100, 109–110
fixation, 103–104
hazardous situations, 95–96, 105
reflexive selection, 68
situation awareness, 101
training, 112–113
visual field, 102–107

O

Observations, research methods, 15–16
Off-the-gas passive response, 244
Older drivers, *see also* Experienced drivers
accident evaluation, 230
adverse weather conditions, 125
alarms, 181
alcohol, 91
car purchasing, 213
cessation of driving, 207
crashes, information acquisition, 169–170
deliberate selection, 71
eye movement recording, 20
heuristic value, ambient-focal framework, 46
information acquisition, 172–175
in-vehicle technologies, 156
licensing, 206–207
"looking but not seeing" accidents, 32
multisensory warning signals, 195
night driving, 90
passenger presence, 91
perceptual and motor processes, 168
poorly designed technologies, 176–177
Olfactory cues, 196
Oncoming traffic, 58
100 car study, 77, 93
Organisation for Economic Co-operation Development, 121
Organizational interventions, 210–212

P

Parameters, visual attention, 80–81
Parents role, graduated licensing, 205
Partial visual occlusion paradigm, 36
Participants, experiment, 236, 238
Passengers
distractions and accidents, 77
mental load, 85
talking with, 191
younger drivers, 91, 93
Passive, off-the-gas response, 244
Passive speed control, 58

Pavement, roadway design, 119–121
PDAs, 93, see also Technologies
Peppermint, 196
Perception
fundamentals, 2–3
performance measurements, 23
testing, 16
Performance
alarms, 177
capacities estimations, 55–57
field of view reduction, 42–43
thinking tasks, 84–85
Peripheral information, 122–123
PET, see Postencroachment time (PET)
Photopic adaptation conditions, 41
Physical parameters, memory, 22
Physical resources, conservation, 53
Pleasure-seeking tendency, 66
Poles, collisions with, 214
Postencroachment time (PET), 134–135, 136
Preattentive selection, 64
Precipitation, 124, see also Weather conditions, adverse
Predictability, alarms, 177
Predicting behavior
hazardous situations, 106–107
preview distance/time, 44–45
road users, 10, 112–113
Preview distance/time
empirical evidence, experiments, 44–45
hazardous situations, 106–107
Primary hazards, 96
Princess Diana, 228
Process, alarms, 177
Psychophysiological techniques, 24
Public lighting, see also Luminance
defined, 128–129
driver workload, 132
dynamic, 130
freeways, 129–130
fundamentals, 128, 141
rural roads, 130–131
Pupils, see also Eyes
measurements, 81, 83
mental load and speed, 85
Purpose, alarms, 178

Q

Questionnaires, self-assessment, 21

R

Radio, 192
Railway tracks, 225
Rain, *see* Weather conditions, adverse
RAPT, *see* Risk awareness and perception training (RAPT)
Reaction time, 22–24
Reading while driving, 93
Real-world application, estimations, 51–52
Recording techniques
drivers' self-assessment, 20–21
eye movement recording, 17–20
measurements, 22–24
memory, 21–22
psychophysiological techniques, 24
reaction time, 22–24
sign recognition, 22
traffic accidents, 22
Records, driving, 21
Redundancy, 5
Reflection
collisions, avoidance, 224
crossmodal information processing, 187
distractions, 75–76
environment, 118
estimations, 51
experience and visual attention, 89
eye allocation, 151–152
information acquisition, 167–168
models, information acquisition, 2
visual requirements, 31–32
Reflexive selection, 65, 67–68
Reliability, driving simulators, 15
Reliance, information acquisition, 176–181
Research methods, 15–17
Response, collision behavioral analysis and experiment
accidents, 229–231
design, 239
discussion, 247–248
evaluation of accidents, 229–231
experimental method and results, 233–247
facility, 234
fundamentals, 227–229, 233–234
hill scenario, 242–247
intersection scenario, 240–242
investigative rationale, 231–233
participants, 236, 238
procedure, 238
recommendations, 248–249
results, 239–247
scenario, 234–236, 240–247
summary, 248–249
Responsibility, attribution of, 77–78
Restricted scanning, 102
Rewards and incentives, 209–210
Risk, adverse weather conditions, 124–125
Risk awareness and perception training (RAPT), 111, 113
Risk compensation, *see* Behavioral adaptation

Road environment and traffic engineering changes
- fundamentals, 213
- hazardous location treatment, 214
- intersection safety, 213–214
- road environment safety, 215
- run-off-road crashes, 214
- special road-user groups, 215
- speed limit reductions, 215

Road environment safety, 215
Road trauma, interventions
- commercial vehicle driver licensing, 208
- driver management, 208
- driver testing, 206
- driver training and education, 205–206
- enforcement programs, 208–209
- fundamentals, 202, 216
- graduated licensing, 204–205
- Haddon matrix, 202–203
- hazardous location treatment, 214
- heavy vehicle fleets, 211–212
- incentives and rewards, 209–210
- intersection safety, 213–214
- licensing, 203–210
- management, 203–210
- motorcycle licensing, 207
- older drivers, 206–207
- organizational interventions, 210–212
- reflection, 201–202
- road environment safety, 213–215
- run-off-road crashes, 214
- small fleets, 212
- special road-user groups, 215
- speed limit reductions, 215
- testing, 206
- traffic engineering changes, 213–215
- training, 203–210
- vehicle changes, 212–213

Road users
- predicting behavior, 10, 112–113
- special groups, 215

Roadway design
- alignment, 123–124
- environment, 122–123
- fundamentals, 118–119, 139–140
- lane width, 120, 140
- lateral clearance, 121
- markings, 121–122
- pavement, 119–121
- roughness, 119–120

Roadway luminance, 40–41, *see also* Luminance; Public lighting
Roadway preview distance/time
- empirical evidence, experiments, 44–45
- hazardous situations, 106–107

Roughness, 119–120
Roundabouts, 214
Rumar model, 9
Run-off-road crashes, 214
Rural roads, 130–131

S

Safety
- freeway lighting, 129
- intersections, 213–214
- lane width, 120, 140
- reduction in speed, 124
- road environment, 215

Salience, 158–159
Scales, self-assessment, 21
Scanning
- and hazards, 93–98
- restricted, 102

Scenarios
- collision response analysis and experiment, 234–236, 240–247
- visual attention, 80

Scotopic adaptation conditions, 41
SDT, *see* Signal detection theory (SDT)
Search, visual attention, 80–81, *see also* Eyes, movement
Secondary hazards, 96
SEEV (salience, effort, expectancy, and value) model, 158–159
Selection with awareness, 64–65
Selection without awareness, 64
Self-assessment, 20–21
Self-explaining roads (SER), 119
Seniors, *see* Older drivers
Sensory conspicuity, 69
Sensory process, 9
Sequence of events, 5
SER, *see* Self-explaining roads (SER)
Shoulders, 214
Signal detection theory (SDT), 177
Signs
- advertising, 69–70, 97–98
- detection theory models, 8–9
- exploratory selection, 69
- eye movement recording, 20
- memory, 21
- performance measurements, 23–24
- recognition, recording techniques, 22
- removal, societal forces, 60
- roadway luminance, 41
- traffic accidents, 22
- visual acuity reduction, 38
- warning, short-term context, 4–5

Simulations
- cell phone use, 101–102
- eye movement training, 109–110
- field of view reduction, 43–44

hands-free phone use, 101–102
hazardous situation training, 112
mobile phones, 190
road environment, 122
road markings, 122
visual acuity reduction, 39
visual demands and driving, 13–15
Situation awareness, 9–11
Slalom course, 38, 42–43
Sleepiness, *see also* Drowsiness; Fatigue
crossmodal information processing, 195–196
deliberate selection, 70
driving errors, 77
in-car entertainment, 192
Small fleets, 212
Snow, *see* Weather conditions, adverse
Social use of car, 91
Societal forces, decisions, 60
Spain, 14, 211
Spanish National Institute for Aerospace Technology, 14
Spatial distribution, fixations, 80–81
Special road-user groups, 215
Speech-based in-vehicle technologies, 85
Speed
adaptation, 25
buildings, 122
control, 56–57, 85–86
driving simulators, 14
enforcement programs, 208–209
estimation performance capacities, 56–57
field of view reduction, 43
housing, 122
lane width, 120
lateral clearance, 121
mental load, 85–86
passive control, movement perception, 58
perception, 56–57, 85–86
peripheral information, 122–123
roughness of road surface, 119–120
same effect as BAC, 215
speed limit reductions, 215
transmission, information theory, 5
transverse markings, 121–122
trees, 122
unsafe reduction in, 124
visual acuity reduction, 39–40
"Split-brain" monkeys, 33
Splitter islands, 214
Standard deviations of lane position index
preview distance/time, roadway, 44
visual acuity reduction, 39
Steering and steering performance/behavior
diagnostic signature, 36–37
driver's field of view, 42–44
gaze direction, 68
heuristic value, ambient-focal framework, 47
measures of, 39
preview distance, 106–107
public lighting, 132
reflexive selection, 68
roadway luminance, 40–41
roadway preview distance/time, 44–45
two-level model, 35–36
visual acuity, 37–40
Steering wheel position, 14
Stevens' power law, 137
Stimuli
collision likelihood, 55
exploratory selection, 69–70
memory, 22
SEEV model, 158–159
selection, perception models, 9
testing, 16–17
Streetlights, 129, *see also* Public lighting
Studded tires, 125, 140
Subconscious selection, 64
Sudden luminance, 67–68, *see also* Luminance; Public lighting
Sunday drivers, 224–225
Sustainable Safety philosophy, 213
Sweden, 211, 215
Swedish Road and Transport Research Institute, 44
Swerving, hill scenario response, 247–248
SWOV, *see* Institute for Road Safety and Research (SWOV)
System capacity, 5

T

Talking, 189–191, *see also* Cell phones
Tau, 56
TCT, *see* Traffic conflict techniques (TCT)
Technologies
alcohol interlock, 212
concierge services, 153
exploratory selection, 69–70
heads-up displays, 69–70
in-vehicle, 152–156
listening to radio, 192
menu navigation, 78
mobile phones, 77, 189–191
navigation systems, 77–78, 152–153
poorly designed, 176
speech-based in-vehicle technologies, 85
younger drivers, 93
Teenagers, *see* Younger drivers
Telematics, 154–156
Testing, 16–17, *see also* Experiments
Theoretical framework, 32–37
Thinking tasks, *see also* Cognitive demands; Decisions; Mental loads

cognitive effort, 83
control, 85–86
distractions, 79
inattentional blindness, 84–85
internal distraction, 83
measurements, mental loads, 83
performance, 84–85
speed, 84–86
testing, 16
vision, 84–85
visual search and performance effects, 84–85
Time-in-lane index, 39
Time-space mathematical relationship, 56
Time-to-collision (TTC), 134–135
Time-to-contact (TTC), 55–56, 232
Time-to-line crossing, 46
Time-to-passage, 232
Tires, studded, 125, 140
TNO Human Factors, 134
Traffic
accident recording techniques, 22
oncoming, movement perception, 58
Traffic conflict techniques (TCT), 134
Traffic engineering, *see* Road environment and traffic engineering changes
Traffic Weather Information Service (TWIS), 126
Training
aerial view training technique, 111
driver, 205–206
eye movements, 109–110
hazardous situations, 112
Mount Cotton driver training course, 42
for novice drivers, 112–113
road trauma, interventions, 203–210
visual search, 107, 108–112
what-happens-next, 113
Transmission speed, 5
Transportation Research Institute (United States), 15
Transport Research Laboratory (United Kingdom), 15
Transverse markings, 58, 121–122, *see also* Markings
Traveler information, 193, *see also* Navigation systems
Trees, 122, 214
Tunnel markings, 58, *see also* Markings
Tunnel vision, 80
TWIS, *see* Traffic Weather Information Service (TWIS)
Two-level model, 35–36

U

UFOV, *see* Useful field of view (UFOV)
Unappreciated alarms, 180
Unconscious selection, 64
Unintentional selection, 64
United Kingdom, 15
United States National Highway Traffic Safety Administration (NHTSA), 53
United States of America, 15, 208
University of Groningen (Holland), 15
University of Leeds (United Kingdom), 15
University of Michigan (United States), 15
University of Monash Accident Research Center (Australia), 15
Useful field of view (UFOV), 71, 170, *see also* Field of view
Utility, alarms, 177

V

Valency, 7
Validity, 15–16
Vehicle changes, 212–213
Velocity, 58, 233
Verbal instructions, 97
Vibrotactile warnings, 68
Videos
conflict quantification, 135–136, 141–142
environment, 133–134
hazardous situations, 94
hazard perception training, 111
modeling, 136–137
police pursuits, 97
Visibility, 125, *see also* Weather conditions, adverse
Vision, thinking tasks, 84–85
Vision Zero philosophy, 213
Visual acuity, 37–40
Visual attention, 79–81, *see also* Eyes, allocation
Visual attention, experience and
accident liability relationship, 89–93
attention, dividing and focusing, 113
behavior prediction, other road users, 112
experienced drivers, 98–102
management, 113
mental model of situation, 113
novice drivers, 98–102, 112–113
reflection, 89
road users, predicting behavior, 112
and scanning, 93–98
training, 108–112
visual field, 102–107
visual search, 98–102, 108–112
Visual demands
adaptation to driving, 24–26
distractions, 79
drivers' self-assessment, 20–21
Endsley's model, 9
eye movement recording, 17–20

fundamentals, 2–3
Gibson's model, 7–8
information theory, 3–7
in-vehicle technologies, 154–156
measurements, 22–24
memory, 21–22
methodologies used, 11–13
models, information acquisition, 3–11
observations, 15–16
psychophysiological techniques, 24
reaction time, 22–24
recording techniques, 17–24
reflections, 2
research methods, 15–17
sign detection theory, 8–9
sign recognition, 22
simulations, 13–15
telematics, 154–156
testing, 16–17
traffic accidents, 22
Visual field, 102–107
Visual requirements, vehicular guidance
ambient (dorsal) vision, 33–35
conclusions, 47–48
diagnostic signature, 36–37
driver's field of view, 42–44
empirical evidence, experiments, 37–45
focal (ventral) vision, 33–35
fundamentals, 32
heuristic value, 45–47
modes of visual processing, 33–35
properties, 34–35
reflection, 31–32
roadway luminance, 40–41
roadway preview distance/time, 44–45
steering behavior, 35–37
theoretical framework, 32–37
two-level model, 35–36
visual acuity, 37–40
Visual search, *see also* Eyes, movement
distractions, 82
experienced *vs.* novice drivers, 98–102
eye movement recording, 20
training, 108–112

W

"Wait and see" strategy, 244
Warning demands, 4–5, 82
Warning signals, 193–195
Weapon focus, 94
Weather conditions, adverse
accident risk, 124–125
behavioral adaptation, 125–127
fundamentals, 140–141
What-happens-next training, 113
Workload, 101, 132, *see also* Cognitive demands; Mental loads; Thinking tasks
Wrong-way incursions, 236

Y

Younger drivers, see also Novice drivers
accident evaluation, 230
alarms, 181
alcohol, 91
crashes, information acquisition, 170–171
distractions, 92
eye movement recording, 20
graduated licensing, 204–205
information acquisition, 172–175
in-vehicle technologies, 156
likelihood of accidents, 89–90
night driving, 90
passenger presence, 91, 93
perceptual and motor processes, 168
poorly designed technologies, 176–177
preview distance/time, roadway, 44
profile of, 93
social use of car, 91